Lecture Notes in Artificial Intelligence   3575

Edited by J. G. Carbonell and J. Siekmann

Subseries of Lecture Notes in Computer Science

T0223706

Stefan Wermter   Günther Palm
Mark Elshaw (Eds.)

Biomimetic Neural Learning for Intelligent Robots

Intelligent Systems, Cognitive Robotics,
and Neuroscience

 Springer

Series Editors

Jaime G. Carbonell, Carnegie Mellon University, Pittsburgh, PA, USA
Jörg Siekmann, University of Saarland, Saarbrücken, Germany

Volume Editors

Stefan Wermter
Mark Elshaw
University of Sunderland St. Peter's Campus
School of Computing and Technology, Centre for Hybrid Intelligent Systems
Sunderland SR6 0DD, UK
E-mail: {stefan.wermter, mark.elshaw}@sunderland.ac.uk

Günther Palm
University of Ulm, Department of Neural Information Processing
89069 Ulm, Germany
E-mail: palm@neuro.informatik.uni-ulm.de

Library of Congress Control Number: 2005928445

CR Subject Classification (1998): I.2, I.2.6, I.2.9, I.5.1

ISSN   0302-9743
ISBN-10   3-540-27440-5 Springer Berlin Heidelberg New York
ISBN-13   978-3-540-27440-7 Springer Berlin Heidelberg New York

This work is subject to copyright. All rights are reserved, whether the whole or part of the material is
concerned, specifically the rights of translation, reprinting, re-use of illustrations, recitation, broadcasting,
reproduction on microfilms or in any other way, and storage in data banks. Duplication of this publication
or parts thereof is permitted only under the provisions of the German Copyright Law of September 9, 1965,
in its current version, and permission for use must always be obtained from Springer. Violations are liable
to prosecution under the German Copyright Law.

Springer is a part of Springer Science+Business Media

springeronline.com

© Springer-Verlag Berlin Heidelberg 2005
Printed in Germany

Typesetting: Camera-ready by author, data conversion by Scientific Publishing Services, Chennai, India
Printed on acid-free paper   SPIN: 11521082   06/3142   5 4 3 2 1 0

Preface

This book presents research performed as part of the EU project on biomimetic multimodal learning in a mirror neuron-based robot (MirrorBot) and contributions presented at the International AI-Workshop on NeuroBotics. The overall aim of the book is to present a broad spectrum of current research into biomimetic neural learning for intelligent autonomous robots. There is a need for a new type of robot which is inspired by nature and so performs in a more flexible learned manner than current robots. This new type of robot is driven by recent new theories and experiments in neuroscience indicating that a biological and neuroscience-oriented approach could lead to new life-like robotic systems.

The book focuses on some of the research progress made in the MirrorBot project which uses concepts from mirror neurons as a basis for the integration of vision, language and action. In this book we show the development of new techniques using cell assemblies, associative neural networks, and Hebbian-type learning in order to associate vision, language and motor concepts. We have developed biomimetic multimodal learning and language instruction in a robot to investigate the task of searching for objects. As well as the research performed in this area for the MirrorBot project, the second part of this book incorporates significant contributions from other research in the field of biomimetic robotics. This second part of the book concentrates on the progress made in neuroscience inspired robotic learning approaches (in short: NeuroBotics).

We hope that this book stimulates and encourages new research in this interesting and dynamic area. We would like to thank all contributors to this book, all the researchers and administrative staff within the MirrorBot project, the reviewers of the chapters and all the participants at the AI-Workshop on NeuroBotics.

Finally, we would like to thank the EU for their support of the MirrorBot project, and Alfred Hofmann and his staff at Springer for their continuing support.

May 2005

Stefan Wermter
Günther Palm
Mark Elshaw

Table of Contents

Part II: Biomimetic Cognitive Behaviour in Robots

Towards Biomimetic Neural Learning for Intelligent Robots

Stefan Wermter[1], Günther Palm[2], Cornelius Weber[1], and Mark Elshaw[1]

[1] Hybrid Intelligent Systems, University of Sunderland,
School of Computing and Technology,
St Peter's Way, Sunderland, SR6 0DD, UK
{Stefan.Wermter, Cornelius.Weber, Mark.Elshaw}@sunderland.ac.uk
www.his.sunderland.ac.uk
[2] Neuroinformatics, University of Ulm,
Oberer Eselsberg, D-89069 Ulm, Germany
palm@neuro.informatik.uni-ulm.de

Abstract. We present a brief overview of the chapters in this book that relate to the development of intelligent robotic systems that are inspired by neuroscience concepts. Firstly, we concentrate on the research of the MirrorBot project which focuses on biomimetic multimodal learning in a mirror neuron-based robot. This project has made significant developments in biologically inspired neural models using inspiration from the mirror neuron system and modular cerebral cortex organisation of actions for use in an intelligent robot within an extended 'pick and place' type scenario. The hypothesis under investigation in the MirrorBot project is whether a mirror neuron-based cell assembly model can produce a life-like perception system for actions. Various models were developed based on principles such as cell assemblies, associative neural networks, and Hebbian-type learning in order to associate vision, language and motor concepts. Furthermore, we introduce the chapters of this book from other researchers who attended our AI-workshop on NeuroBotics.

1 Introduction

Many classical robot systems ignore biological inspiration and so do not perform in a robust learned manner. This is reflected in most of the conventional approaches to the programming of (semi-) autonomous robots: Many details of the program have to be reprogrammed and fine-tuned by hand even for slight changes in the application, which is time consuming and error-prone. Hence, there is a need for a new type of computation that is able to take its inspiration from neuroscience and perform in an intelligent adaptive manner to create biomimetic robotic systems. Many researchers including the contributors to this book hold that by taking inspiration from biological systems that would allow the development of autonomous robots with more robust functionality than is possible with current robots. An additional benefit of biomimetic robots is that

S. Wermter et al. (Eds.): Biomimetic Neural Learning, LNAI 3575, pp. 1–18, 2005.
© Springer-Verlag Berlin Heidelberg 2005

they can provide an indication of how the biological systems actually could work in order to provide feedback to neuroscientists. In order to indicate the progress made towards Biomimetic Neural Learning for Intelligent Robots this book is split into two parts.

In the first part we present some of the research findings from the biomimetic multimodal learning in a mirror neuron-based robot (MirrorBot) project. The aim of this EU FET project was to develop models for biomimetic multimodal learning using a mirror neuron-based robot to investigate an extended 'pick and place' scenario. This task involves the search for objects and integrates multimodal sensory inputs to plan and guide behaviour. These perceptual processes are examined using models of cortical assemblies and mirror neurons to explore the emergence of semantic representations of actions, percepts, language and concepts in a MirrorBot, a biologically-inspired neural robot. The hypothesis under investigation and focus of this book is whether a mirror neuron-based cell assembly model will produce a life-like perception system for actions. The MirrorBot project combines leading researchers in the areas of neuroscience and computational modeling from the University of Sunderland, Parma and Ulm, INRIA Lorraine/LORIA-CNRS and Cognition and Brain Sciences Unit, Cambridge. The findings of the neuroscience partners form the basis of the computational models that are used in the development of the robotic system. The neuroscience partners concentrate on two cerebral cortex systems by examining how humans process and represent different word categories and the mirror neuron system.

The extended 'pick and place' scenario involves the MirrorBot neural robot assistant being positioned between two tables that have multiple objects positioned on them and is required to perform various behaviours on objects based on a human verbal instruction. The robot takes in three or four word instructions that contain an actor, action and object such as 'bot pick plum' or 'bot show brown nut'. The instructional grammar developed for the MirrorBot contains approximately 50 words with the actor being the 'bot'. The actions that are performed are divided into those that are performed by the hand, leg or head. For instance, the action performed by the hand include 'pick', 'put' and 'lift', the leg actions include 'go' and 'move' and the head actions include 'show' and 'turn-head'. The objects include natural objects such as 'orange', 'nut' and artefact objects such as 'ball' and 'cup'. In order to perform the appropriate behaviours the robot assistant using neural learning must perform such diverse activities as language recognition, object localization, object recognition, attention, grasping actions, docking, table localization, navigation, wandering and camera positioning.

In the second part of the book we provide chapters from researchers in the field of biomimetic robotic neural learning systems who attended the AI-Workshop on NeuroBotics. The aim of this workshop and hence of this book is to contribute to robotic systems which use methods of learning or artificial neural networks and/or are inspired by observations and results in neuroscience and animal behaviour. These chapters were selected to give an indication of the diversity of the research that is being performed into biomimetic robotic learning and to provide a broader perspective on neural robotics. For instance, chapters will consider the development of a virtual platform for modeling biomimetic robots, a robotic arm, robot recognition in RoboCup, sensory motor control of robot limbs and navigation. These models are utilised by both robot simulators and actual robots and make use of neural approaches that are both supervised and unsupervised.

2   Modular Cerebral Cortex Organisation of Actions: Neurocognitive Evidence for the MirrorBot Project

Neuroscience evidence reflected in the development of the MirrorBot biomimetic robotic systems comes from research at Cambridge related to how words are processed and represented in the cerebral cortex based on neurocognitive experiments of Pulvermüller. Accordingly, words are represented and processed using Hebbian learning, synfire chains and by use of semantic features. Hebbian learning supports the basis of higher cognitive behaviour through a simple synaptic approach based on cell assemblies for cortical processing [27, 30, 28, 29]. Cell assemblies rely on a connectivity structure between neurons that support one another's firing and hence have a greater probability of being co-activated in a reliable fashion [43, 41, 47, 23]. Synfire chains are formed from the spatiotemporal firing patterns of different associated cell assemblies and rely on the activation of one or more cell assemblies to activate the next assembly in the chain [27, 41, 18]. Hence, neurocognitive evidence on word representation and processing in the cerebral cortex suggests that cognitive representations are distributed among cortical neuronal populations [29, 33, 27]. The word meaning is critical for determining the cortical populations that are activated for the cognitive representation task.

When looking at the web of cell assemblies which process and represent particular word types Pulvermüller [27] notes that activation is found in both hemispheres of the cerebral cortex for content words. Semantic word categories elicit different activity patterns in the fronto-central areas of the cortex, in the areas where body actions are known to be processed [40, 11]. Perception words are represented by assemblies in the perisylvian cortex and posterior cortex [27, 31] and nouns related to animals activate the inferior temporal or occipital cortices [28, 27, 29].

Emotional words are felt to activate the amygdala and cells in the limbic system more than words associated with tools and their manipulation [26]. The link between the assemblies in these two regions is achieved through the amygdala and frontal septum [27]. For action words that involve moving ones own body the perisylvian cell assembly is also associated with assemblies in the motor, premotor and prefrontal cortices [27, 30]. For content words the semantic features that influence the cell assemblies come from various modalities and include the complexity of activity performed, facial expression or sound, the type and number of muscles involved, the colour of the stimulus, the object complexity and movement involved, the tool used, and whether the person can see itself doing this activity. The combination of these characteristics into a single depiction is produced by pathways linking sensory information from diverse modalities to the same neurons. For objects the semantic features represented by cell assemblies typically relate to their colour, smell and shape. If a word is repeatedly presented with a stimulus the depiction of this stimulus is incorporated into the one for the word to produce a new semantic feature. In general, words are depicted via regions historically known as language regions and additional regions connected with the words semantics.

Concerning a division between action related and non-action related words [33], Pulvermüller states that there is a finer-grained grounding of language instruction in actions. This produces a division of the representation in the cerebral cortex based on the part of the body that performs that action between leg, head and hand [11, 29, 30, 27, 12]. It is well known that there is a division in the motor cortex between the regions that perform head/face, hand/arm and leg actions [25]. For instance, the region of the motor cortex that controls face movement is found in the inferior precentral gyrus, hand and arm in the middle region of the precentral gyrus and the leg actions are located in the dorsomedial area [29, 30]. Given the difference in the regions of the cortex that are responsible for performing actions it is also stated by Pulvermüller that a similar difference can be identified when representing action verbs and so grounding language instructions in actions based on the part of the body that performs the action [29].

Pulvermüller and his colleagues have performed various experiments [28, 12, 11, 13, 29, 40] on cerebral cortex processing of action verbs to test their hypothesis on the representation of action verbs based on the body part that performs. These include experiments where (i) different groups of subjects are given leg-, arm- and face-related action verbs and pseudo-words and asked to state whether they are a word; (ii) subjects are asked to use a rating system to answer questions on the cognitive processes a word arouses; (iii) subjects rank words based on whether they are leg-, arm- or head-related; and (iv) there is a comparison between hearing walk- and talk-type verbs. In these experiments EEG electrodes are positioned at various points along the scalp to produce recordings of cerebral cortex activation. From these experiments areas are identified where the activation is the same for all action verbs and more importantly are different depending on the action verbs based on the body parts they relate to.

Differences between the three types of action verbs based on the body parts were observed by Pulvermüller and his colleagues. They found a greater activation for face-words in the frontal-lateral regions of the left hemisphere close to the premotor cortex associated with face and head. For face- and leg-related action verbs there are different regions along the motor strip that are identified to process verbs from these two verb categories. Leg-type words produce greater activation in the cortical region of the cerebral cortex used to produce leg actions and for the face-words there is greater activation in the inferior regions near to the face region of the motor cortex [32]. It is found that hand-related words are located in more lateral regions of the cortex than leg-words. Consistent with the somatotopy of the motor and premotor cortex [25], leg-words elicited greater activation in the central cerebral cortex region around the vertex, with face-words activating the inferior-frontal areas, thereby suggesting that the relevant body part representations are differentially activated when action words are being comprehended.

In addition the average response time for lexical decisions is faster for face-associated words than for arm-associated words, and the arm-associated words are faster than leg ones. There is also greater activation in the right parieto-occipital areas for arm- and leg-words relative to head words. The evidence of these experiments points to the word semantics being represented in different parts of the cerebral cortex in a systematic way. Particularly the representation of the word is related to the actual motor and premotor regions of the cerebral cortex that perform the action.

3   Mirror Neuron System Inspiration for MirrorBot Project

Research at Parma has provided a great deal of evidence on the mirror neuron system that inspired the robotic research for the MirrorBot project. Rizzolatti and co-workers [35, 8] found that neurons located in the rostral region of a primate's inferior frontal cortex area, the F5 area, are activated by the movement of the hand, mouth or both. These neurons fire as a result of the action, but not of the isolated movements that make up this action. The recognition of motor actions comes from the presence of a goal and so the motor system does not solely control movements [9, 37]. Hence, what turns a set of movement into an action is the goal and holding the belief that performing the movements will achieve a specific goal [1]. The F5 neurons are organised into diverse categories based on the actions that cause them to fire, which are 'grasping', 'holding' and 'tearing' [34, 9].

Certain grasping-related neurons fire when grasping an object whether it be performed by the hand, mouth or both [7]. This supports both the view that these neurons do not represent the motor action but the actual goal of performing the grasping task. Within area F5 there are two types of neuron: the first known as canonical neurons only respond to the performing of the action and the second mirror neurons that respond not only when performing an action but also when seeing or hearing the action performed [17, 36, 34]. Hence, the mirror neuron system produces a neural representation that is identical for the performance and recognition of the action [1].

These mirror neurons are typically found in area F5c and do not fire in response to the presence of the object or mimicking of the action. Mirror neurons required the action to interact with the actual object. They respond not only to the aim of the action but also how the action is carried out [44]. However, as shown by Umilta et al. 2001 [44] an understanding of an invisible present object causes the activation of the mirror neurons if the hand reaches for the object in the appropriate manner. This is achieved when they are first shown the action being performed completely visible and then with the hand-object interaction hidden. As the performance and recognition of an action causes activation in the premotor areas which is responsible for the hand movements when observing the action there is a set of mechanisms that prevent the behaviour being mimicked. The mirror neuron system indicates that the motor cortex is not only involved in the production of actions but in the action understanding from perceptual information [36] and so the observer has the same internal representation of action as the actor [44].

In this book Gallese [6] considers an exciting extension to findings related to the mirror neuron system based on the system providing an understanding of the emotional state of the performer for the observer. We do not exist independent of the actions, emotions and sensations of others as we understand the intentions of others. This indicates that as well as recognising an action the mirror neuron system has a role in predicting the consequences of what is being performed. Furthermore by allocating intentions to the actions monkeys and humans are able to use the mirror neurons to aid social interactions. This is achieved through the mirror neuron system providing intention to the motor sequence to identify further goals from this sequence. Hence, Gallese [6] in this chapter notes that we do not just see and recognise an action using the mirror neuron system but by using this system we also associate emotions and sensations to this observed behaviour. This occurs as if the observer is performing a similar action and feeling the same feelings and sensations. This offers a form of mind reading by the observed by attaching intentions to the behaviour. Hence, this points to the ability through embodied simulation to gain insight into the minds of others. Although this does not account for all social cognition.

It is observed that mirror neurons in humans are also excited by both the performance and observation of an action [9]. The F5 area in primates corresponds to various cortical areas in humans including the left superior temporal sulcus, the left inferior parietal lobule and the anterior region of Broca's area. The association of mirror neurons with Broca's area in humans and F5 in primates provides an indication that mirror neurons might have evolved in humans into the language system [34]. The role of the mirror neuron system in language can be seen from the findings of Pulvermüller [30, 11] in that processing and representation of words includes the activation of some of the same regions as those that are found to perform the action. The ability in the first instance to recognise an action is required for the development of a communication system between members of a group and finally for an elaborate language system [17, 34]. The concept of the mirror neuron system being the foundation of the language system directs the multimodal models developed as part of the MirrorBot project.

4   Computational Models in the MirrorBot Project

The neuroscience findings related to word processing and representation based on the findings of research at Cambridge and the mirror neuron system from research at Parma formed the basis for various computational robotic models in the MirrorBot project. In this book we incorporate chapters that provide some of these models that are able to perform language sequence detection [16, 18], spatial visual attention [46, 4], auditory processing [22], navigation [3] and language based multimodal input integration for robot control and multimodal information fusion [18, 4, 20, 48, 24]. These models are a significant contribution to the field of biomimetic neural learning for intelligent robots as they offer brain-inspired robotic performance that is able to produce behaviour based on the fusion of sensory data from multiple sources.

The research of Knoblauch and Pulvermüller [16] described in this book relates to the use of a computational system to consider if word sequences are grammatically correct and so perform sequence detection. A particularly important feature of language is its syntactic structure. For a robot to be able to perform language processing in a biomimetic manner it should be able to distinguish between grammatically correct and incorrect word sequences, categorise words into syntactic classes and produce rules. The model of Knoblauch and Pulvermüller [16] incorporates a biologically realistic element in that it uses numerous sequence detectors to show that associative Hebb-like learning is able to identify word sequences, produces neural representations of grammatical structures, and links sequence detectors into neural assemblies that provides a biological basis of syntactic rule knowledge. The approach consists of two populations of neuron, the WW population for word webs and population SD for sequence detectors. Each neuron is based on leaky-integrate units. The model was found to create auto-associative substitution learning and generalized sequential order to new examples and achieves the learning of putative neural correlate of syntactical rules.

Vitay et al. [46] have developed a distributed model that allows to sequentially focus salient targets on a real-world image. This computational model relies on dynamical lateral-inhibition interactions within different neural maps organized according to a biologically-inspired architecture. Despite the localized computations, the global emergent behaviour mimics the serial mechanism of attention-switching in the visual domain. Attention is understood here as the ability to focus a given stimulus despite noise and distractors, what is represented here by a localized group of activated neurons. The task is to sequentially move this bubble of activity on the different salient targets in the image. They use a preprocessed representation of the image according to task requirements: objects potentially interesting are made artificially salient by enhancing their visual representation. Three mechanisms are involved to achieve that task: first a mechanism allowing to focus a given stimulus despite noise and competing stimuli; second a switching mechanism that can inhibit at demand the currently focused target to let the first mechanism focus another location; third a working memory system that remembers previously focused locations to avoid coming back to a previously inspected object. The cooperation between these different mechanisms is not sequential but totally dynamic and distributed: no need for a central executive that would control the timing between the functionally different systems, the inner dynamics of the neurons make the work. This model has been successfully implemented on a robotic platform.

The chapter by Murray et al. [22] offers a biomimetically inspired hybrid architecture that uses cross-correlation and recurrent neural networks for acoustic tracking in robots. This research is motivated by gaining an understanding of how the mammalian auditory system tracks sound sources and how to model the mechanisms of the auditory cortex to enable acoustic localisation. The system is based on certain concepts from the auditory system and by using a recurrent neural network can dynamically track a sound as it changes azimuthally in the environment. The first stage in the model determines the azimuth of the sound source from within the environment by using Cross-Correlation and provides this angle to the neural predictor to predict the next angle in the sequence with the use of the recurrent neural network. To train the recurrent neural network to recognize various speeds, a separate training sub-group was created for each individual speed. This is to ensure that the network learns the correct temporal sequence it needs to recognize and provide prediction for the speeds. It has been shown that within the brain there is short term memory to perform such prediction tasks and in order to forecast the trajectory of an object it is required that previous positions are remembered to establish predicitons.

The navigation approach of Chokshi et al. [3] is based on modeling the place cells by using self-organising maps. The overall aim of this approach is to localise a robot using two locations based on visual stimulus. An internal representation of the world is produced by the robot in an unsupervised manner. The self-organising maps receive visual images that are used to produce internal representation of the environment that act like place codes using landmarks. Localisation by the robot is achieved by a particular position being associated with a specific active neuron. An overall architecture has been developed that uses modules to do diverse operations such as visual information derivation and motor control. The localisation was performed using a Khepera robot in an environment divided into 4 sections. These sections were also divided into squares that were used to determine the place cell error. Landmarks of different coloured cubes and pyramids were positioned at the edge of the environment. Each square represented a place cell, with the training and testing involving the images associated with each cell. An interesting finding of this study was that as the robot approached a specific landmark it was found that the appropriate place cell in the self-organising map output layer had great activation and once this robot leaves the landmark the activation reduces until it reaches 0. Clustering was also seen for landmarks that were close together and distinct landmarks were positioned further apart on the self-organising map.

The first language model based on multimodal inputs from the MirrorBot project considered in this book, is that of Markert et al. [18] who have developed an approach that through associative memories and sparse distributed representations can associate words with objects and characteristics of the object and actions. The approach enables a robot to process language instructions in a manner that is neurobiologically inspired using cell assemblies. The fundamental concept behind the cortical model developed is the use of cortical regions for different properties of an entity. Hence, a feature of the cortical model is the combining of visual, tactile, auditory language, goal and motor regions. Each of the regions is implemented using a spike counter architecture. Looking at regional activations it is possible to describe how inputs from multiple modalities are combined to create an action.

This language model using multimodal inputs is also used by Fay et al [4] who developed a neurobiologically plausible robotic system that combines visual attention, object recognition, language and action processing using neural associative memory. This involves finding and pointing with the camera at a specific fruit or object in a complex scene based on a spoken instruction. This requires the understanding of the instruction, relating the noun with a specific object that is recognised using the camera and coordinating motor output with planning and sensory information processing. The kinds of spoken commands that the robot is able to parse include 'bot show plum and 'bot put apple to yellow cup'. In the architecture preprocessing involves extracting features from the auditory and the visual input chosen by attention control. The cortical model used to perform speech recognition, language processing, action planning and object recognition consists of various neural networks such as radial basis networks and associator networks. A speech recogniser is used to receive the language instruction which is checked for syntactic consistency. This work is closely related to the work by Knoblauch and Pulvermüller described above. If the word sequence is syntactically correct the global goal is divided into a sequence of subgoals whose solution fulfills the overall goal. Object recognition is performed by a hierarchy of radial-basis-function networks which divide a complex recognition task into various less complex tasks. The model is able to associate a restricted set of objects with sentence like language instructions by associating the noun with properties of the object such as colour and actions. The model is of particular significance to this book as it shows how data from diverse modalities of language and vision can be brought together to perform actions.

Wermter et al. [48] produce two architectures that are able to successfully combine the three input modalities of high-level vision, language instructions and motor directions to produce simulated robot behaviour. The flat multimodal architecture uses a Helmholtz machine and receives the three modalities at the same time to learn to perform and recognise the three behaviours of 'go, 'pick' and 'lift' . The hierarchical architecture at the lower-level uses a Helmholtz machine and at the upper-level a SOM to perform feature binding. These multimodal architectures are neuroscience-inspired by using concepts from action verb processing based on neurocognitive evidence of Pulvermüller and specifically features of the mirror neuron system. The architectures are able to display certain features of the mirror neuron system which is a valuable development as the activation patterns for both the performance of the action and its recognition are close. Furthermore, we are able to indicate the role played by the mirror neuron system in language processing. A particular interesting finding obtained with the hierarchical architecture is that certain neurons in the Helmholtz machine layer of the hierarchical model respond only to the motor input and so act like the canonical neurons in area F5 and others to the visual stimulus and so are analogous to the mirror neurons. Furthermore, the lower-level Helmholtz machine is analogous to area F5 of the primate cortex and the SOM area to area F6. The F5 area contains neurons that can produce a number of different grasping activities. F6 performs as a switch, facilitating or suppressing the effects of F5 unit activations so that only the required units in the F5 region are activated to perform or recognise the required action.

An additional multimodal model from the MirrorBot project described in this book is that of Ménard et al. [20] which is a self-organising approach inspired by cortical maps. When an input is given to the map, a distributed activation pattern appears. From this a small group of units is selected by mutual inhibition that contains the most active units. Unlike Kohonens self-organising maps where this decision is made based on a global winner-takes-all approach, the decision is based on a numerical distribution process. The fundamental processing element of the Biologically-Inspired Joint Associative MAps (BIJAMA) model is a disk-shaped map consisting of identical processing units. The global behaviour of the map incorporates adaptive matching procedures and competitive learning. Ménard et al. [20] use several self-organising maps that are linked to one another to achieve cooperate processing by achieving word-action association. This multi-associative model is used to associate multimodalities of language and action and is able to deal with an association between modalities that is not one-one. The main issue associated with this approach is that learning in a map is dependent on the other maps, so that the inter-map connectivity biases the convergence to a certain state. The model is able to organise word representation in such a way that for instance a body word is associated with a body action.

Panchev [24] also developed a multimodal language processing model related to the MirrorBot project. This model uses a spiking neural model that is able to recognise a human instruction and then produce robot actions. Learning is achieved through leaky Integrate-And-Fire neurons that have active dendrites and dynamic synapses. Using spiking neurons the overall aim of this research is to model the primary sensory regions, higher cognitive functional regions and motor control regions. Hence, in this architecture there are modules that are able to recognise single word instructions, recognise objects based on colour and shape and a control system for navigation. The model uses a working memory that is based on oscillatory activity of neural assemblies from the diverse modalities. As this is one of the first robot control models that is based on spiking neurons it offers the opportunity to consider new behaviours and computational experiments that could be compared with the activity identified in the brain. This research is a significant contribution to the MirrorBot project as it shows the use of spiking neurons for spatiotemporal data processing. It is felt that as this model is able to approximate current neuroscience evidence it could directly address future neuroscience studies in the area of multimodal language processing. A future direction of this work is to consider the incorporation of goal behaviours by having certain objects more attractive than others.

5   Biomimetic Cognitive Behaviour in Neural Robots

We now turn to the chapters in the book that relate to research into biomimetic robots outside the MirrorBot project. The chapters summarised in this section show the diversity necessary to build biomimetic intelligent systems.

Reinforcement and reward based learning has proved a successful technique in biomimetic robots. In this book there are four chapter related to this technique from Jasso and Triesch [15], Hafner and Kaplan [10], Sung et al. [42] and Sheynikhovich et al. [39]. Jasso and Triesch [15] who consider the development of a virtual reality platform that is useful for biomimetic robots as it can be used to model cognitive behaviour. This environment allows the examination how cognitive skills are developed as it is now understood that these skills can be learned through the interaction with the parent and the environment in changeable social settings. This learning relates to visual activities such as gaze and point following and shared attention skills. Usually the models incorporate a single child and a parent. The environment is a room that contains objects and furniture and virtual agents that receive images from their camera. These images are processed and used to influence the behaviour of the agent. The chapter shows that the platform is suitable for modeling how gaze following emerges through infant-parent interactions. Gaze following is the capacity to alter ones own attention to an object that is the attention of another person. The environment consists of a living room containing toys and furniture and contains a parent agent and child agents. The child agents use reinforcement learning to alter its gaze to that of the parent based on a reward.

Hafner and Kaplan [10] in this book present research on biomimetic robots that are able to learn and understand pointing gestures from each other. Using a simple feature-based neural approach it is possible to achieve discrimination between left and right pointing gestures. The model is based on reward mechanisms and is implemented on two AIBO robot dogs. The adult robot is positioned pointing to an object using its left or right front leg and the child robot is positioned watching it. From the pointing gesture, the child robot learns to guess the direction of the object the adult robot is attending to and starts searching for it. The experiment used different viewing angles and distances between the two robots as well as different lighting conditions. This is a first step in order to bootstrap shared communication systems between robots by attention detection and manipualtion.

A further learning approach for biomimetic robots is that of Sung et al. [42] who use grid based function approximators for reinforcement learning. The approach uses techniques gained from active learning to achieve active data acquisition and make use of Q-learning methods that incorporate piecewise linear grid-based approximators. A feature of active learning is active data acquisition with algorithms being developed to reduce the effort required to produce training data. The learning algorithm that relates piecewise linear based approximators to reinforcement learning consists of two components. The first component is used for data acquisition for learning and the second carries out the learning. The suitability of the approach is tested on the 'Mountain-Car' problem which is typically used to evaluate reinforcement learning approaches. When doing so it was found that the number of required state transitions during learning was reduced. It is anticipated that this approach will be applicable to reinforcement learning for a real-world problem.

As navigation is such an important activity for biomimetic robots this book includes a second chapter on this by Sheynikhovich et al. [39]. Their biologically inspired model is based on rodent navigation. The model is multimodal in that it combines visual inputs from images and odometer readings to produce firing in artificial place cells using Hebbian synapses. By using reinforcement type learning the model is able to recreate behaviours and actual neurophysiological readings from the rodent. Although the model starts with no knowledge of the environment learning occurs through populations of place cells as the artificial rodent interacts with the environment. In the model visual information is correlated with odometer data related to the rotation and displacement using Hebbian learning in order that ambiguity in the visual data is resolved using the odometer readings. The model was implemented on a simulator and a mobile Khephera robot and is able to achieve similar performance to animals. This was seen when the robot learned the navigational task of reaching a hidden platform from random positions in the environment. In training the robot was given reward each time it found the platform and was able overtime to reduce the number of steps required to find the platform.

With regards to biomimetic learning the approach of Bach [2] looks at using distributed and localised representations to achieve learning and planning. To perform plan-based control there is a need to have a localist representation of the objects and events in the model of the world. In this approach compositional hierarchies implemented using MicroPsi node nets are used as a form of executable semantic networks that are seen as knowledge-based artificial neural networks. By using MicroPsi node nets it is possible to achieve backpropagation learning and symbolic plan representations. MicroPsi agents have a group of motivational variables that determine demands that direct how the agent performs. In the first instance, the agent does not know what actions will fulfill their desires and so performs trial-and-error actions. When the action is felt to have a positive impact on the demand a link is established between them. Experiments are performed by using a simulated environment that provides the agents with resources and dangers.

Folgheraiter and Gini [5] have developed an artificial robotic arm that replicates the functionality and structure of the human arm. In order to test the control system architecture the arm was developed with a spherical joint with 3 degrees of freedom, and an elbow with 1 degree of freedom. The control is arranged in a modular-hierarchical manner that has three levels: the lower level replicates the spinal reflexes that is used to control artificial muscle activities; the middle level produces the required arm movement trajectories; and at the higher level the circuits in the cerebral cortex and the cerebellum are found to control the path generator operation. The model uses a multilayer perceptron in order to solve the inverse kinematics problem as it is possible to train the network from actual readings from the human arm. In order to model the reflex behaviours a simplified model of the human spinal cord was used that concentrates on modeling membrane potential instead of spiking behaviour of the neurons. By using light materials it is possible to include this arm into a humanoid robot.

The next chapter in this book includes a model of human-like controlling of a biped robot [38]. Humanoid robots require complex design and complicated mathematical models. Scarfogliero et al. [38] demonstrate that a Light Adaptive-Reactive biped is simple, cheap and effective and is able to model the human lower limb. By use of this model it is possible to understand how humans walk and how this might be incorporated into a robot. The model is able to alter joint stiffness for position-control by use of servo motors. Scarfogliero et al. [38] devised a motor device based on torsional spring and damper to recreate the elastic characteristics of muscles and tendons. By using this approach it is possible to have good shock resistance and to determine the external load. As the robot uses location and velocity feedback it is able to perform a fine position operation even though it has no a-priori knowledge of external load.

The chapter in this book by Meng and Lee [21] considers the production of a biologically plausible novelty habituation model based on topological mapping for sensory-motor learning. Meng and Lee [21] examine embedded developmental learning algorithms by using a robotic system made up of two arms that are fitted with two-fingered gripper and a pan/tilt head that includes a colour camera. This robot is built in such away that it recreates the positioning of the head and arms of a child. Novelty and habituation are fundamental for early learning by assisting a system to examine new events/places while still monitoring the current state to gain experience for the full environment. The chapter considers the problem of sensory-motor control of limbs based on a childs arm movements during the first three months of life. Learning is achieved through a hierarchical mapping arrangement which consists of fields of diverse sizes and overlap at diverse mapping levels. The fields include local information such as sensory data and data on movement from the motor and stimulus data. The maps contain units known as fields that are receptive regions. In the experiments various parameters are considered such as the condition of the environment, field extent and habituation variables. By using the concepts from novelty and habitation as the basis of early robot learning it is possible to learn sensory-motor coordination skills in the critical areas first, before going onto the less critical areas. Hence, the robot uses novelty to learn to coordinate motor behaviour with sensory feedback.

Robot control using visual information was performed by Hermann et al. [14] to examine modular learning using neural networks for biomimetic robots. This chapter describes a modular architecture that is used to control the position/orientation of a robot manipulator by feedback from the visual system. The outlined modular approach is felt to overcome some of the main limitations associated with neural networks. Using modular learning is a useful approach for robots as there is limited data for training, robots must function in real-time in a real-time environment. A single neural network may not be sufficient to perform a complex task, however by using a modular sequential and bidirectional arranges of neural modules solutions can be found. Motivated by biological modularity Hermann et al. [14] use extended Kohonen maps that combine a self-organising map with ADA-Line networks (SOM-LLM) in the modular model. This neural network approach was selected as the SOM-LLM is simple and offers a topological representation. To test the modular architecture it was used to control a robot arm using two cameras that are positioned on a robot head. This chapter is able to offer an approach that can combine a set of neural modules that can converge and so learn complex systems.

In the area of RoboCup Mayer et al. [19] have developed a neural detection system that achieves colour-based attention control and neural object recognition to determine whether another robot is observed. The ability to recognise teammate robots and opposition robots is fundamental for robot soccer games. Problems associated with self-localisation and communication between robot teammates has lead to the need for an approach that is able to detect teammates and the opposition in a reliable manner based on vision. The approach identified in Mayer et al. [19] in this book uses the following steps (i) identify region of interest; (ii) gain features from this region; (iii) classify the features using neural networks; and (iv) arbitrate the classification outcome to establish if a robot is recognised and whether it is part of the own or opposition team. Regions of interest are typically determined using blob search using a segmented and colour-indexed picture. The types of features that are used in the robot are width, how much black is in the image and an orientation histogram. Once the features are established they are passed into two multilayer preceptron neural networks that are used to classify the features. One network was used to process the simple features and the other for the orientated histogram levels. These two networks produce a probability value stating if a robot is present. A decision as to whether a robot is present is based on whether the joint probablity from these two neural networks is greater than a threshold. The team the robot belongs to depends on the recognition of a colour marker. This approach has proved to give very good performance when classifing the presence of a robot and whether it belongs to the opposition or the own team.

Two approaches for visual homing using a descriptor that characterises local image sections in a scale invariant fashion are considered by Vardy and Oppacher [45] in this book. Visual homing is returning to a location by contrasting the current image with the one at the goal. This approach is based on the behaviour of insects like bees or ants. The two homing approaches rely on edges being extracted from the input images using a Sobel filter. The first approach uses the common technique of corresponding descriptors among images and the second approach establishes a home vector by determining the local image regions which are most similar between the two images, and assuming that these correspond to the foci of expansion and contraction. This second approach makes use of the structure of the motion field for pure translation. The second method found a home vector more directly using the stationary local image region closest from the two images. The first approach was able to out-perform the warping method, while the second performs equivalently to the warping method.

6   Conclusion

As can be seen from this book there is much research being carried out towards biomimetic robots. In particular, the MirrorBot project has contributed to the development of biomimetic robots by taking neuroscience evidence and producing neural models, but also several other joint European projects have worked in the same direction (SpikeFORCE, BIOLOCH, CIRCE, CYBREHAND, MIRROR). The breadth of the international research community sharing the research goal of biomimetic robotics can be seen not only from the contributions to this workshop, but also from many closely related conferences that have been organised in recent years. The chapters included in this book show that the MirrorBot project has successfully developed models that are able to check the syntactic consistency of word sequences, visually explore scenes and integrate multiple inputs to produce sophisticated robotics systems. This shows that we can overcome the present limitations of robotics and improve on some of the progress made by basing robots on biological inspiration such as the mirror neuron concept and modular cerebral cortex organisation of actions. The second part of the book shows the diversity of the research in the field of biomimetic neural robot learning. Although this research produces different approaches to diverse sets of robot function they are all connected by performance, flexibility and reliability that can be achieved by those based on biological systems. The biological systems thereby act as a common guideline for these diverse, cooperative, cooperating and competing approaches. Hence, there is a need to base robotic systems on biological concepts to achieve robust intelligent systems. As shown in this book the current progress in biomimetic robotics is significant, however more time is needed before we see it in full operation showing fully autonomous biomimetic robots.

References

1. M. Arbib. From monkey-like action recognition to human language: An evolutionary framework for neurolinguistics. *Behavioral and Brain Science*, pages 1–9, 2004.
2. J. Bach. Representations for a complex world: Combining distributed and localist representations for learning and planning. In S. Wermter, G. Palm, and M. Elshaw, editors, *Biomimetic Neural Learning for Intelligent Robots*. Springer-Verlag, Heidelberg, Germany, 2005.
3. K. Chokshi, S. Wermter, P. Panchev, and K. Burn. Image invariant robot navigation based on self organising neural place codes. In S. Wermter, G. Palm, and M. Elshaw, editors, *Biomimetic Neural Learning for Intelligent Robots*. Springer-Verlag, Heidelberg, Germany, 2005.
4. R. Fay, U. Kaufmann, A. Knoblauch, H. Markert, and G. Palm. Combining visual attention, object recognition and associative information processing in a neurobotic system. In S. Wermter, G. Palm, and M. Elshaw, editors, *Biomimetic Neural Learning for Intelligent Robots*. Springer-Verlag, Heidelberg, Germany, 2005.

5. M. Folgheraiter and G. Gini. Maximumone: an anthropomorphic arm with bio-inspired control system. In S. Wermter, G. Palm, and M. Elshaw, editors, *Biomimetic Neural Learning for Intelligent Robots*. Springer-Verlag, Heidelberg, Germany, 2005.

6. V. Gallese. The intentional attunement hypothesis. the mirror neuron system and its role in interpersonal relations. In S. Wermter, G. Palm, and M. Elshaw, editors, *Biomimetic Neural Learning for Intelligent Robots*. Springer-Verlag, Heidelberg, Germany, 2005.

7. V. Gallese, L. Escola, I. Intskiveli, M. Umilta, M. Rochat, and G. Rizzolatti. Goal-relatedness in area F5 of the macaque monkey during tool use. Technical Report 17, MirrorBot, 2003.

8. V. Gallese, L. Fadiga, L. Fogassi, and G. Rizzolatti. Action recognition in the premotor cortex. *Current Opinion in Neurobiology*, 119:593–609, 1996.

9. V. Gallese and A. Goldman. Mirror neurons and the simulation theory of mind-reading. *Trends in Cognitive Science*, 2(12):493–501, 1998.

10. V. Hafner and F. Kaplan. Learning to interpret pointing gestures: Experiments with four-legged autonomous robots. In S. Wermter, G. Palm, and M. Elshaw, editors, *Biomimetic Neural Learning for Intelligent Robots*. Springer-Verlag, Heidelberg, Germany, 2005.

11. O. Hauk, I. Johnsrude, and F. Pulvermüller. Somatotopic representation of action of action words in human motor and premotor cortex. *Neuron*, 41:301–307, 2004.

12. O. Hauk and F. Pulvermüller. Neurophysiological distinction of action words in the frontal-central cortex. Technical Report 7, MirrorBot, 2003.

13. O. Hauk and F. Pulvermüller. Neurophysiological distinction of action words in the frontal lobe: An ERP study using minimum current estimates. *European Journal of Neuroscience*, 21:1–10, 2004.

14. G. Hermann, P. Wira, and J-P Urban. Modular learning schemes for visual robot control. In S. Wermter, G. Palm, and M. Elshaw, editors, *Biomimetic Neural Learning for Intelligent Robots*. Springer-Verlag, Heidelberg, Germany, 2005.

15. H. Jasso and J. Triesch. A virtual reality platform for modeling cognitive development. In S. Wermter, G. Palm, and M. Elshaw, editors, *Biomimetic Neural Learning for Intelligent Robots*. Springer-Verlag, Heidelberg, Germany, 2005.

16. A. Knoblauch and F. Pulvermüller. Sequence detector networks and associative learning of grammatical categories. In S. Wermter, G. Palm, and M. Elshaw, editors, *Biomimetic Neural Learning for Intelligent Robots*. Springer-Verlag, Heidelberg, Germany, 2005.

17. E. Kohler, C. Keysers, M. Umilta, L. Fogassi, V. Gallese, and G. Rizzolatti. Hearing sounds, understanding actions: Action representation in mirror neurons. *Science*, 297:846–848, 2002.

18. H. Markert, A. Knoblauch, and G. Palm. Detecting sequences and understanding language with neural associative memories and cell assemblies. In S. Wermter, G. Palm, and M. Elshaw, editors, *Biomimetic Neural Learning for Intelligent Robots*. Springer-Verlag, Heidelberg, Germany, 2005.

19. G. Mayer, U. Kaufmann, G. Kraetzschmar, and Palm G. Neural robot detection in robocup. In S. Wermter, G. Palm, and M. Elshaw, editors, *Biomimetic Neural Learning for Intelligent Robots*. Springer-Verlag, Heidelberg, Germany, 2005.

20. O. Menard, F. Alexandre, and H. Frezza-Buet. Towards word semantics from multi-modal acoustico-motor integration: Application of the bijama model to the setting of action-dependant phonetic representations. In S. Wermter, G. Palm, and M. Elshaw, editors, *Biomimetic Neural Learning for Intelligent Robots*. Springer-Verlag, Heidelberg, Germany, 2005.

21. Q. Meng and M. Lee. Novelty and habituation: The driving force in early stage learning for developmental robotics. In S. Wermter, G. Palm, and M. Elshaw, editors, *Biomimetic Neural Learning for Intelligent Robots*. Springer-Verlag, Heidelberg, Germany, 2005.

22. J. Murray, H. Erwin, and S. Wermter. A hybrid architecture using cross-correlation and recurrent neural networks for acoustic tracking in robots. In S. Wermter, G. Palm, and M. Elshaw, editors, *Biomimetic Neural Learning for Intelligent Robots*. Springer-Verlag, Heidelberg, Germany, 2005.

23. G. Palm. *Neural Assemblies. An Alternative Approach to Artificial Intelligen e*. Springer-Verlag, 1982.

24. C. Panchev. A spiking neural network model of multi-modal language processing of robot instructions. In S. Wermter, G. Palm, and M. Elshaw, editors, *Biomimetic Neural Learning for Intelligent Robots*. Springer-Verlag, Heidelberg, Germany, 2005.

25. W. Penfield and T. Rasmussen. *The cerebral cortex of man*. Macmillan, Cambridge, MA, 1950.

26. D. Perani, S. Cappa, T. Schnur, M. Tettamanti, S. Collina, M. Rosa, and F. Fazio. The neural correlates of verbs and noun processing a PET study. *Brain*, 122:2337–2344, 1999.

27. F. Pulvermüller. Words in the brain's language. *Behavioral and Brain Sciences*, 22(2):253–336, 1999.

28. F. Pulvermüller. Brain reflections of words and their meaning. *Trends in Cognitive Neuroscience*, 5(12):517–524, 2001.

29. F. Pulvermüller. A brain perspective on language mechanisms: from discrete neuronal ensembles to serial order. *Progress in Neurobiology*, 67:85–111, 2002.

30. F. Pulvermüller. *The Neuroscience of Language: On Bain Circuits of Words and*. Cambridge Press, Cambridge, UK, 2003.

31. F. Pulvermüller, R. Assadollahi, and T. Elbert. Neuromagnetic evidence for early semantic access in word recognition. *European Journal of Neuroscience*, 13:201–205, 2001.

32. F. Pulvermüller, M. Häre, and F. Hummel. Neurophysiological distinction of verb categories. *Cognitive Neuroscience*, 11(12):2789–2793, 2000a.

33. F. Pulvermüller, B. Mohr, and H. Schleichert. Semantic or lexico-syntactic factors: What determines word class specific activity in the human brain? *Neuroscience Letters*, 275(81-84):2789–2793, 1999.

34. G. Rizzolatti and M. Arbib. Language within our grasp. *Trends in Neuroscience*, 21(5):188–194, 1998.

35. G. Rizzolatti, L. Fadiga, V. Gallese, and L. Fogassi. The mirror system, imitation, and the evolution of language. *Cognitive Brain Research*, 3:131–141, 1996.

36. G. Rizzolatti, L. Fogassi, and V. Gallese. Neurophysiological mechanisms underlying the understanding and imitation of action. *Nature Review*, 2:661–670, 2001.

37. G. Rizzolatti, L. Fogassi, and V. Gallese. Motor and cognitive functions of the ventral premotor cortex. *Current Opinion in Neurobiology*, 12:149–154, 2002.

38. U. Scarfogliero, M. Folgheraiter, and G. Gini. Larp, biped robotics conceived as human modelling. In S. Wermter, G. Palm, and M. Elshaw, editors, *Biomimetic Neural Learning for Intelligent Robots*. Springer-Verlag, Heidelberg, Germany, 2005.

39. D. Sheynikhovich, R. Chavarriaga, T. Strösslin, and W. Gerstner. Spatial representation and navigation in a bio-inspired robot. In S. Wermter, G. Palm, and M. Elshaw, editors, *Biomimetic Neural Learning for Intelligent Robots*. Springer-Verlag, Heidelberg, Germany, 2005.

40. O. Shtyrov, Y. Hauk and F. Pulvermüller. Distributed neuronal networks for encoding category-specific semantic information: the mismatch negativity to action words. *European Journal of Neuroscience*, 19:1–10, 2004.

41. M. Spitzer. *The Mind Within the Net: Models of Learning, Thinking and Acting.* MIT Press, Cambridge, MA, 1999.

42. A. Sung, A. Merke, and M. Riedmiller. Reinforcement learning using a grid based function approximator. In S. Wermter, G. Palm, and M. Elshaw, editors, *Biomimetic Neural Learning for Intelligent Robots.* Springer-Verlag, Heidelberg, Germany, 2005.

43. A. Treves and E. Rolls. Computational analysis of the role of the hippocampus in memory. *Hippocampus*, 4(3):374–391, 1994.

44. M. Umilta, E. Kohler, V. Gallese, L. Fogassi, L. Fadiga, and G. Keysers, C.and Rizzolatti. I know what you are doing: A neurophysical study. *Neuron*, 31:155–165, 2001.

45. A. Vardy and F. Oppacher. A scale invariant local image descriptor for visual homing. In S. Wermter, G. Palm, and M. Elshaw, editors, *Biomimetic Neural Learning for Intelligent Robots.* Springer-Verlag, Heidelberg, Germany, 2005.

46. J. Vitay, N. Rougier, and F. Alexandre. A distributed model of spatial visual attention. In S. Wermter, G. Palm, and M. Elshaw, editors, *Biomimetic Neural Learning for Intelligent Robots.* Springer-Verlag, Heidelberg, Germany, 2005.

47. S. Wermter, J. Austin, D. Willshaw, and M. Elshaw. Towards novel neuroscience-inspired computing. In S. Wermter, J. Austin, and D. Willshaw, editors, *Emergent Neural Computational Architectures based on Neuroscience*, pages 1–19. Springer-Verlag, Heidelberg, Germany, 2001.

48. S. Wermter, C. Weber, V. Gallese, and F. Pulvermüller. Neural grounding robot language in action. In S. Wermter, G. Palm, and M. Elshaw, editors, *Biomimetic Neural Learning for Intelligent Robots.* Springer-Verlag, Heidelberg, Germany, 2005.

The Intentional Attunement Hypothesis
The Mirror Neuron System and Its Role in
Interpersonal Relations

Vittorio Gallese

Dept. of Neuroscience, University of Parma, via Volturno, 39,
I-43100 Parma, ITALY
`vittorio.gallese@unipr.it`
`http://www.unipr.it/arpa/mirror/english/staff/gallese.htm`

Abstract. Neuroscientific research has unveiled neural mechanisms mediating between the personal experiential knowledge we hold of our lived body, and the implicit certainties we simultaneously hold about others. Such personal, body-related experiential knowledge enables our intentional attunement with others, which in turn constitutes a shared manifold of intersubjectivity. This we-centric space allows us to personally characterize and provide experiential understanding to the actions performed by others, and the emotions and sensations they experience. A direct form of "experiential understanding" is achieved by modeling the behavior of other individuals as intentional experience on the basis of the equivalence between what the others do and feel and what we do and feel. This parsimonious modeling mechanism is embodied simulation. The mirror neuron system is likely a neural correlate of this mechanism. This account shades some light on too often sidelined aspects of social cognition. More generally, it emphasizes the role played in cognition by neural sensory-motor integration.

1 Introduction

The dominant view in cognitive science is to put most efforts in clarifying what are the formal rules structuring a solipsistic, representational mind. Much less investigated is what triggers the sense of social identity that we experience with the multiplicity of "other selves" populating our social world. Is the solipsistic type of analysis inspired by folk-psychology. The exclusive explanatory approach to social cognition? In particular, is it doing full justice to the phenomenal aspects of our social intentional relations? My answer is no to both questions.

At difference with Mr. Spock, the famous alien character of the Star Trek saga, our social mental skills are not confined to a declarative, conceptualized, and objective perspective. Normally, we are not alienated from the actions, emotions and sensations of others, because we are *attuned to the intentional relations of others*. By means of intentional attunement, "the others" are much more than being different representational systems; they become *persons*, like us.

S. Wermter et al. (Eds.): Biomimetic Neural Learning, LNAI 3575, pp. 19–30, 2005.
© Springer-Verlag Berlin Heidelberg 2005

In the present paper I will show that the same neural circuits involved in action control and in the first person experience of emotions and sensations are also active when witnessing the same actions, emotions and sensations of others, respectively. I will posit that the mirror neuron systems, together with other mirroring neural clusters outside the motor domain, constitute the neural underpinnings of embodied simulation, the functional mechanism at the basis of intentional attunement.

This paper is exclusively focused on the relationships between the mirror neuron system, embodied simulation and the experiential aspects of social cognition. A longer and more elaborate version will appear soon in Phenomenology and the Cognitive Sciences [15]. For sake of concision, many other issues related to mirror neurons and simulation will not be addressed here. The vast literature on the mirror neuron system in humans and its relevance for theory of mind, imitation and the evolution of language is reviewed and discussed in several papers [17, 43, 45, 12, 42, 44, 20, 4]. For an analysis of the role played by embodied simulation in conceptual structure and content, see [18].

2   The Mirror Neuron System for Actions in Monkeys and Humans: Empirical Evidence

About ten years ago we discovered in the macaque monkey brain a class of premotor neurons that discharge not only when the monkey executes goal-related hand actions like grasping objects, but also when observing other individuals (monkeys or humans) executing similar actions. We called them "mirror neurons" [21, 46]. Neurons with similar properties were later discovered in a sector of the posterior parietal cortex reciprocally connected with area F5 (PF mirror neurons, see [45, 22]).

The observation of an object-related hand action leads to the activation of the same neural network active during its actual execution. Action observation causes in the observer the automatic activation of the same neural mechanism triggered by action execution. We proposed that this mechanism could be at the basis of a direct form of action understanding [21, 46] see also [10, 11, 12, 13, 14, 20, 45, 44].

Further studies carried out in our lab corroborated and extended our original hypothesis. We showed that F5 mirror neurons are also activated when the final critical part of the observed action, that is, the hand-object interaction, is hidden [49]. In a second study we showed that a particular class of F5 mirror neurons, "audio-visual mirror neurons" can be driven not only by action execution and observation, but also by the sound produced by the same action [41]. "Audio-visual mirror neurons" respond to the sound of actions and discriminate between the sounds of different actions, but do not respond to other similarly interesting sounds such as arousing noises, or monkeys' and other animals' vocalizations. It doesn't significantly differ at all for the activity of this neural network if a peanut being broken, is specified at the motor, visual or auditory level. Such neural mechanism enables to represent the consequences of an action, thus its goal, in a way that is in principle also open to misrepresentation (e.g. neurons responding to a sound different from that produced by the action coded by them when executed or observed). Furthermore, the same conceptual content ("the goal of action A") results from a multiplicity of states subsuming it, namely, differently triggered patterns of activations within a population of "audio-visual mirror neurons". These neurons instantiate sameness of informational content at a quite "abstract" level. If the different mode of presentation of events as intrinsically different as sounds, images, or voluntary body actions, is nevertheless bound together within the same neural substrate, what we have is a mechanism instantiating a form of conceptualization. This perspective can be extended to other parts of the sensory-motor system (see [18]).

Furthermore, this perspective seems to suggest that thought may not be entirely separate from what animals can do, because it directly uses sensory-motor bodily mechanisms — the same ones used by non-human primates to function in their environments. According to this hypothesis, thought is at least partly an exaptation of the neural activities underpinning the operations of our bodies.

We recently explored the most lateral part of area F5 where we described a population of mirror neurons related to the execution/observation of mouth actions [8]. The majority of these neurons discharge when the monkey executes and observes transitive, object-related ingestive actions, such as grasping, biting, or licking. However, a small percentage of mouth-related mirror neurons discharge during the observation of intransitive, communicative facial actions performed by the experimenter in front of the monkey, such as lip protrusion and lips-smacking [8].

Several studies using different experimental methodologies and techniques have demonstrated also in the human brain the existence of a mirror neuron system matching action perception and execution. During action observation there is a strong activation of premotor and parietal areas, the likely human homologue of the monkey areas in which mirror neurons were originally described (for a review, see [45, 12, 44, 14]. Furthermore, the mirror neuron matching system for actions in humans is somatotopically organized, with distinct cortical regions within the premotor and posterior parietal cortices being activated by the observation/execution of mouth, hand, and foot related actions [5].

A recent brain imaging study, in which human participants observed communicative mouth actions performed by humans, monkeys and dogs showed that the observation of communicative mouth actions led to the activation of different cortical foci according to the different observed species. The observation of human silent speech activated the pars opercularis of the left inferior frontal gyrus, a sector of Broca's region. The observation of monkey lip-smacking activated a smaller part of the same region bilaterally. Finally, the observation of the barking dog activated only exstrastriate visual areas. Actions belonging to the motor repertoire of the observer (e.g., biting and speech reading) or very closely related to it (e.g. monkey's lip-smacking) are mapped on the observer's motor system. Actions that do not belong to this repertoire (e.g., barking) are mapped and henceforth categorized on the basis of their visual properties [6].

The involvement of the motor system during observation of communicative mouth actions is also testified by the results of a TMS study by Watkins et al. (2003) [50], in which they showed that the observation of communicative, speech-related mouth actions, facilitate the excitability of the motor system involved in the production of the same actions.

3 Action Observation as Action Simulation

The mirror neuron system for action is activated both by transitive, object-related and intransitive, communicative actions, regardless of the effectors performing them. When a given action is planned, its expected motor consequences are forecast. This means that when we are going to execute a given action we can also predict its consequences. The action model enables this prediction. Given the shared sub-personal neural mapping between what is acted and what is perceived – constituted by mirror neurons – the action model can also be used to predict the consequences of actions performed by others. Both predictions (of our actions and of others' actions) are instantiations of embodied simulation, that is, modeling processes.

The same functional logic that presides over self-modeling is employed also to model the behavior of others: to perceive an action is equivalent to internally simulating it. This enables the observer to use her/his own resources to experientially penetrate the world of the other by means of a direct, automatic, and unconscious process of simulation.

Embodied simulation automatically establishes a direct experiential link between agent and observer, in that both are mapped in a neutral fashion. The stimuli whose observation activates mirror neurons, like a grasping hand, its predicted outcome, and the sound it produces, all consist of the specific interaction between an agent and a target. It is the agentive relational specification to trigger the mirror neurons' response. The mere observation of an object not acted upon indeed does not evoke any response. Furthermore, the effector-target interaction must be successful. Mirror neurons respond if and only if an agentive relation is practically instantiated by an acting agent, regardless of its being the observer or the observed. The agent parameter must be filled. Which kind of agent is underspecified, but not *unspecified*. Indeed, not all kinds of agents will do. The abovementioned brain imaging experiment on communicative actions shows that only stimuli consistent with or closely related to the observer's behavioral repertoire are effective in activating the mirror neuron system for actions [6].

To summarize, action observation constitutes a form of embodied simulation of action. This, however, is different from the simulation processes occurring during motor imagery. The main difference is what triggers the simulation process: an internal event – a deliberate act of will –in the case of motor imagery, and an external event, in the case of action observation. This difference leads to slightly different and non-overlapping patterns of brain activation (see [12, 13]). However, both conditions share a common mechanism: the simulation of actions by means of the activation of parietal and premotor cortical networks. I submit that this simulation process also constitutes a basic level of experiential understanding, a level that does not entail the explicit use of any theory or declarative representation.

4 The Mirror Neuron System and the Understanding of Intentions

According to my hypothesis, "intentional attunement" is a basic requisite for social identity. In that respect, I think that monkeys may exploit the mirror neuron system to optimize their social interactions. At least, the evidence we have collected so far seems to suggest that the mirror neuron system for actions is enough sophisticated to enable its exploitation for social purposes. Recent results by Cisek and Kalaska (2004) [7] show that neurons in the dorsal premotor cortex of the macaque monkey can covertly simulate observed behaviors of others, like a cursor moved to a target on a computer screen, even when the relation between the observed sensory event and the unseen motor behavior producing it is learned through stimulus-response associations. My hypothesis is that monkeys might entertain a rudimentary form of "teleological stance", a likely precursor of a full-blown intentional stance. This hypothesis extends to the phylogenetic domain the ontogenetic scenario proposed by Gergely and Csibra (2003) [23] for human infants. But monkeys certainly do not entertain full-blown mentalization. Thus, what makes humans different? First of all, from a behavioral point of view human infants for years heavily rely on interactions with their caregivers and with other individuals to learn how to cope with the world. This is an important difference between humans and other species that may play a major role in bootstrapping more sophisticated cognitive social skills.

At present we can only make hypotheses about the relevant neural mechanisms underpinning the mentalizing abilities of humans that are still poorly understood from a functional point of view. In particular, we do not have a clear neuroscientific model of how humans can understand the intentions promoting the actions of others they observe. A given action can be originated by very different intentions. Suppose one sees someone else grasping a cup. Mirror neurons for grasping will most likely be activated in the observer's brain. A simple motor equivalence between the observed action and its motor representation in the observer's brain, however, can only tell us *what* the action is (it's a grasp) and not *why* the action occurred. Determining why action A (grasping the cup) was executed, that is, determining its intention, can be equivalent to detecting the goal of the still not executed and impending subsequent action (say, drink from the cup). In a recent fMRI study [38] we showed that premotor mirror neurons-related areas not only code the "what" of an action but also its "why", that is, the intention promoting it. Detecting the intention of Action A is equivalent to predict its distal goal, that is, the goal of the subsequent Action B.

These results are consistent with those of another recently published fMRI study. Schubotz and von Cramon (2004) [47] contrasted the observation of biological hand actions with that of abstract motion (movements of geometric shapes). In both conditions 50% of the stimuli failed to attain the normally predictable end-state. The task of participants was to indicate whether the action was performed in a goal-directed manner (button "yes") or not (button "no"). After abstract motion observation, participants had to indicate whether the object sequence was regular until the end of presentation (button yes) or not (button no). Results showed that both conditions elicited significant activation within the ventral premotor cortex. In addition, the prediction of biological actions also activated BA 44/45, which is part of the mirror neuron system. Schubotz and von Cramon (2004) [47] concluded that their findings point to a basic premotor contribution to the representation or processing of sequentially structured events, supplemented by different sets of areas in the context of either biological or non biological cues.

The statistical frequency of action sequences (the detection of what most frequently follows what) as they are habitually performed or observed in the social environment, can therefore constrain preferential paths of inferences/predictions. It can be hypothesized that this can be accomplished by chaining different populations of mirror neurons coding not only the observed motor act, but also those that in a given context would normally follow. Ascribing intentions would therefore consist in predicting a forthcoming new goal. If this is true, it follows that one important difference between humans and monkeys could be the level of recursivity attained by the mirror neuron system in our species. A similar proposal has been recently put forward in relation to the faculty of language (see [36, 9]). According to this perspective, action prediction and the ascription of intentions are related phenomena, underpinned by the same functional mechanism. In contrast with what mainstream cognitive science would maintain, action prediction and the ascription of intentions therefore do not belong to different cognitive realms, but both pertain to embodied simulations underpinned by the activation of chains of logically-related mirror neurons.

Many scholars are exclusively focusing on clarifying differences between humans and other primates with respect to the use of propositional attitudes. According to this mainstream view, humans have Theory of Mind, non-human primates don't. This in my opinion squares to a "Ptolemaic Paradigm", with a very strong anthropocentric aftertaste. This paradigm so pervasive in contemporary cognitive science is too quick in establishing a direct and nomological link between our use of propositional attitudes and their supposed neural correlates. No one can deny that we use propositional attitudes, unless embracing a radical eliminativism (which is not my case). But it is perfectly possible that we will never find boxes in our brain containing the neural correlates of beliefs, desires and intentions as such. Such a search is the real reductionism.

As pointed out by Allen and Bekoff (1997) [1], this "all-or-nothing" approach to social cognition, this desperate search for a "mental Rubicon" (the wider the better) is strongly arguable. When trying to account for our cognitive abilities we forget that they are the result of a long evolutionary process. It is reasonable to hypothesize that this evolutionary process proceeded along a line of continuity (see [17, 22]).

It is perhaps more fruitful to establish to which extent different cognitive strategies may be underpinned by similar functional mechanisms, which in the course of evolution acquire increasing complexity. The empirical data briefly reviewed in this chapter are an instantiation of this strategy of investigation. The data on mirror neurons in monkeys and mirroring circuits in the human brain seem to suggest that the ease with which we are capable to "mirror" in the behavior of others, and recognize them as similar to us, - in other words, our "Intentional Attunement" with others - may rely on a series of matching mechanisms that we have just started to uncover.

5 Mirroring Emotions and Sensations

Emotions constitute one of the earliest ways available to the individual to acquire knowledge about its situation, thus enabling to reorganize this knowledge on the basis of the outcome of the relations entertained with others. The coordinated activity of sensory-motor and affective neural systems results in the simplification and automatization of the behavioral responses that living organisms are supposed to produce in order to survive. The integrity of the sensory-motor system indeed appears to be critical for the recognition of emotions displayed by others (see [2, 3]), because the sensory-motor system appears to support the reconstruction of what it would feel like to be in a particular emotion, by means of simulation of the related body state.

We recently published an fMRI study showing that experiencing disgust and witnessing the same emotion expressed by the facial mimicry of someone else, both activate the same neural structure – the anterior insula – at the same overlapping location [51]. This suggests, at least for the emotion of disgust, that the first- and third-person experiences of a given emotion are underpinned by the activity of a shared neural substrate. When I see the facial expression of someone else, and this perception leads me to experience *that* expression as a particular affective state, I do not accomplish this type of understanding through an argument by analogy. The other's emotion is constituted, experienced and therefore directly understood by means of an embodied simulation producing a shared body state. It is the activation of a neural mechanism shared by the observer and the observed to enable direct experiential understanding. A similar simulation-based mechanism has been proposed by Goldman and Sripada (2004) [30] as "unmediated resonance".

Let us focus now on somatic sensations as the target of our social perception. As repeatedly emphasized by phenomenology, touch has a privileged status in making possible the social attribution of lived personhood to others. "Let's be in touch" is a common clause in everyday language, which metaphorically describes the wish of being related, being in contact with someone else. Such examples show how the tactile dimension be intimately related to the interpersonal dimension.

New empirical evidence suggests that the first-person experience of being touched on one's body activates the same neural networks activated by observing the body of someone else being touched [40]. Within SII-PV, a multimodal cortical region, there is a localized neural network similarly activated by the self-experienced sensation of being touched, and the perception of an external tactile relation. This double pattern of activation of the same brain region suggests that our capacity to experience and directly understand the tactile experience of others could be mediated by embodied simulation, that is, by the externally triggered activation of *some* of the same neural networks presiding over our own tactile sensations. A similar mechanism likely underpins our experience of the painful sensations of others (see [35, 48]).

6 The Many Sides of Simulation

The notion of simulation is employed in many different domains, often with different, not necessarily overlapping meanings. Simulation is a functional process that possesses a certain representational content, typically focusing on possible states of its target object. For example, in motor control theory simulation is characterized as the mechanism employed by forward models to predict the sensory consequences of impending actions. According to this view, the predicted consequences are the simulated ones.

In philosophy of mind, on the other hand, the notion of simulation has been used by the proponents of Simulation Theory of mind reading to characterize the pretend state adopted by the attributer in order to understand others' behavior (see [31, 32, 33, 34, 24, 25, 26, 27, 28, 17, 29]).

I employ the term "embodied simulation" as an automatic[1], nonconscious, and pre-reflexive functional mechanism, whose function is the modeling of objects, agents, and events. Simulation, as conceived of in the present paper, is therefore not necessarily the result of a willed and conscious cognitive effort, aimed at interpreting the intentions hidden in the overt behavior of others, but rather a basic functional mechanism of our brain. However, because it also generates representational content, this functional mechanism seems to play a major role in our epistemic approach to the world. It represents the outcome of possible actions, emotions, or sensations one could take or experience, and serves to attribute this outcome to another organism as a real goal-state it is trying to bring about, or as a real emotion or sensation it is experiencing.

Successful perception requires the capacity of predicting upcoming sensory events. Similarly, successful action requires the capacity of predicting the expected consequences of action. As suggested by an impressive and coherent amount of neuroscientific data (for a review, see [12, 18]), both types of predictions seem to depend on the results of unconscious and automatically driven neural states, functionally describable as simulation processes.

To which extent embodied simulation is a motor phenomenon? According to the use I make of this notion, embodied simulation *is not conceived of as being exclusively confined to the domain of motor control*, but rather as a more general and basic endowment of our brain. It applies not only to actions or emotions, where the motor or viscero-motor components may predominate, but also to sensations like vision and touch. It is mental because it has content. It is embodied not only because it is neurally realized, but also because it uses a pre-existing body-model in the brain realized by the sensory-motor system, and therefore involves a non-propositional form of self-representation.

7 Conclusions

We have discovered some of the neural mechanisms mediating between the multi level experiential knowledge we hold of our lived body, and the *implicit certainties* we simultaneously hold about others. Such body-related experiential knowledge enables us to directly understand some of the actions performed by others, and to decode the emotions and sensations they experience. Our seemingly effortless capacity to conceive of the acting bodies inhabiting our social world as *goal-oriented persons* like us depends on the constitution of a "we-centric" shared meaningful interpersonal space. I propose that this shared manifold space (see [11, 12, 13, 14])

---

[1] It is "automatic" in the sense that it is obligatory.

can be characterized at the functional level as embodied simulation, a specific mechanism, likely constituting a basic functional feature by means of which our brain/body system models its interactions with the world.

The mirror neuron matching systems and the other non-motor mirroring neural clusters represent *one particular* sub-personal instantiation of embodied simulation. With this mechanism we do not just "see" an action, an emotion, or a sensation. Side by side with the sensory description of the observed social stimuli, internal representations of the body states associated with these actions, emotions, and sensations are evoked in the observer, 'as if' he/she would be doing a similar action or experiencing a similar emotion or sensation. This proposal also opens new interesting perspectives for the study of the neural underpinnings of psychopathological states and psychotherapeutic relations (see [19]), and of aesthetic experiences.

In contrast with what argued by Jacob and Jeannerod (2004, forthcoming) [39], social cognition is not *only* explicitly reasoning about the contents of someone else's mind. Our brains, and those of other primates, appear to have developed a basic functional mechanism, embodied simulation, which gives us an experiential insight of other minds. The shareability of the phenomenal content of the intentional relations of others, by means of the shared neural underpinnings, produces intentional attunement. Intentional attunement, in turn, by collapsing the others' intentions into the observer's ones, produces the peculiar quality of familiarity we entertain with other individuals. This is what "being empathic" is about. By means of a shared neural state realized in two different bodies that nevertheless obey to the same functional rules, the "objectual other" becomes "another self". Furthermore, the mirror neuron system for actions in humans appear to be suitable for the detection of the intentions promoting the behavior of others. Thus, as previously hypothesized ([17]), the mirror neuron system could be at the basis of basic forms of mind reading.

This of course doesn't account for all of our social cognitive skills. Our most sophisticated mind reading abilities likely require the activation of large regions of our brain, certainly larger than a putative domain-specific Theory of Mind Module. As correctly pointed out by Jacob and Jeannerod (2004, forthcoming) [39], the same actions performed by others in different contexts can lead the observer to radically different interpretations. Thus, social stimuli are also understood on the basis of the explicit cognitive elaboration of their contextual aspects and of previous information.

The point is that these two mechanisms are not mutually exclusive. Embodied simulation is experience-based, while the second mechanism is a cognitive description of an external state of affairs. Embodied simulation scaffolds the propositional, more cognitively sophisticated mind reading abilities. When the former mechanism is not present or malfunctioning, as perhaps in autism (see [16, 19]), the latter one can provide only a pale, detached account of the social experiences of others. It is an empirical issue to determine how much of social cognition, language included, can be explained by embodied simulation and its neural underpinnings.

References

1. Allen, C., and Bekoff, M.: Species of Mind. Cambridge, MIT Press (1997)
2. Adolphs. R.: Cognitive neuroscience of human social behaviour. Nat Rev Neurosci, 4(3) (2003) 165-178
3. Adolphs, R., Damasio, H., Tranel, D., Cooper, G., Damasio, A.R.: A role for somatosensory cortices in the visual recognition of emotion as revealed by three-dimensional lesion mapping. J. Neurosci, 20 (2000) 2683-2690
4. Arbib, M.: The Mirror System Hypothesis. Linking Language to Theory of Mind. http://www.interdisciplines.org/coevolution/papers/11 (2004)
5. Buccino, G., Binkofski, F., Fink, G.R., Fadiga, L., Fogassi, L., Gallese, V., Seitz, R.J., Zilles, K., Rizzolatti, G., Freund, H.-J.: Action observation activates premotor and parietal areas in a somatotopic manner: an fMRI study. European Journal of Neuroscience, 13 (2001) 400-404
6. Buccino, G., Lui, F., Canessa, N., Patteri, I., Lagravinese, G., Benuzzi, F., Porro, C.A., Rizzolatti, G.: Neural circuits involved in the recognition of actions performed by nonconspecifics: An fMRI study. J Cogn. Neurosci. 16 (2004) 114-126
7. Cisek, P., Kalaska, J.: Neural correlates of mental rehearsal in dorsal premotor cortex. Nature 431 (2004) 993-996
8. Ferrari P.F., Gallese V., Rizzolatti G., Fogassi L.: Mirror neurons responding to the observation of ingestive and communicative mouth actions in the monkey ventral premotor cortex. European Journal of Neuroscience 17 (2003) 1703-1714
9. Fitch, WT and Hauser, MD: (2004) Computational constraints on syntactic processing in a non human primate. Science 303: 377-380.
10. Gallese, V.: The acting subject: towards the neural basis of social cognition. In Metzinger, T. (Ed.), Neural Correlates of Consciousness. Empirical and Conceptual Questions. Cambridge, MA. MIT Press (2000) 325-333
11. Gallese, V.: The "Shared Manifold" Hypothesis: from mirror neurons to empathy. Journal of Consciousness Studies: 8(5-7) (2001) 33-50
12. Gallese, V.: The manifold nature of interpersonal relations: The quest for a common mechanism. Phil. Trans. Royal Soc. London, 358 (2003a) 517-528
13. Gallese, V.: The roots of empathy: The shared manifold hypothesis and the neural basis of intersubjectivity. Psychopatology, 36(4) (2003b) 171-180
14. Gallese V.: "Being like me": Self-other identity, mirror neurons and empathy. In: Perspectives on Imitation: From Cognitive Neuroscience to Social Science. S. Hurley and N. Chater (Eds). Boston, MA: MIT Press (2004) in press
15. Gallese, V.: Embodied simulation: From neurons to phenomenal experience. Phenomenology and the Cognitive Sciences (2005a) in press.
16. Gallese V.: La Molteplicità Condivisa: Dai Neuroni Mirror all'Intersoggettività. (2005b) In preparation.
17. Gallese, V., Goldman, A.: Mirror neurons and the simulation theory of mind-reading. Trends in Cognitive Sciences 12 (1998) 493-501.
18. Gallese, V. Lakoff, G.: The brain's concepts: The Role of the Sensory-Motor System in Reason and Language. Cognitive Neuropsychology (2005) in press.
19. Gallese, V., Migone, P.: Intentional attunement: Mirror neurons and the neural underpinnings of interpersonal relations (2005) In preparation
20. Gallese, V., Keysers, C., Rizzolatti, G.: A unifying view of the basis of social cognition. Trends in Cognitive Sciences 8 (2004) 396-403

21. Gallese, V., Fadiga, L., Fogassi, L., Rizzolatti, G.: Action recognition in the premotor cortex. Brain 119 (1996) 593-609
22. Gallese, V., Fadiga, Fogassi, L., Rizzolatti, G.: Action representation and the inferior parietal lobule. In Prinz, W., and Hommel, B. (Eds.) Common Mechanisms in Perception and Action: Attention and Performance, Vol. XIX. Oxford: Oxford University Press (2002) 247-266
23. Gergely, G. and Csibra, G.: Teleological reasoning in infancy: the naive theory of rational action. TICS 7 (2003) 287-292.
24. Goldman, A.: Interpretation Psychologized Mind and Language 4, (1989) 161-185
25. Goldman, A.: The psychology of folk psychology. Behavioral Brain Sciences: 16 (1993a) 15-28
26. Goldman, A.: Philosophical Applications of Cognitive Science. Boulder, Colo., Westview Press (1993b)
27. Goldman, A.: The Mentalizing Folk. In Metarepresentation (Sperber, D., Ed.), London, Oxford University Press (2000).
28. Goldman, A.: Imitation, Mindreading, and Simulation. In S. Hurley and N. Chater (Eds.), *Perspectives* on Imitation: From Cognitive Neuroscience to Social Science, Cambridge, MA: MIT Press (2004) in press
29. Goldman, A., Gallese, V. Reply to Schulkin. Trends in Cognitive Sciences 4 (2000) 255-256
30. Goldman, A., Sripada, C.S.: Simulationist Models of Face-based Emotion Recognition. Cognition (2004) in press
31. Gordon, R.: Folk psychology as simulation. Mind and Language 1 (1986) 158-171.
32. Gordon, R.: Simulation without introspection or inference from me to you. In Mental Simulation, Davies, M. and Stone, T., (Eds.) Blackwell (1995) 53-67
33. Gordon, R.: 'Radical' Simulationism. In P. Carruthers and P. Smith (Eds.), Theories of Theories of Mind, Cambridge, UK: Cambridge University Press (1996) 11-21
34. Gordon, R.: Intentional Agents Like Myself. In S. Hurley and N. Chater (Eds.), Perspectives on Imitation: From Cognitive Neuroscience to Social Scienc*e*, Cambridge, MA, MIT Press (2004) in press
35. Hutchison, W.D., Davis, K.D., Lozano, A.M., Tasker, R.R., Dostrovsky, J.O.: Pain related neurons in the human cingulate cortex. Nature Neuroscience, 2 (1999) 403-405.
36. Hauser, MD, Chomsky, N., Fitch, WT.: The faculty of language: What is it, Who has it, and How did it evolve? Science 298 (2002) 1569-1579
37. Hauser, MD and Fitch, WT.: Computational constraints on sysntactic processing in a non human primate. Science 303 (2004) 377-380
38. Iacoboni, M., Molnar-Szakacs, I., Gallese, V., Buccino, G., Mazziotta, J., Rizzolatti, G.: Grasping intentions with mirror neurons. Soc. for Neurosci Abs., (2004), 254 11
39. Jacob, P., and Jeannerod, M. (2004) The Motor theory of social cognition. TICS, in press.
40. Keysers, C., Wickers, B., Gazzola, V., Anton, J-L., Fogassi, L., Gallese, V.: A Touching Sight: SII/PV Activation during the Observation and Experience of Touch. Neuron 42, April 22 (2004) 1-20
41. Kohler, E., Keysers, C., Umiltà, M.A., Fogassi, L., Gallese, V., Rizzolatti, G.: Hearing sounds, understanding actions: Action representation in mirror neurons. Science 297 (2002) 846-848
42. Metzinger, T., Gallese, V.: The emergence of a shared action ontology: Building blocks for a theory. Consciousness and Cognition, 12 (2003) 549-571
43. Rizzolatti, G., Arbib, M.: Language within our grasp. Trends Neurosci. 21 (1998) 188-192

44. Rizzolatti, G., Craighero, L.: The mirror neuron system. Ann. Rev. Neurosci. 27 (2004) 169-192
45. Rizzolatti, G., Fogassi, L., Gallese, V.: Neurophysiological mechanisms underlying the understanding and imitation of action. Nature Neuroscience Reviews 2 (2001) 661-670
46. Rizzolatti, G., Fadiga, L., Gallese, V., Fogassi, L.: Premotor cortex and the recognition of motor actions. Cog. Brain Res. **3** (1996) 131-141
47. Schubotz, RI, von Cramon, DY: Sequences of Abstract Nonbiological Stimuli Share Ventral Premotor Cortex with Action Observation and Imagery. J Neurosci. 24 (2004) 5467-5474
48. Singer, T., Seymour, B., O'Doherty, J., Kaube, H., Dolan, R.J., Frith, C.F.: Empathy for pain involves the affective but not the sensory components of pain. Science 303 (2004) 1157-1162
49. Umiltà, M.A., Kohler, E., Gallese, V., Fogassi, L., Fadiga, L., Keysers, C., Rizzolatti, G."I know what you are doing": A neurophysiologycal study. Neuron 32 (2001) 91-101
50. Watkins KE, Strafella AP, Paus T.: Seeing and hearing speech excites the motor system involved in speech production. Neuropsychologia 41(8) (2003) 989-94
51. Wicker, B., Keysers, C., Plailly, J., Royet, J-P., Gallese, V., Rizzolatti, G.: Both of us disgusted in my insula: The common neural basis of seeing and feeling disgust. Neuron, 40 (2003) 655-664

Sequence Detector Networks and Associative Learning of Grammatical Categories

Andreas Knoblauch[1,2] and Friedemann Pulvermüller[1]

[1] MRC Cognition and Brain Sciences Unit,
15 Chaucer Road, Cambridge CB2 2EF, England
Tel: (+44)-1223-355294; Fax: (+44)-1223-359062
andreas.knoblauch@mrc-cbu.cam.ac.uk
friedemann.pulvermuller@mrc-cbu.cam.ac.uk
[2] Abteilung Neuroinformatik, Fakultät für Informatik,
Universität Ulm, Oberer Eselsberg, D-89069 Ulm, Germany

Abstract. A fundamental prerequisite for language is the ability to distinguish word sequences that are grammatically well-formed from ungrammatical word strings and to generalise rules of syntactic serial order to new strings of constituents. In this work, we extend a neural model of syntactic brain mechanisms that is based on syntactic sequence detectors (SDs). Elementary SDs are neural units that specifically respond to a sequence of constituent words AB, but not (or much less) to the reverse sequence BA. We discuss limitations of the original version of the SD model (Pulvermüller, Theory in Biosciences, 2003) and suggest optimal model variants taking advantage of optimised neuronal response functions, non-linear interaction between inputs, and leaky integration of neuronal input accumulating over time. A biologically more realistic model variant including a network of several SDs is used to demonstrate that associative Hebb-like synaptic plasticity leads to learning of word sequences, formation of neural representations of grammatical categories, and linking of sequence detectors into neuronal assemblies that may provide a biological basis of syntactic rule knowledge. We propose that these syntactic neuronal assemblies (SNAs) underlie generalisation of syntactic regularities from already encountered strings to new grammatical word sequences.

1 Introduction

A fundamental feature of all languages is syntax. The specifically human syntactic capability includes the abilities

- to distinguish learned word sequences that are grammatically well-formed from new ungrammatical word strings,
- to categorise words into lexical categories, such as noun, verb, adjective etc.,
- to generalise rules of syntactic serial order to new strings of constituents, which one has never encountered before but are in accordance with the rules of syntax.

S. Wermter et al. (Eds.): Biomimetic Neural Learning, LNAI 3575, pp. 31–53, 2005.
© Springer-Verlag Berlin Heidelberg 2005

It is undeniable that these abilities have a neural implementation in the human brain. However, researchers disagree to what degree the abilities to learn complex sequences, to form lexical categories, and to generalise syntactic rules or regularities are based on genetically determined and pre-wired circuits [1, 2] or emerge in a universal learning device as a consequence of associative learning [3, 4]. We here attempt to address this dispute by demonstrating that, given a pre-structured network shaped according to known neuroanatomical and neurophysiological properties of the cortex, sequence learning and lexical category formation are a necessary consequence of associative Hebb-like synaptic learning. We also suggest that rule generalisation may be based on learned neural links between the learned lexical category networks.

Before addressing these questions we analyze and extend a recently proposed cortical model for language related sequence detection [5, 6]. In contrast to alternative models of nonlinear sequence detection, e.g. based on delay lines [7, 8] or multiplicative low-pass filters [9, 10], our model is linear and can be interpreted in terms of cortical microcircuitry and neurophysiological observations of working-memory related sustained activity [11]. The basic idea of the model is that a sequence detector (SD) for the word sequence AB receives *weak* inputs from the representation of the *first* word A (via long-range axons ending on distal apical dendritic sites), and *strong* inputs from the representation of the *second* word B (via axon collaterals of nearby neurons ending on basal dendritic sites).

In section 2 we work out some limitations of the original model with respect to the limited range of word delays and the limited amplitude difference between its critical sequence AB and inverse sequence BA. Then we suggest optimized model variants which make predictions on the time course of word web activations in the brain. Finally we develop a biologically more realistic network model variant which consists of many individual SD units. In order to reconcile the two views mentioned above we demonstrate in section 3 how simple Hebbian learning in our model can lead to generalisation of sequences and the emergence of grammatical categories.

2   Sequence Detectors (SDs)

2.1   Simple Linear SD Model

A *sequence detector* (SD) for the sequence (A, B) is a cell, or a set of cells[1], which receives weak input of strength $w > 0$ from a cell set α representing A, and strong input of strength $s > w$ from another cell set β representing B [5, 6]. The two input sets are also referred to as *word webs* (WWs) and represent the occurrences of sequence set elements by output activities

$$y_\alpha(t) = \sum_i \exp(-(t - t_{A,i})/\tau) \cdot H(t - t_{A,i}) \qquad (1)$$

$$y_\beta(t) = \sum_i \exp(-(t - t_{B,i})/\tau) \cdot H(t - t_{B,i}), \qquad (2)$$

---

[1] We have the idea that a single SD is a cell group consisting of many neurons.

where $H(x)$ is the Heaviside function ($H(x) = 1$ for $x \geq 0$, $H(x) = 0$ otherwise), and $t_{A,i}$ is the i-th occurrence of string symbol A and similarly for $t_{B,i}$. Thus, the word webs α and β are modeled as linear first-order low-pass filters with time constant $\tau > 0$.

If word webs α and β are activated once at times $t_{A,1} = 0$ and $t_{B,1} = \Delta > 0$, respectively, (i.e., β follows α by a delay of Δ), then the *membrane potentials* $x_{\alpha,\beta}$ and $x_{\beta,\alpha}$ of the *critical sequence detector* (α, β) and the *inverse sequence detector* (β, α) can be written as the sum of their inputs,

$$x_{\alpha,\beta}(t) = w \cdot \exp(-t/\tau) \cdot H(t) + s \cdot \exp(-(t - \Delta)/\tau) \cdot H(t - \Delta) \tag{3}$$
$$x_{\beta,\alpha}(t) = s \cdot \exp(-t/\tau) \cdot H(t) + w \cdot \exp(-(t - \Delta)/\tau) \cdot H(t - \Delta). \tag{4}$$

For the *output* of the sequence detectors we write

$$y_{\alpha,\beta}(t) = f(x_{\alpha,\beta}(t)) \tag{5}$$
$$y_{\beta,\alpha}(t) = f(x_{\beta,\alpha}(t)), \tag{6}$$

where the activation function f is assumed to be non-linear and sigmoidal, for example $f(x) = H(x - \Theta)$ for a threshold Θ.

Although "on average", i.e., when integrating the inputs over time, there is no difference between critical and inverse SDs, the *peak* value of $x_{\alpha,\beta}(t)$ can exceed the peak of $x_{\beta,\alpha}(t)$ (see Fig. 1). For $w < s$, the *peak values* are

$$P_{\alpha,\beta} := \max_t x_{\alpha,\beta}(t) = s + w \cdot e^{-\Delta/\tau} \tag{7}$$
$$P_{\beta,\alpha} := \max_t x_{\beta,\alpha}(t) = \max(s, w + s \cdot e^{-\Delta/\tau}). \tag{8}$$

If we can find a threshold Θ with $0 < P_{\beta,\alpha} < \Theta < P_{\alpha,\beta}$ then for $f(x) = H(x - \Theta)$ only the *critical* sequence detector (α, β), but not the *inverse* detector (β, α) gets activated.

When do we obtain $P_{\beta,\alpha} < P_{\alpha,\beta}$, i.e., when can the SD distinguish between critical and inverse input? The condition is obviously fulfilled if $P_{\beta,\alpha} = s$, i.e., if the first input peak for the inverse SD (β, α) is larger than or equal the second input peak. If the second peak is larger then we have to require $s + w \cdot e^{-\Delta/\tau} > w + s \cdot e^{-\Delta/\tau}$, or equivalently, $(s - w) \cdot (1 - e^{-\Delta/\tau}) > 0$. This is obviously true for any $s > w$ and $\Delta > 0$.

Maximal Peak Difference. We want to maximize the peak difference

$$d_P := P_{\alpha,\beta} - P_{\beta,\alpha}. \tag{9}$$

First we determine the optimal delay Δ for fixed w and s. In order to resolve the max-expression in eq. 8, we have to find out when the first peak of the inverse SD (β, α) is larger than the second peak. For this we have to resolve $s > w + s \cdot e^{-\Delta/\tau}$ for Δ. With

$$\Delta_P := -\tau \cdot \ln(1 - w/s), \tag{10}$$

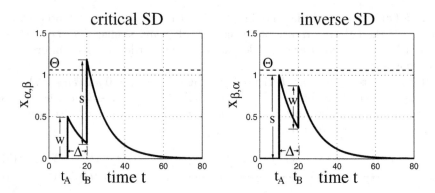

Fig. 1. Working principle of linear sequence detectors (SDs) according to [5]. The plots show the summed inputs of two SDs obtained from two input units α and β representing items A and B which occur at times t_A and $t_B > t_A$, respectively. **Left:** The critical SD (α, β) obtains first weak input w from α and then strong input s from β. The summed input $x_{\alpha,\beta}$ can exceed threshold Θ at time t_B. **Right:** The inverse SD (β, α) receives first strong input and then weak input such that the summed input $x_{\beta,\alpha}$ does not exceed Θ. Parameters are $\tau = 10$, $w = 0.5$, $s = 1$, $t_A = 10$, $t_B = 20$

we find that the first peak is larger for $\Delta > \Delta_P$, the second peak is larger for $\Delta < \Delta_P$, and the peak values are the same for $\Delta = \Delta_P$. For $\Delta \geq \Delta_P$, the peak difference is $d_P = w \cdot e^{-\Delta/\tau}$ which *decreases* monotonically in Δ. For $\Delta \leq \Delta_P$, the peak difference is $d_P = (s - w) \cdot (1 - e^{-\Delta/\tau})$, which *increases* monotonically in Δ. Together, d_P is maximal for $\Delta = \Delta_P$,

$$d_P \leq d_P|_{\Delta=\Delta_P} = \frac{w \cdot (s - w)}{s} \qquad (11)$$

Next, we determine optimal w for given s. For this it is sufficient to maximize $w \cdot (s - w) = -w^2 + sw$. The maximum occurs for $-2w + s = 0$ and the maximal peak difference is

$$d_P \leq d_P|_{\Delta=\Delta_P, w=s/2} = \frac{s}{4}. \qquad (12)$$

Thus, the peak height of a sequence detector can exceed the peak height of the inverse sequence detector by no more than 25 percent.

Maximal Delay Range. Now we maximize the range of valid delays $\Delta \in [\Delta_{\min}; \Delta_{\max}]$ for which a sequence detector can be above threshold, while its inverse counterpart remains well below threshold. For fixed threshold Θ and noise parameter η we require

$$P_{\alpha,\beta}(\Delta) > \Theta \qquad (13)$$

$$P_{\beta,\alpha}(\Delta) < \Theta - \eta \qquad (14)$$

for $0 \leq \eta < \Theta/5$ and all $\Delta \in [\Delta_{\min}; \Delta_{\max}]$. A positive η guarantees that the inverse SD remains below threshold even if primed by noise or previous activation.

Obviously, we have to require $0 < w < s < \Theta - \eta \le \Theta < w + s$. Thus, the first peak of the inverse detector is already below threshold, and it is sufficient to require

$$s + w \cdot e^{-\Delta/\tau} > \Theta \Leftrightarrow \Delta < \Delta_{\max} := \tau \cdot \ln \frac{w}{\Theta - s} \tag{15}$$

$$w + s \cdot e^{-\Delta/\tau} < \Theta - \eta \Leftrightarrow \Delta > \Delta_{\min} := \tau \cdot \ln \frac{s}{\Theta - \eta - w}. \tag{16}$$

Note that $\Delta_{\min} \to \Delta_P$ for $s \to \Theta - \eta$. We have to require that the interval length

$$d_\Delta := \Delta_{\max} - \Delta_{\min} = \tau \cdot \ln \frac{w \cdot (\Theta - \eta - w)}{s \cdot (\Theta - s)}. \tag{17}$$

is positive, $w \cdot (\Theta - \eta - w) > s \cdot (\Theta - s)$. For $s > w$ this is equivalent to

$$s > h(w) := \frac{\Theta + \sqrt{\Theta^2 - 4w \cdot (\Theta - \eta - w)}}{2}. \tag{18}$$

Note that $\Delta_{\min}$ may be quite large, e.g. larger than τ. Thus we may want to impose an upper bound on $\Delta_{\min}$. For $q > 0$ we require

$$\frac{\Delta_{\min}}{\tau} \le q \Leftrightarrow s \le e^q \cdot (\Theta - \eta - w). \tag{19}$$

Now we can maximize the interval length d_Δ with respect to the input strengths w and s. Since $\ln(x)$ is monotonically increasing, it is sufficient to maximize its argument in eq. 17, i.e., the function

$$f(w, s) := \frac{w \cdot (\Theta - \eta - w)}{s \cdot (\Theta - s)} \text{ for } (w, s) \in \tag{20}$$

$$R := \{(w, s) : 0 < w < h(w) < s < \min(\Theta - \eta, e^q \cdot (\Theta - \eta - w))\}. \tag{21}$$

The partial derivatives of f are

$$\frac{\partial f}{\partial w} = \frac{\Theta - \eta - 2w}{s \cdot (\Theta - s)} \tag{22}$$

$$\frac{\partial f}{\partial s} = \frac{w \cdot (\Theta - \eta - w) \cdot (2s - \Theta)}{s^2 \cdot (\Theta - s)^2} \tag{23}$$

From eq. 22 we get $\partial f / \partial w = 0$ only for $w = w_0 := (\Theta - \eta)/2$. Since $\partial f / \partial^2 w < 0$, for any fixed $s < \Theta$ we obtain maximal $f(w)$ at $w = w_0$. On the other hand, since $\partial f / \partial s > 0$ for $(w, s) \in R$ we have to choose s as large as possible. The situation is illustrated in Fig. 2.

As long as Q_1 lies on the right side of X, or equivalently,

$$q \ge q_2 := \ln 2 \approx 0.7, \tag{24}$$

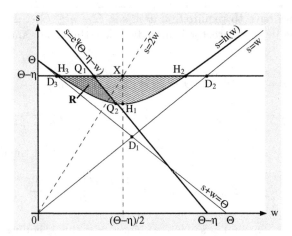

Fig. 2. The valid range R (see eq. 21) for weak and strong input strengths (w, s) corresponds to the double hatched region (Q_1, Q_2, H_3). Without the constraint eq. 19 on $\Delta_{\min}$ (i.e., $q \to \infty$) R expands to the single hatched region (H_1, H_2, H_3). If additionally $\eta \to 0$ (i.e., no noise) R expands to the triangle (D_1, D_2, D_3). Outside R the SD does not work properly. Optimal w, s are in point X or Q_1 (see text). **Coordinates** (w,s) are $\mathbf{X}((\Theta - \eta)/2, \Theta - \eta)$, $\mathbf{D_1}(\Theta/2, \Theta/2)$, $\mathbf{D_2}(\Theta - \eta, \Theta - \eta)$, $\mathbf{D_3}(\eta, \Theta - \eta)$, $\mathbf{H_1}((\Theta - \eta)/2, (\Theta + \sqrt{2\Theta\eta - \eta^2})/2)$, $\mathbf{H_{2/3}}((\Theta - \eta \pm \sqrt{\Theta^2 - 6\Theta\eta + 5\eta^2})/2, \Theta - \eta)$, $\mathbf{Q_1}((\Theta - \eta) \cdot (1 - e^{-q}), \Theta - \eta)$, $\mathbf{Q_2}((\Theta - \eta - e^{-q}\Theta)/(1 - e^{-2q}), (\Theta - e^{-q}(\Theta - \eta))/(1 - e^{-2q}))$

the optimal choice will be in X, i.e., $s = \Theta - \eta$ and $w = s/2$. For smaller q with Q_1 still being on the right side of $H_3(\eta, \Theta - \eta)$, or equivalently,

$$q_2 > q \geq q_1 := -\ln\left(\frac{1}{2} + \sqrt{\frac{1}{4} - \frac{\eta}{\Theta - \eta}}\right) \tag{25}$$

$$\geq -\ln(1 - \frac{\eta}{\Theta - \eta}) \geq \frac{\eta}{\Theta - \eta}, \quad ' = ' \text{ for } \eta \to 0. \tag{26}$$

the optimum must lie somewhere on $s = e^q \cdot (\Theta - \eta - w)$. Thus we have to maximize $f_q(w) := f(w, e^q \cdot (\Theta - \eta - w))$. We have

$$f_q(w) := \frac{e^{-q} \cdot w}{\Theta - e^q \cdot (\Theta - \eta - w)}, \tag{27}$$

$$f_q'(w) = \frac{e^{-q} \cdot (\Theta - e^q \cdot (\Theta - \eta))}{(\Theta - e^q \cdot (\Theta - \eta - w))^2}. \tag{28}$$

Note that $f_q'(w)$ cannot change its sign since the counter is independent of w and the denominator is $(\Theta - s)^2 > 0$. Also, the counter of f_q' cannot become zero for any considered q, since from $\Theta = e^q \cdot (\Theta - \eta)$ it follows $q = -\ln(1 - \eta/\Theta) \leq q_1$. Thus, $f_q'(w) \leq 0$ for all considered w and q. Thus, the optimum is in Q_1.

Maximal Peak Height. Finally we maximize the peak height $P_{\alpha,\beta}$ of the critical SD under the constraint $(w, s) \in R$ (see eqs. 7,21). Since $P_{\alpha,\beta}$ is maximal for minimal Δ it is sufficient to maximize

$$P(w, s) := P_{\alpha,\beta}|_{\Delta = \Delta_{\min}} = s + \frac{w \cdot (\Theta - \eta - w)}{s} \tag{29}$$

with respect to w and s for $(w, s) \in R$ (see Fig. 2). The partial derivatives are

$$\frac{\partial P}{\partial w} = \frac{\Theta - \eta - 2w}{s}, \tag{30}$$

$$\frac{\partial P}{\partial s} = 1 - \frac{w(\Theta - \eta - w)}{s}. \tag{31}$$

Thus, for fixed s we obtain maximal P for $w_0 = (\Theta - \eta)/2$, and P decreases for w smaller or larger than w_0. Further, $\partial P/\partial s = 0$ for $s = s_0 = \sqrt{w(\Theta - \eta - w)}$, and P increases for s smaller or larger than s_0. Since $s_0(w)$ is maximal for $w = w_0$ we have $s_0 \leq (\Theta - \eta)/2$ and thus $(w, s_0) \notin R$. Thus, for fixed w we have to choose s as large as possible. Together, the maximum of P occurs in point X of Fig. 2 if $X \in R$. If $X \notin R$ the maximum must be somewhere on $s = e^q \cdot (\Theta - \eta - w)$. Inserting the equivalent expression $w = \Theta - \eta - s \cdot e^{-q}$ into eq. 29 we then have to maximize

$$P_q(s) := P(\Theta - \eta - s \cdot e^{-q}, s) = s(1 - e^{-2q}) + e^{-q}(\Theta - \eta). \tag{32}$$

Since $P_q(s)$ increases with s we have to choose s as large as possible. Thus the maximum occurs in point Q_1 of Fig. 2. In summary, both maximal P and maximal d_Δ are obtained for the same values of input strengths w and s. The maxima are in Q_1 for $q_1 < q < q_2$ and in X for $q > q_2 = \ln 2$.

Examples. Assume threshold $\Theta = 1$, noise parameter $\eta = 0.1$, and a decay time constant $\tau = 2\,\mathrm{sec}$ for the word web input activity. Correspondingly, we find $q_1 \approx 0.136$ and $q_2 \approx 0.693$. We want to find input strengths w and s such that the delay interval length d_Δ, the peak height $P_{\alpha,\beta}$, and the peak difference d_P are all maximal.

If we do not constrain the minimal word web delay ($q > q_2$) we can choose point X in Fig. 2, i.e., $s = 0.9$ and $w = 0.45$. We obtain a minimal word web delay $\Delta_{\min} \approx 1.39\,\mathrm{sec}$, a maximal delay $\Delta_{\max} \approx 3\,\mathrm{sec}$, and thus a maximal delay range of $d_\Delta \approx 1.61\,\mathrm{sec}$. For optimal delay $\Delta_P \approx 1.39\,\mathrm{sec}$ we obtain a maximal critical peak height $P_{\alpha,\beta} = 1.125$ (versus $P_{\beta,\alpha} = 0.9$), and the maximal possible peak difference $d_P = s/4 = 0.225$.

However, a minimal word delays $\Delta_{\min} \approx 1.39\,\mathrm{sec}$ may be too large for processing of word sequences. Let us require $\Delta_{\min} \leq 0.5\,\mathrm{sec}$ which corresponds to $q = 0.5\,\mathrm{sec}\,/\tau = 0.25$. Since $q_1 < q < q_2$ the maxima can be found in Q_1. Thus we choose $w \approx 0.199$ and $s = 0.9$. We obtain $\Delta_{\min} = 0.5\,\mathrm{sec}$, but only $\Delta_{\max} \approx 1.37\,\mathrm{sec}$, and thus maximal $d_\Delta \approx 0.877\,\mathrm{sec}$. For optimal delay $\Delta_P = 0.5\,\mathrm{sec}$ we obtain maximal $P_{\alpha,\beta} \approx 1.06$ (versus $P_{\beta,\alpha} = 0.9$), but a peak difference of only $d_P \approx 0.16 < s/4$.

For word web delays $\Delta < q_1 \cdot \tau \approx 0.272\,\mathrm{sec}$ the SD will not be able to work appropriately at all.

Summary so far. We have analysed the linear SD model with exponentially decaying word web activity [6] with respect to the maximal peak difference d_P between critical and inverse SD (eqs. 11,12), the maximal above-threshold peak height $P_{\alpha,\beta}$ of the critical SD (eq. 29), and the maximal interval length d_Δ (eq. 17) of valid word web delays $\Delta \in [\Delta_{\min}; \Delta_{\max}]$.

We have found parameters where the model works quite well (see Fig. 2). However, the above examples and the previous analysis reveal some limitations of the original model: The peak difference d_P is limited to at most 25 percent of the inverse SD's peak height, and d_P can be even much smaller for non-optimal word web delay $\Delta \neq \Delta_P$. Another problem is the limited range of valid word web delays for plausible values τ of the decay constant of word web activity. The examples reveal that $\Delta_{\min}$ and d_Δ are too small for processing complex sentences where delays Δ between key words (e.g., subject noun and predicate verb) may vary between some hundred milliseconds and many seconds. We will discuss possible solutions to these problems in the following section.

2.2   Perspectives on Improving the Linear SD Model

The analysis of the previous section reveals some limitations of the simple linear SD model: It is not very robust because the maximal peak or amplitude difference between critical and inverse SD cannot exceed 25 percent (eq. 12), and for reasonable noise parameter η there can be severe limitations on minimal and maximal possible word web delays $\Delta_{\min}$ and $\Delta_{\max}$.

One particularity of the model is the assumption that the word web activity decays exponentially with $\exp(-t)$, based on some experimental findings of sustained neural activity related to working memory [11]. It is possible to relieve the restrictions on Δ considerably by using alternative decay functions $k(t)$, for example such that $k(t) < \exp(-t)$ for small t, but $k(t) > \exp(-t)$ for larger t (see [12]). However, it is easy to show that the 25 percent limitation on the maximal amplitude difference holds for any linear summative SD model.

A possible solution would be a non-linear multiplicative SD model, perhaps based on NMDA receptor mediated currents [13]. In such a model, the "weak" input fibres would make synapses predominantly endowed with NMDA receptors, and the "strong" input fibre synapses would be endowed predominantly with AMPA receptors. Since NMDA currents decay much slower than AMPA currents (typically $\tau_{\mathrm{NMDA}} = 150\mathrm{msec}$ versus $\tau_{\mathrm{AMPA}} = 5\mathrm{msec}$) and because NMDA receptors mediate excitation only when the neuron is already excited (by AMPA input) the SD unit would be activated strongly only if it first receives NMDA and secondly AMPA input (cf. [13, 14]).

We will not explain this possibility here for space constraints, but a more detailed description of the improved model variants will be given in a forthcoming paper [12]. In the following we will discuss further biologically more realistic network variants of the simple linear SD model with exponential word web decay that can most easily be implemented and analyzed.

2.3   "Leaky-Integrate" SD Model

In the linear model variant of sequence detection discussed above, the SD's postsynaptic activity state at time t was considered to be influenced only by the activity state of its presynaptic neural units at the *same time t*. This is a simplification, because postsynaptic activation is known to be the cumulative result of inputs that arrived during a longer past and that, like the presynaptic activity itself, decay over time (e.g., see [15]).The temporal decay of the postsynaptic potential x can be described with a so-called "leaky-integrator" time constant τ_x. To make our linear model of sequence detection more realistic neurobiologically, we introduce leaky integration and therefore rewrite eqs. 3,4 as differential equations

$$\tau_x \cdot \frac{d}{dt}x_{\alpha,\beta}(t) = -x_{\alpha,\beta}(t) + w \cdot y_\alpha(t) + s \cdot y_\beta(t), \tag{33}$$

$$\tau_x \cdot \frac{d}{dt}x_{\beta,\alpha}(t) = -x_{\beta,\alpha}(t) + s \cdot y_\alpha(t) + w \cdot y_\beta(t), \tag{34}$$

where $y_\alpha(t) = \exp(-t/\tau) \cdot H(t)$ and $y_\beta(t) = \exp(-(t-\Delta)/\tau) \cdot H(t-\Delta)$. We can find solutions separately for the time intervals $t \in [0; \Delta)$ and $t \in [\Delta; \infty)$ by first solving the simpler differential equation

$$\tau_x \cdot \frac{dx}{dt} = -x + c \cdot e^{-t/\tau} \tag{35}$$

with initial value $x(0) = \nu$. With eq. 59, the solution of eq. 35 is

$$x(t) = \nu \cdot e^{-t/\tau_x} + \frac{c}{1 - \tau_x/\tau} \cdot \alpha_{\tau,\tau_x}(t), \tag{36}$$

where $\alpha_{\tau,\tau_x}(t) := \exp(-t/\tau) - \exp(-t/\tau_x)$ (see appendix A.1). With eqs. 52,53 we can determine the maximum

$$t_{\max}(c,\nu) = \frac{\tau \cdot \tau_x}{\tau - \tau_x} \cdot \ln\left(\frac{\nu}{c} + \frac{\tau}{\tau_x} \cdot \left(1 - \frac{\nu}{c}\right)\right) \tag{37}$$

$$x_{\max}(c,\nu) = \frac{c}{1 - \frac{\tau_x}{\tau}} \cdot \left(\frac{\nu}{c} + \frac{\tau}{\tau_x} \cdot \left(1 - \frac{\nu}{c}\right)\right)^{-\frac{\tau_x}{\tau - \tau_x}}$$

$$- \left(\frac{c}{1 - \frac{\tau_x}{\tau}} - \nu\right) \cdot \left(\frac{\nu}{c} + \frac{\tau}{\tau_x} \cdot \left(1 - \frac{\nu}{c}\right)\right)^{-\frac{\tau}{\tau - \tau_x}} \tag{38}$$

Now we can solve the original differential eqs. 33,34 for $x_{\alpha,\beta}$ and $x_{\beta,\alpha}$. For $x_{\alpha,\beta}(t)$, $t \in [0; \Delta]$ we can apply eq. 36 using $c = w$ and initial value $\nu = x_{\alpha,\beta}(0) = 0$. For $t \in [0; \Delta]$ we use $c = s + w \cdot e^{-\Delta/\tau}$ and $\nu = x_{\alpha,\beta}(\Delta) = w \cdot \alpha_{\tau,\tau_x}(\Delta)/(1 - \tau_x/\tau)$. This and a similar procedure for the inverse detector's potential $x_{\beta,\alpha}$ finally yields

$$x_{\alpha,\beta}(t) = \frac{1}{1 - \frac{\tau_x}{\tau}} \cdot \begin{cases} w \cdot \alpha(t) & ,0 \le t \le \Delta \\ w \cdot \alpha(\Delta) \cdot e^{-\frac{t-\Delta}{\tau_x}} + (s + we^{-\frac{\Delta}{\tau}}) \cdot \alpha(t-\Delta), & t \ge \Delta \end{cases} \tag{39}$$

$$x_{\beta,\alpha}(t) = \frac{1}{1 - \frac{\tau_x}{\tau}} \cdot \begin{cases} s \cdot \alpha(t) & ,0 \le t \le \Delta \\ s \cdot \alpha(\Delta) \cdot e^{-\frac{t-\Delta}{\tau_x}} + (w + se^{-\frac{\Delta}{\tau}}) \cdot \alpha(t-\Delta), & t \ge \Delta \end{cases} \tag{40}$$

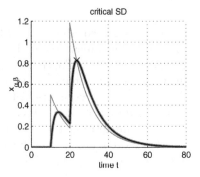
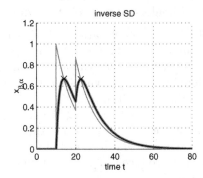

Fig. 3. Membrane potential of the critical SD (left) and the inverse SD (right) for the "leaky-integrate" model as computed from eqs. 39,40 (black line). Word webs are activated at $t_A = 10$ and $t_B = 20$. Parameters are $\tau_{WW} = 10$, $\tau_{WW} = 10$, $w = 0.5$, $s = 1$. The thin gray line represents the response of the original non-leaky model (cf. Fig. 1). Also shown are simulation results (thick gray line) and the peak maxima (crosses) as computed from eqs. 41,42

where for brevity we write $\alpha(t)$ instead of $\alpha_{\tau,\tau_x}(t)$. The peak amplitude of the critical SD is the peak of the "second" alpha function and can therefore be obtained from eq. 38 by substituting $\nu = w \cdot \alpha(\Delta)/(1 - \tau_x/\tau)$ and $c = s + w \cdot e^{-\Delta/\tau}$, while the peak amplitude of the inverse SD is the peak of either the first or the second alpha function,

$$P_{\alpha,\beta} = x_{\max}\left(s + w \cdot e^{-\Delta/\tau}, \frac{w \cdot \alpha(\Delta)}{1 - \tau_x/\tau}\right), \tag{41}$$

$$P_{\beta,\alpha} = \max\left(x_{\max}(s, 0), x_{\max}\left(w + s \cdot e^{-\Delta/\tau}, \frac{s \cdot \alpha(\Delta)}{1 - \tau_x/\tau}\right)\right). \tag{42}$$

Fig. 3 illustrates and verifies the shown formulas eqs. 39-42. The potential of the leaky-integrate SDs is essentially a smoothed version of the original SDs' potential (cf. Fig. 1).

Numerical simulations as shown in Fig. 4 suggest the following preliminary results: (1) The maximal amplitude difference occurs for an inter-stimulus delay $\Delta \approx \Delta_P$ such that the two peaks of the inverse SD have similar height. It appears that Δ_P increases monotonically with τ_x, in particular, $\Delta_P \geq \ln 2 \approx 0.7$ for $\tau_x > 0$. This implies that in biologically more realistic models the minimal word web delay $\Delta_{\min}$ will be even larger than discussed in section 2.1. (2) For $\Delta \approx \Delta_P$ the maximal amplitude difference occurs also for $s = 2w$, the same as for the original SD model. For $\Delta \neq \Delta_P$ the optimum occurs at $s \neq 2w$. The maximal amplitude difference is < 25 percent, slightly smaller than for the original model. However, the loss appears to be small (only a few percent) for reasonable parameters.

In the following simulations we will use $\tau \gg \tau_x$, for example, $\tau = 100$ and $\tau_x = 1$. In this case the behavior of a single SD is still very similar to the original model (see appendix A.1, Fig. 8).

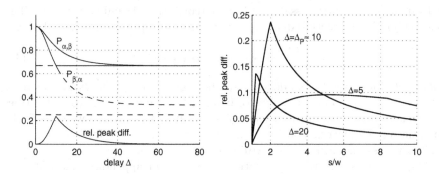

Fig. 4. Left: Peak height of the critical $(P_{\alpha,\beta})$ and inverse SD $(P_{\beta,\alpha}$, solid line; the dashed lines correspond to the smaller one of the two activity peaks, cf. eq. 42) and their relative difference as functions of stimulus delay Δ for otherwise same parameters as in Fig. 3. The optimum for relative peak difference occurs for larger $\Delta_P \approx 10$ than for the non-leaky model ($\Delta_P = \ln 2 \cdot \tau \approx 7$) which implies also higher $\Delta_{\min}$ for the leaky model. Note that Δ is near the optimum at the discontinuity, i.e., when the first and second peak of the inverse detector have nearly equal height. **Right**: Relative peak difference as a function of s/w for optimal $\Delta = \Delta_P \approx 10$ (solid) compared to $\Delta = 5$ and $\Delta = 20$ (dashed), and otherwise same parameters as for the left panel. For $\Delta = \Delta_P$, the optimum still occurs for $s = 2w$ but is (slightly) below the 25 percent bound. For Δ smaller (larger) than Δ_P, the optimum occurs for s/w larger (smaller) than 2

2.4 Assembly SD Model

So far we have analyzed only a single SD unit. As we will see in the next section, learning and generalization of grammatical rules and categories can be accomplished if several related single SD units group into a functional unit called an *assembly SD*. Figure 5 illustrates an assembly SD consisting of $n = 5$ individual SD units which are connected with each other and receive strong and weak inputs from two word webs α and β.

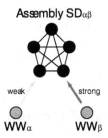

Fig. 5. An assembly SD consisting of $k = 5$ mutually connected individual SD units receiving inputs from two word webs α and β

In analogy to eq. 35 the assembly SD model consisting of n units can be described by a system of n differential equations for the membrane potentials x_i ($i = 1, ..., n$),

$$\tau_x \cdot \frac{d\mathbf{x}}{dt} = -\mathbf{x} + \mathbf{A}\mathbf{x} + \mathbf{b} + \mathbf{c} \cdot e^{-t/\tau}, \tag{43}$$

where bold lowercase letters denote vectors of dimension n and bold uppercase letters represent $n \times n$ matrices. Vector $\mathbf{b}$ is constant input, and vector $\mathbf{c}$ describes the strength of the exponentially decaying input from the word webs. $\mathbf{A}$ is the synaptic weight matrix. For simplicity we assume $A_{ij} = a$ for $i, j = 1, ..., n$. Further we assume initial values $\mathbf{x}(0) = \nu = (\nu_1, ..., \nu_n)$.

The system of differential equations is solved in appendix A.4. With the eigenvalue $\lambda := na - 1$ and $\tau_\lambda := -\tau_x/\lambda$ the unique solution is

$$x_i(t) = (\nu_i - \bar{\nu}) \cdot e^{-t/\tau_x} + \bar{\nu} \cdot e^{-t/\tau_\lambda}$$

$$+ (b_i - \bar{b}) \cdot (1 - e^{-t/\tau_x}) + \bar{b} \cdot \frac{1 - e^{-t/\tau_\lambda}}{-\lambda}$$

$$+ (c_i - \bar{c}) \cdot \frac{e^{-t/\tau} - e^{-t/\tau_x}}{1 - \tau_x/\tau} + \bar{c} \cdot \frac{e^{-t/\tau} - e^{-t/\tau_\lambda}}{-\lambda - \tau_x/\tau} \tag{44}$$

$$\bar{x}(t) = \bar{\nu} \cdot e^{-t/\tau_\lambda} + \bar{b} \cdot \frac{1 - e^{-t/\tau_\lambda}}{-\lambda} + \bar{c} \cdot \frac{e^{-t/\tau} - e^{-t/\tau_\lambda}}{-\lambda - \tau_x/\tau}. \tag{45}$$

for $i = 1, ..., n$ and means $\bar{\nu} := (\sum_{i=1}^n \nu_i)/n$, $\bar{b} := (\sum_{i=1}^n b_i)/n$, $\bar{c} := (\sum_{i=1}^n c_i)/n$, and $\bar{x} := (\sum_{i=1}^n x_i)/n$. The system is stable for a negative eigenvalue $\lambda = na - 1 < 0$ and then converges with $x_i \to b_i - \bar{b} + \bar{b}/ - \lambda$ for $t \to \infty$.

Note that the solution for the mean potential $\bar{x}$ (eq. 45) of the assembly SD has the same form as the solution eq. 36 for an isolated leaky-integrate SD (set $\bar{b} = 0$ and substitute $\nu = \bar{\nu}$, $c = \bar{c} \cdot (1 - \tau_x/\tau)/(-\lambda - \tau_x/\tau)$ in eq. 36). Thus the mean activity $\bar{x}$ of an assembly SD receiving input $\mathbf{w} \cdot y_\alpha(t) + \mathbf{s} \cdot y_\beta(t)$ behaves exactly the same way as an individual leaky-integrate SD with time constant τ_x' and weak and strong inputs w' and s',

$$\tau_x' := \tau_\lambda, \quad w' := \frac{-\lambda\tau - 1}{\tau - \tau_\lambda} \cdot \bar{w}, \quad s' := \frac{-\lambda\tau - 1}{\tau - \tau_\lambda} \cdot \bar{s}, \tag{46}$$

where y_α, y_β are as in section 2.3 and $\bar{w}$, $\bar{s}$ are the mean component values of the input strength vectors $\mathbf{w}$, $\mathbf{s}$. The most prominent effect of grouping elementary SDs into an assembly will be a slower time course of the membrane potentials during sequence processing since $\tau_x' \gg \tau_x$ for $na \to 1$. This is because of the strong internal connections within the assembly, which lead to full information exchange between elementary SDs.

3   Associative Learning of Grammatical Categories

3.1   Network and Learning Model

Figure 6 illustrates our simple network model. It contains two neuron populations, population WW for word webs, and population SD for elementary sequence

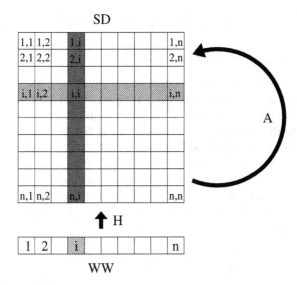

Fig. 6. Network model for learning grammatical categories. The Word Web population WW projects onto the sequence detector layer SD via hetero-associative connections H. Initially, the in-column connections from WW i to layer SD are strong (dark gray), the in-row connections are weak (light gray), and all remaining connections are very weak. Thus, SD ij is selective for word i followed by word j. Additionally, there are initially very weak auto-associative recurrent connections A within population SD

detectors. Each neuron is modelled similar to the leaky-integrate units described in section 2.3. For m word webs, population SD contains m^2 units, one neuron for each possible combination of word webs. Each word web makes two types of forward connections onto related sequence detectors: (i) weak connections to those SDs that detect the word web as first element, and (ii) strong connections onto those SDs that detect the word web as second element. In addition there are plastic recurrent connections between all pairs of SDs which are initially weak and unspecific. One may describe our network as auto-associative memory - the SD population or "layer" - connected in an hetero-associative fashion with a layer for word webs (cf. [16, 17, 18]).

Each word web i is modelled as a simple leaky-integrate unit with the potential $x_{\mathrm{WW},i}$ obeying

$$\frac{dx_{\mathrm{WW},i}}{dt} = -\frac{1}{\tau} \cdot x_{\mathrm{WW},i} + \sum_k \delta(t - t_{ik}), \qquad (47)$$

where τ is the decay time constant of the word web's sustained activity, $\delta(t)$ is the Dirac impulse ($\delta(t) =$ "∞" for $t = 0$, and $\delta(t) = 0$ for $t \neq 0$), and t_{ik} is the time of the k-th activation of WW i. Word web output $y_{\mathrm{WW},i} = x_{\mathrm{WW},i}$ is simply linear. Similarly, potential $x_{\mathrm{SD},ij}$ and output $y_{\mathrm{SD},ij}$ of sequence detector ij can be described by

$$\tau_x \cdot \frac{dx_{\mathrm{SD},ij}}{dt} = -x_{\mathrm{SD},ij} + \sum_k H_{kij} \cdot y_{\mathrm{WW},k} + \sum_{k,l} a_{klij} \cdot y_{\mathrm{SD},kl}, \qquad (48)$$

$$y_{\mathrm{SD},ij} = \max(0, \min(\Theta, x_{\mathrm{SD},ij})), \qquad (49)$$

where τ_x is the membrane time constant of the sequence detectors, H_{kij} represents the strength of the connection from WW k to SD ij, and A_{klij} represents the strength of the connection from SD kl to SD ij. Initially, the "in-row" hetero-associative synaptic strengths were "weak" with $H_{kkj} = w$, the "in-column" hetero-associative connections were "strong" with $H_{kik} = s$ for $i \neq k$, the "diagonal" connections were "very strong" with $H_{kkk} = s_1$, and all remaining connections were "very weak" with $H_{kij} = w_1$ for $i, j \neq k$ and $A_{klij} = w_A$.

For synaptic learning we used a basic Hebbian coincidence rule with decay. The weight $\omega_{ij} \in [0; \omega_{\max}]$ of the synapse between two neurons i and j follows the differential equation

$$\frac{d\omega_{ij}}{dt} = -d + f_{\mathrm{pre}}(y_i) \cdot f_{\mathrm{post}}(y_j) \qquad (50)$$

where d is the decay term, and f_{pre} and f_{post} are positive sigmoid functions of pre- and postsynaptic activities y_i and y_j, respectively.

The anatomical interpretation of our model is that different WWs are located at different cortical sites, and that SD ij is located in the vicinity of WW j [5, 6]. Therefore the *short-range* "in-column" connections from WW to SD are "strong" and the remaining *long-range* connections are "weak" or "very weak". The "diagonal" SD ii can be interpreted as being part of word-related neuronal assemblies (with similar properties as WW i).

3.2   Generalizing Sequences and Learning of Grammatical Categories

What is the data set a learning mechanism for serial order in language can operate on? A typical situation in language use is that very common word strings are frequently being encountered and that one of the words in the string is occasionally replaced by another one. If this happens repeatedly, it is possible to define classes of string elements that are frequently being substituted with each other, and these can be considered the basis of so-called lexical categories or grammatical word classes. Think, for example of nouns such as boys, whales, fish, lobsters and verbs such as swim, sleep, jump, talk etc. Each member of one of these lexical or lexico-semantic categories can, in principle, co-occur with any member of the respective other category. The child learns this, although in real life it usually does not encounter all possible pairings. Some of the words of one class co-occur with some members of the other class, and the resulting word substitutions between strings are sufficient to generalise a rule that links all members of class one to all members of class two. The basis of this learning would be the substitution pattern of words between strings. The most elementary case where this putative auto-associative substitution learning [6] could be explored is the case where word strings $(1, 3), (1, 4), (2, 3)$ are being presented to a network and it is investigated whether, as a result of this learning, the network would process the not encountered string that completes the substitution pattern, namely $(2,4)$, in the same way as the learned word sequences (see [19, 6, 5]). An example would be a child who encounters "boys swim", "boys sleep", "whales swim" and generalises the acquired sequence knowledge to generate (or accept) the new string "whales sleep". The important question is whether a network of the properties detailed so far in this paper would explain this substitution learning at the neuronal level.

To control for the relevance of substitutions of string elements and to test the specificity of the generalisation behavior of the network to those word webs that were activated during string substitutions, we added a few items that only occurred in one context. In the following we illustrate some simulations of our model with $m = 7$ WWs and $m^2 = 49$ SDs (see Fig. 7). The network develops in three phases: (1) *Training phase*: Initially SD ij receives weak input from WW i and strong input from WW j and only very weak input from other sources. Thus SD ij responds strongly only to its critical sequence (i,j) as described in section 2.3. When sequence (i,j) is presented then the three SDs ii, ij, and jj respond strongly at the same time. For this the corresponding recurrent synapses within layer SD get strengthened by Hebbian learning and therefore the corresponding "triplet cliques" are established in the network. In our example of Fig. 7 we have presented the sequences $(1,3)$, $(1,4)$, $(2,3)$, and $(5,6)$. (2) *Replay phase*:

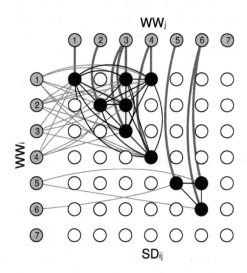

Fig. 7. SD network after training with sequences $(1,3)$, $(1,4)$, $(2,3)$, and $(5,6)$. Learning results in two discrete assemblies representing the productive rule $\{1,2\} \rightarrow \{3,4\}$ and the associative learning of $\{5\} \rightarrow \{6\}$, respectively. Thin gray edges indicate the "weak" hetero-associative connections from WW to SD, thick gray edges indicate "strong" connections. Black connections inside layer SD indicate auto-associative connections constituting the cell assemblies made up of SDs. SD: sequence detector; WW: word web

While in phase 1 feed-forward connections are dominant, in the replay phase the network dynamics are dominated by the recurrent connections within layer SD while the same sequences from phase 1 are presented again for several times. The resulting strengthening of recurrent connections gives rise to point attractors in the state space which correspond to cell assemblies representing grammatical rules. For example, the replay of sequence $(1, 3)$ will activate a whole assembly of seven SDs 13, 14, 23, 11, 22, 33, and 44. The result is that the connections from WWs 1 and 3 to *all* seven SDs of the assembly get strengthened heteroassociatively. Additionally also the autoassociative connections between the seven SDs get strengthened. A neuronal assembly forms which is no longer selective for one sequence of WW activations, but rather for a sequence of *any member* of a set of WWs followed by the activation of any member of a second set. (3) *Testing phase*: Finally, the network is brought back to a feed-forward dominated regime. Due to the learned hetero-associative connections now the network is capable of generalizing word sequences such as $(2, 4)$ that never occurred before but that are supported by the substitution pattern of sequence constituents. Now the network can be said to represent two discrete "rules" $\{1, 2\} \rightarrow \{3, 4\}$ and $\{5\} \rightarrow \{6\}$, where the word sets $\{1, 2\}$ and $\{3, 4\}$ constitute two distinct "grammatical categories".

Our simulations demonstrate (a) that a variant of the sequence detector model produces auto-associative substitution learning and generalises sequential order information to new, never encountered strings, and (b) that this generalisation is specific to those items that participated in substitutions. We submit that the proposed mechanisms can be the basis of learning of both grammatical categories and grammatical rules of syntactic serial order. It would be the elementary sequence detectors strongly connected to each other that form these abstract higher order Syntactic Neuronal Assemblies (SNAs).

4   Summary

In this work we have analysed different versions of linear summative sequence detector models [5, 6] and their potential use in sequence generalisation and the development of grammatical categories.

We have shown that a putative neural correlate of syntactical rules can be learned by an auto-associative memory on the basis of the pattern of substitutions observable in language use. String elements, which we call "words" here for simplicity, can be classified according to the frequent contexts (preceding and following neighbor words) in the string. We demonstrate here that an auto-associative memory can learn *discrete* representations which can be considered a neural equivalent of links between syntactic categories and can explain specific syntactic generalization.

Clearly our current simulations (see Fig. 7) account only for very simple grammatical rules and categories (such as noun and verb) where each word can be subject to at most one rule such that the resulting cell assemblies have no overlaps. If a word belonged to several different categories the resulting cell assemblies would overlap which leads to the requirement of appropriate threshold control, for example by inhibitory interneurons [18, 20]. The application of our model to real language data will be discussed in greater detail in a forthcoming paper [21].

Acknowledgements. This work was supported in part by the MirrorBot project of the European Union (IST-2001-35282).

References

1. Chomsky, N.: Syntactic structures. Mouton, The Hague (1957)
2. Hauser, M., Chomsky, N., Fitch, W.: The faculty of language: what is it, who has it, and how did it evolve? Science **298(5598)** (2002) 1569–1579
3. Elman, J.: Finding structure in time. Cognitive Science **14** (1990) 179–211
4. Elman, J., Bates, L., Johnson, M., Karmiloff-Smith, A., Parisi, D., Plunkett, K.: Rethinking innateness. A connectionist perspective on development. MIT Press, Cambridge, MA (1996)
5. Pulvermüller, F.: The neuroscience of language: on brain circuits of words and serial order. Cambridge University Press, Cambridge, UK (2003)
6. Pulvermüller, F.: Sequence detectors as a basis of grammar in the brain. Theory in Bioscience **122** (2003) 87–103
7. Kleene, S.: Representation of events in nerve nets and finite automata. In Shannon, C., McCarthy, J., eds.: Automata studies. Princeton University Press, Princeton, NJ (1956) 3–41
8. Braitenberg, V., Heck, D., Sultan, F.: The detection and generation of sequences as a key to cerebellar function: experiments and theory. Behavioral and Brain Sciences **20** (1997) 229–245
9. Reichardt, W., Varju, D.: Übertragungseigenschaften im Auswertesystem für das Bewegungssehen. Zeitschrift für Naturforschung **14b** (1959) 674–689
10. Egelhaaf, M., Borst, A., Reichardt, W.: Computational structure of a biological motion-detection system as revealed by local detector analysis in the fly's nervous system. Journal of the Optical Society of America (A) **6** (1989) 1070–1087
11. Fuster, J.: Memory in the cerebral cortex. MIT Press, Cambridge, MA (1999)
12. Knoblauch, A., Pulvermüller, F.: Associative learning of discrete grammatical categories and rules. in preparation (2005)
13. Destexhe, A., Mainen, Z., Sejnowski, T.: Kinetic models of synaptic transmission. [15] chapter 1 1–25
14. Knoblauch, A., Wennekers, T., Sommer, F.: Is voltage-dependent synaptic transmission in NMDA receptors a robust mechanism for working memory? Neurocomputing **44-46** (2002) 19–24
15. Koch, C., Segev, I., eds.: Methods in neuronal modeling. MIT Press, Cambridge, Massachusetts (1998)
16. Palm, G.: On associative memories. Biological Cybernetics **36** (1980) 19–31
17. Palm, G.: Neural Assemblies. An Alternative Approach to Artificial Intelligence. Springer, Berlin (1982)
18. Knoblauch, A., Palm, G.: Pattern separation and synchronization in spiking associative memories and visual areas. Neural Networks **14** (2001) 763–780

19. Pulvermüller, F.: A brain perspective on language mechanisms: from discrete neuronal ensembles to serial order. Progress in Neurobiology **67** (2002) 85–111
20. Knoblauch, A.: Synchronization and pattern separation in spiking associative memory and visual cortical areas. PhD thesis, Department of Neural Information Processing, University of Ulm, Germany (2003)
21. Pulvermüller, F., Knoblauch, A.: Emergence of discrete combinatorial rules in universal grammar networks. in preparation (2005)

A   Mathematical Appendix

A.1   Properties of Alpha Functions

The difference $\alpha_{\tau_2,\tau_1}(t)$ of two exponentials with time constants $\tau_1 < \tau_2$ is called *alpha function* and has the following maximum $\alpha_{\max}$ at $t = t_{\max}$ (see Fig. 8),

$$\alpha_{\tau_2,\tau_1}(t) := e^{-t/\tau_2} - e^{-t/\tau_1} \tag{51}$$

$$\alpha_{\tau_2,\tau_1}'(t) = 0 \Leftrightarrow t = t_{\max} := \frac{\tau_1 \cdot \tau_2}{\tau_2 - \tau_1} \cdot \ln \frac{\tau_2}{\tau_1} \tag{52}$$

$$\alpha_{\max} := \alpha_{\tau_2,\tau_1}(t_{\max}) = \left(\frac{\tau_2}{\tau_1}\right)^{-\frac{\tau_1}{\tau_2-\tau_1}} - \left(\frac{\tau_2}{\tau_1}\right)^{-\frac{\tau_2}{\tau_2-\tau_1}}. \tag{53}$$

For brevity we will simply write $\alpha(t)$ in the following. We have $\alpha(0) = 0$ and $\alpha(t) \to 0$ for $t \to \infty$, and a single *maximum* $\alpha_{\max}$ occurs at $t_{\max}$. With $c := \tau_2/\tau_1$ the maximum can be written as

$$\frac{t_{\max}}{\tau_1} = \frac{c}{c-1} \cdot \ln c, \quad \frac{t_{\max}}{\tau_2} = \frac{\ln c}{c-1}, \quad \alpha_{\max} = c^{-\frac{1}{c-1}} - c^{-\frac{c}{c-1}}. \tag{54}$$

For $c \to 1$ we have $t_{\max} \to \tau_1$ and $\alpha_{\max} \to 0$. For $c \to \infty$ we have $t_{\max}/\tau_1 \to \infty$, $t_{\max}/\tau_2 \to 0$, and $\alpha_{\max} \to 1$. Fig. 8(right) shows $t_{\max}/\tau_2$ and $1 - \alpha_{\max}$ as functions of c on a logarithmic scale.

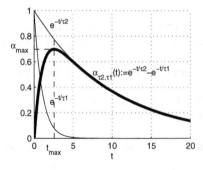
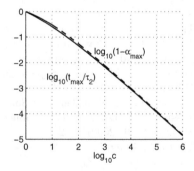

Fig. 8. Left: The alpha function $\alpha_{\tau_2,\tau_1}(t)$ for $\tau_2 = 10$, $\tau_1 = 10$. A single maximum $\alpha_{\max}$ occurs at $t = t_{\max}$. Right: $t_{\max}/\tau_2$ (solid) and $1 - \alpha_{\max}$ (dashed) as functions of $c := \tau_2/\tau_1$ on logarithmic scales (base 10)

A.2   Linear Differential Equations

For analysing the "leaky-integrate" SD model it is necessary to solve the initial value problem for a single linear inhomogenous differential equation of the type

$$\dot{x} = a \cdot x + b. \tag{55}$$

where a, b, and x are functions of t, and $x(t_0) = \nu$. Assume that h is the solution for the corresponding homogenous problem $\dot{x} = a\dot{x}$ with $h(t_0) = 1$. Then it is easy to verify that the unique solution for the inhomogenous equation with $x(t_0) = \nu$ is

$$x(t) = \nu \cdot h(t) + h(t) \cdot \int_{t_0}^{t} \frac{b(s)}{h(s)} ds. \tag{56}$$

For our purposes it will be necessary to solve the following differential equation,

$$\dot{x} = ax + b + c \cdot e^{dt}. \tag{57}$$

for constant a, b, c, and d. The solution of the corresponding homogenous equation is $h(t) = e^{at}$. Thus the unique solution for the inhomogenous equation with initial value $x(0) = \nu$ is

$$x(t) = \nu \cdot e^{at} + e^{at} \cdot \int_{t_0}^{t} \frac{b + ce^{ds}}{e^{as}} ds \tag{58}$$

$$= \nu \cdot e^{at} - \frac{b}{a} \cdot (1 - e^{at}) + \frac{c}{d-a} \cdot (e^{dt} - e^{at}). \tag{59}$$

Note that eq. 59 contains an alpha function, $e^{dt} - e^{at} = \alpha_{-1/d,-1/a}(t)$.

A.3   Diagonalization and Systems of Differential Equations

Differential equation *systems* of the type $\dot{\mathbf{x}} = \mathbf{A}\mathbf{x}$ can be solved most easily if $\mathbf{A}$ is a *diagonal* $n \times n$ matrix. Then each equation can be solved independently of the others, e.g., using eq. 59. If $\mathbf{A}$ is not diagonal we can still try to *diagonalise* $\mathbf{A}$ by finding an appropriate linear coordinate transform $\mathbf{z} := \mathbf{C}^{-1} \cdot \mathbf{x}$ for an invertible matrix $\mathbf{C}$ such that $\mathbf{D} = \mathbf{C}^{-1} \cdot \mathbf{A} \cdot \mathbf{C}$ is diagonal. Then it is easy to solve the equivalent system in z-coordinates, $\dot{\mathbf{z}} = \mathbf{D}\mathbf{z}$. Finally, the solution in x-coordinates can be obtained by the inverse transform $\mathbf{x} := \mathbf{C} \cdot \mathbf{z}$. Unfortunately, it is not possible to diagonalise *any* matrix $\mathbf{A}$. (Anyhow, it is always possible to find the so-called Jordan matrix of $\mathbf{A}$ where only *two*) diagonals are non-zero.)

If $\mathbf{A}$ has a simple form we can try to apply combined *row and column transformations* on $\mathbf{A}$ in order to eliminate non-diagonal matrix components. Row and column transformations can be described using Kronecker's symbol δ_{ij}, where $\delta_{ij} = 1$ for $i = j$ and 0 otherwise. We define

$$\delta^{(k,l)} := ((\delta_{ik} \cdot \delta_{jl})_{ij})^{n \times n} \tag{60}$$

as the matrix where all components are zero except a single one-entry at row k, column l. Verify that $\delta^{(k,l)} \cdot \mathbf{A}$ is the $n \times n$-matrix which at line k has a copy of the l-th line of $\mathbf{A}$ and is 0 otherwise. Similarly, $\mathbf{A} \cdot \delta^{(k,l)}$ is the matrix which at column l has a copy of the k-th column of $\mathbf{A}$. With the unity matrix $\mathbf{I}$ we can do the following rows or columns operations on matrix $\mathbf{A}$,

- $(\mathbf{I} + c \cdot \delta^{(k,l)}) \cdot \mathbf{A}$ adds c times the l-th row to the k-th row.
- $\mathbf{A} \cdot (\mathbf{I} + c \cdot \delta^{(k,l)})$ adds c times the k-th column to the l-th column.

Further verify that

$$(\mathbf{I} + c \cdot \delta^{(k,l)})^{-1} = \mathbf{I} - c \cdot \delta^{(k,l)} \qquad (61)$$

Before solving the assembly SD system eq. 43 (see appendix A.4) we now diagonalise the $n \times n$ matrix

$$\mathbf{A_h} := (\mathbf{A} - \mathbf{I})/\tau_x = \frac{1}{\tau_x} \cdot \begin{pmatrix} a-1 & a & \cdots & a & a \\ a & a-1 & \cdots & a & a \\ \vdots & \vdots & \ddots & \vdots & \vdots \\ a & a & \cdots & a-1 & a \\ a & a & \cdots & a & a-1 \end{pmatrix}. \qquad (62)$$

where $\mathbf{A}$, τ_x, and a are as in section 2.4. We have to find $\mathbf{C}$ such that $\mathbf{D} = \mathbf{C}^{-1} \cdot \mathbf{A_h} \cdot \mathbf{C}$ is diagonal. This can be achieved by applying adequate row operations on $\mathbf{A}$. However, note that whenever we add row i to row j, at the same time we have to add column j to row i, and vice versa. First we subtract the n-th row from rows $1, 2, ..., n-1$ which yields

$$\mathbf{C_1} \cdot \mathbf{A_h} = \frac{1}{\tau_x} \cdot \begin{pmatrix} -1 & 0 & \cdots & 0 & 1 \\ 0 & -1 & \cdots & 0 & 1 \\ \vdots & \vdots & \ddots & \vdots & \vdots \\ 0 & 0 & \cdots & -1 & 1 \\ a & a & \cdots & a & a-1 \end{pmatrix}, \text{for } \mathbf{C_1} = \begin{pmatrix} 1 & 0 & \cdots & 0 & -1 \\ 0 & 1 & \cdots & 0 & -1 \\ \vdots & \vdots & \ddots & \vdots & \vdots \\ 0 & 0 & \cdots & 1 & -1 \\ 0 & 0 & \cdots & 0 & 1 \end{pmatrix}, \qquad (63)$$

where $\mathbf{C_1} := \prod_{i=n-1}^{1}(\mathbf{I} - \delta^{(i,n)}) = (\mathbf{I} - \delta^{(n-1,n)}) \cdot (\mathbf{I} - \delta^{(n-2,n)}) \cdot .... \cdot (\mathbf{I} - \delta^{(1,n)})$ corresponds to the row operations. Since on the left and right side of $\mathbf{A_h}$ there must be inverse matrices we have also to apply the corresponding column transformations, i.e., we add the $1,2,...,n-1$-th column to the n-th column. With $\mathbf{C_1}^{-1} := \prod_{i=1}^{n-1}(\mathbf{I} + \delta^{(i,n)})$ we obtain

$$\mathbf{C_1} \mathbf{A_h} \mathbf{C_1}^{-1} = \frac{1}{\tau_x} \begin{pmatrix} -1 & 0 & \cdots & 0 & 0 \\ 0 & -1 & \cdots & 0 & 0 \\ \vdots & \vdots & \ddots & \vdots & \vdots \\ 0 & 0 & \cdots & -1 & 0 \\ a & a & \cdots & a & na-1 \end{pmatrix} \text{ for } \mathbf{C_1}^{-1} = \begin{pmatrix} 1 & 0 & \cdots & 0 & 1 \\ 0 & 1 & \cdots & 0 & 1 \\ \vdots & \vdots & \ddots & \vdots & \vdots \\ 0 & 0 & \cdots & 1 & 1 \\ 0 & 0 & \cdots & 0 & 1 \end{pmatrix}. \qquad (64)$$

Secondly, we can add c times the $1, 2, ..., n-1$-th row to the n-th row, and subsequently $-c$ times the n-th column to the $1,2,...,n-1$-th column. In order to eliminate the non-diagonal elements of row n we have to choose c such that $a - c - c(na - 1) = 0$. This is true for a factor $c = 1/n$. Thus with $\mathbf{C_2} := \prod_{i=n-1}^{1} \mathbf{I} + \frac{1}{n} \cdot \delta^{(n,i)}$ and $\mathbf{C_2}^{-1} := \prod_{i=1}^{n-1} \mathbf{I} - \frac{1}{n} \cdot \delta^{(n,i)}$ we have

$$
\mathbf{C_2} = \frac{1}{n} \cdot \begin{pmatrix} n & 0 & \cdots & 0 & 0 \\ 0 & n & \cdots & 0 & 0 \\ \vdots & \vdots & \ddots & \vdots & \vdots \\ 0 & 0 & \cdots & n & 0 \\ 1 & 1 & \cdots & 1 & n \end{pmatrix}, \quad \mathbf{C_2}^{-1} = \frac{1}{n} \cdot \begin{pmatrix} n & 0 & \cdots & 0 & 0 \\ 0 & n & \cdots & 0 & 0 \\ \vdots & \vdots & \ddots & \vdots & \vdots \\ 0 & 0 & \cdots & n & 0 \\ -1 & -1 & \cdots & -1 & n \end{pmatrix}. \tag{65}
$$

With $\mathbf{C}^{-1} := \mathbf{C_2} \cdot \mathbf{C_1}$ and $\mathbf{C} := \mathbf{C_1}^{-1} \cdot \mathbf{C_2}^{-1}$ we finally obtain

$$
\mathbf{C}^{-1} = \frac{1}{n} \cdot \begin{pmatrix} n & 0 & \cdots & 0 & -n \\ 0 & n & \cdots & 0 & -n \\ \vdots & \vdots & \ddots & \vdots & \vdots \\ 0 & 0 & \cdots & n & -n \\ 1 & 1 & \cdots & 1 & 1 \end{pmatrix}, \quad \mathbf{C} = \frac{1}{n} \cdot \begin{pmatrix} n-1 & -1 & \cdots & -1 & n \\ -1 & n-1 & \cdots & -1 & n \\ \vdots & \vdots & \ddots & \vdots & \vdots \\ -1 & -1 & \cdots & n-1 & n \\ -1 & -1 & \cdots & -1 & n \end{pmatrix}, \tag{66}
$$

$$
\mathbf{D} = \mathbf{C}^{-1} \cdot \mathbf{A_h} \cdot \mathbf{C} = \frac{1}{\tau_x} \cdot \begin{pmatrix} -1 & 0 & \cdots & 0 & 0 \\ 0 & -1 & \cdots & 0 & 0 \\ \vdots & \vdots & \ddots & \vdots & \vdots \\ 0 & 0 & \cdots & -1 & 0 \\ 0 & 0 & \cdots & 0 & na-1 \end{pmatrix}. \tag{67}
$$

A.4 Solution of the Assembly SD System

The assembly SD system eq. 43 in section 2.4 can equivalently be written as

$$
\frac{d\mathbf{x}}{dt} = \mathbf{A_h} \cdot \mathbf{x} + \mathbf{u} + \mathbf{v} \cdot e^{p \cdot t}, \tag{68}
$$

with $\mathbf{u} := \mathbf{b}/\tau_x$, $\mathbf{v} := \mathbf{c}/\tau_x$, $p := -1/\tau$, and $\mathbf{A_h}$ as in eq. 62. The solution becomes easy if we diagonalise $\mathbf{A_h}$, i.e. if we have an invertible matrix $\mathbf{C}$ such that $\mathbf{D} = \mathbf{C}^{-1} \cdot \mathbf{A_h} \cdot \mathbf{C}$ is diagonal (see appendix A.3). With the linear coordinate transform $\mathbf{z} := \mathbf{C}^{-1} \cdot \mathbf{x}$ we have to solve the equivalent system in $\mathbf{z}$-coordinates,

$$
\frac{d\mathbf{z}}{dt} = \mathbf{C}^{-1} \cdot \mathbf{A_h} \cdot \mathbf{C} \cdot \mathbf{z} + \mathbf{C}^{-1} \cdot (\mathbf{u} + \mathbf{v} \cdot e^{p \cdot t}), \tag{69}
$$

where the coordinate transform matrices $\mathbf{C}$, $\mathbf{C}^{-1}$ and diagonal matrix $\mathbf{D} := \mathbf{C}^{-1} \cdot \mathbf{A_h} \cdot \mathbf{C}$ are given by eqs. 66-67. Thus, it is sufficient to solve the independent equations

$$
\dot{z}_i = qz_i + (u_i - u_n) + (v_i - v_n) \cdot e^{pt}, \quad i = 1, ..., n-1 \tag{70}
$$
$$
\dot{z}_n = rz_n + \bar{u} + \bar{v} \cdot e^{pt} \tag{71}
$$

with $q := -1/\tau_x$, $r := (na - 1)/\tau_x$, $\bar{u} = (\sum_{j=1}^{n} u_i)/n$, and $\bar{v} = (\sum_{j=1}^{n} v_i)/n$. The initial values for $\mathbf{z}$ are

$$
\mathbf{z}(0) = \mathbf{C}^{-1} \cdot \nu = (\nu_1 - \nu_n, \cdots, \nu_{n-1} - \nu_n, \bar{\nu})^T \tag{72}
$$

With eq. 59 in appendix A.2 we obtain for each equation a unique solution,

$$z_i(t) = (\nu_i - \nu_n) \cdot e^{qt} - \frac{u_i - u_n}{q} \cdot (1 - e^{qt}) + \frac{v_i - v_n}{p - q} \cdot (e^{pt} - e^{qt}), \quad (73)$$

$$z_n(t) = \bar{v} \cdot e^{rt} - \frac{\bar{u}}{r} \cdot (1 - e^{rt}) + \frac{\bar{v}}{p - r} \cdot (e^{pt} - e^{rt}), \quad (74)$$

for $i = 1, ..., n - 1$. Transforming back with $\mathbf{x} = \mathbf{C} \cdot \mathbf{z}$ yields the unique solution in $\mathbf{x}$-coordinates,

$$x_i(t) = \nu_i \cdot e^{qt} + \bar{v} \cdot (e^{rt} - e^{qt})$$
$$- \frac{u_i}{q} \cdot (1 - e^{qt}) - \bar{u} \cdot \left(\frac{1 - e^{rt}}{r} - \frac{1 - e^{qt}}{q} \right)$$
$$+ \frac{v_i}{p - q} \cdot (e^{pt} - e^{qt}) + \bar{v} \cdot \left(\frac{e^{pt} - e^{rt}}{p - r} - \frac{e^{pt} - e^{qt}}{p - q} \right) \quad (75)$$

for $i = 1, ..., n$. Resubstituting the original symbols yields finally the unique solution eq. 44.

B   List of Symbols and Abbreviations

WW	word web
SD	sequence detector
SNA	syntactic neuronal assembly
t	time
$H(t)$	Heaviside function, see text following eq. 2
$\alpha_{\tau_2, \tau_1}(t)$ or $\alpha(t)$	alpha function with time constants τ_1 and τ_2, see eq. 52
$\delta(t)$	Dirac impulse function, see text following eq. 47
A, B, also $1,2,3,...$	word symbols or word numbers
(A, B), AB, also $12,23,...$	word sequences
α, β	symbols for WWs representing A and B, respectively
(α, β) or $\alpha\beta$	critical SD representing the word sequence $A \rightarrow B$
(β, α) or $\beta\alpha$	inverse SD representing the word sequence $B \rightarrow A$
y_α, y_β	output activity of WWs α and β, respectively
$x_{\alpha,\beta}$, $x_{\beta,\alpha}$	membrane potentials of SDs (α, β) and (β, α)
$y_{\alpha,\beta}$, $y_{\beta,\alpha}$	output activity of SDs (α, β) and (β, α)
w, s	weak and strong strength of input from WWs to SDs
Θ	SD threshold
η	noise parameter of SD, see eq. 14
τ	decay time constant for WW activity
τ_x	decay time constant of a leaky-integrate SD's potential
τ_λ	time constant for assembly SD system
λ	eigenvalue of assembly SD system
$P_{\alpha,\beta}$, $P_{\beta,\alpha}$	peak potential of critical and inverse SD, respectively
d_P	peak difference $P_{\alpha,\beta} - P_{\beta,\alpha}$
Δ	delay between first and second word in a word sequence
Δ_P	delay Δ such that inverse SD's peaks have equal height
$\Delta_{\min}$, $\Delta_{\max}$	minimal/maximal Δ such that the SD works appropriately
d_Δ	interval length $\Delta_{\max} - \Delta_{\min}$ of valid delays

$h(w)$	constraint eq. 18 equivalent to $d_\Delta > 0$ (cf. Fig. 2)
q	constraint eq. 19 on minimal word web delay $\Delta_{\min}$
q_1, q_2	limiting constants for q defined in eqs. 25,24
R	set of valid tuples (w, s) (see eq. 21)
n	assembly size
m	number of WWs
$\mathbf{H}$	hetero-associative connections from layer WW to layer SD
$\mathbf{A}, a$	auto-associative recurrent connections within layer SD

A Distributed Model of Spatial Visual Attention

Julien Vitay, Nicolas P. Rougier, and Frédéric Alexandre

Loria laboratory, Campus Scientifique, B.P. 239,
54506 Vandœuvre-lès-Nancy Cedex, France
{vitay, rougier, falex}@loria.fr

Abstract. Although biomimetic autonomous robotics relies on the massively parallel architecture of the brain, a key issue for designers is to temporally organize behaviour. The distributed representation of the sensory information has to be coherently processed to generate relevant actions. In the visuomotor domain, we propose here a model of visual exploration of a scene by the means of localized computations in neural populations whose architecture allows the emergence of a coherent behaviour of sequential scanning of salient stimuli. It has been implemented on a real robotic platform exploring a moving and noisy scene including several identical targets.

1   Introduction

The brain, in both humans and animals, is classically presented as a widely distributed and massively parallel architecture dedicated to information processing whose activity is centered around both perception and action. One the one hand, it includes multiple sensory poles able to integrate the huge sensory information through multiple pathways in order to offer the brain a coherent and highly integrated view of the world. On the other hand, it also includes several motor poles able to coordinate the whole range of body effectors, from head to toes or from muscles of the neck to muscles of the last knuckle of the left little toe.

Despite this huge amount of information to be processed, we are able to play the piano (at least some of us) with both left and right hand while reading the partition, tapping the rhythm with our feet, listening to the flute accompanying us and possibly singing the theme song. Most evidently, the brain is a well organized structure able to easily perform those kind of parallel performances.

Nonetheless, real brain performance does not lie in the parallel execution of some uncorrelated motor programs, hoping they could ultimately express some useful behaviour. Any motor program is generally linked to other motor programs through perception because we, as a body, are an indivisible entity where any action draws consequence on the whole body. If I'm walking in the street and suddenly decide to turn my head, then I will have to adapt my walking program in order to compensate for the subtle change in the shape of my body. In other words, the apparent parallelism of our actions is quite an illusion and requires de facto a high degree of coordination of motor programs. But even more striking is the required serialization for every action like for example grasping an object:

S. Wermter et al. (Eds.): Biomimetic Neural Learning, LNAI 3575, pp. 54–72, 2005.
© Springer-Verlag Berlin Heidelberg 2005

I cannot pretend to grasp the orange standing ahead of me without first walking to the table where it is currently lying.

This is quite paradoxical: behaviour is carried out by a massive parallel structure whose goal is finally to coordinate and serialize several elementary action programs. This is the key issue about the kind of performances that are presently identified as the most challenging in biomimetic robotics. The goal of this domain is to develop new computational models, inspired from brain functioning and to embed them in robots to endow them with strong capacities in perception, action and reasoning. The goal is to exploit the robot as a validation platform of brain models, but also to adapt it to natural interactions with humans, for example for helping disabled persons. These strategic orientations have been chosen, for example, in the Mirrorbot european project, gathering teams from neurosciences and computer science. Peoplebot robotic platforms are instructed, via a biologically oriented architecture, to localize objects in a room, reach them and grasp them. Fruits have been chosen to enable simple language oriented instructions using color, shape and size hints.

To build such technological platforms, fundamental research must be done, particularly in computational neurosciences. The most important topic is certainly that of multimodal integration. Various perceptual flows are received by sensors, preprocessed and sent to associative areas where they are merged in an internal representation. The principle of internal representation is fundamental in this neuronal approach. The robot learns by experience to extract in each perceptual modality the most discriminant features together with the conditional probabilities in the multimodal domain of occurrence of these features, one with regard to the other, possibly in a different modality.

In a natural environment, features have to be extracted in very numerous dimensions like for example, in the visual domain, motion, shape, color, texture, etc. Multimodal learning will result in a high number of scattered representations. As an illustration, one can think of learning the consequences of eye or body movement on the position of an item in the visual scene, learning the correlations between some classes of words (e.g. colors, objects) and some visual modalities (e.g. color, shape), learning to merge the proprioception of one's hand and its visual representation to anticipate key events in a grasping task, etc. It is clear that in autonomous robotics, all these abilities in the perceptual, multimodal and sensorimotor domains are fundamental prerequisite and, accordingly, a large amount of modeling work has been devoted to them in the past and are still developed today.

In this paper, we wish to lay emphasis on another important aspect, presently emerging in our domain. Nowadays, tasks to be performed by the robot are increasingly complex and are no longer purely associative tasks. As an illustration, in the Mirrorbot project, we are interested in giving language instructions to the robot like "grasp the red apple". Then, the robot has to observe its environment, select red targets, differentiate the apple, move toward it and end by grasping it. To tell it more technically, one thing is to have at disposal elementary behaviors, another more complicated thing is to know when to trigger the most appropriate and inhibit the others, particularly in a real world including many distractors.

In the framework of brain understanding and multimodal application, we investigated further the nature of the numerical computations required to implement a selective attention mechanism that would be robust against both noise and distractors. This kind of mechanism is an essential part of any robotic system since it allows to recruit available computational power on a restricted area of the perception space, allowing further processing on the interesting stimuli. The resulting model we introduce in this paper is a widely distributed architecture able to focus on a visual stimulus in the presence of a high level of noise or distractors. Furthermore, its parallel and competitive nature gives us some precious hints concerning the paradox of brain, behaviour and machine.

2   The Critical Role of Attention in Behaviour

Despite the massively parallel architecture of the brain, it appears that its processing capacities are limited in several domains: sensory discrimination, motor learning, working memory, etc. Several neuropsychological experiments have pinpointed this limitation. In the visual perception domain, the fundamental experiment by Treisman and Gelade [1] has drawn the distinction between two modes of visual search: when an object has characteristics sufficiently different from its background or other objects, it litterally "pops-out" from the scene and the search for it is very quick and independent from the number of other objects; oppositely, when this object shares some features with distracting objects or when it does not differ enough from its background, the search is very difficult and the time needed for it increases linearly in average with the number of distractors. These two search behaviours are then respectively called "parallel search" and "serial search". In the MirrorBot scenario, the parallel search could be useful when the robot has to find an orange among other non-orange fruits: the "orange-color" feature is sufficient for the robot to find its target. On the contrary, if one asks the robot to find a small green lemon among big green apples and small yellow lemons, the "green-colour" and "small size" features are not sufficient by themselves to dicriminate the green lemon: a conjunction of the two features is needed to perform the task. With respect to the results of Treisman and Gelade, the search would have to be serial, which means that the small and/or green objects have to be scanned sequentially until the green lemon is found.

Why such a limitation in the brain? Ungerleider and Mishkin [2] described the organization of the visual cortex as being composed of two major pathways: the ventral pathway (labelled as the "what" pathway because of its involvement in visual recognition) and the dorsal pathway (labelled as the "where" or "how" pathway because of its involvement in spatial representation and visuomotor transformation). Areas in the ventral pathway (composed by areas from V1 to V2 to V4 to TEO to TE) are specific for certain visual attributes with increasing receptive fields along this pathway: from $0.2°$ in V1 to $25°$ in TE. The complexity of the visual attributes encoded in these areas also increases throughout this pathway: V1 encodes simple features like orientation or luminance in a on-center off-surround fashion, V4 mainly encodes colour and inferotemporal areas (IT, comprising TEO and TE) respond to complex shapes and features. This description corresponds to a feed-forward hierarchical structure of the ventral pathway where low-level areas encode local specific features and high-level areas encode complex objects in a distributed and non-spatial manner. This approach raises several problems: although it is computationally interesting for working memory or language purposes to have a non-spatial representation of a visual object, what happens to this representation when several identical objects are present at the same time in the scene? As this high-level representation in IT is supposed to be highly distributed to avoid the "grandmother neuron" issue [3], how can the representation of several different objects be coherent and under-standable by prefrontal cortex (for example)? Moreover, the loss of the spatial information is a problem when the recognition of a given object has to evoke a motor response, e.g. an ocular saccade. The ventral stream can only detect the presence of a given object, not its position, what would instead be the role of the dorsal pathway (or occipito-parietal pathway). How is the coherence be-tween these two pathways ensured? These problems are known as the "binding problem". Reynolds and Desimone [4] state that attention is a key mechanism to solve that problem.

Visual attention can be seen as a mechanism enhancing the processing of interesting (understood as behaviourally relevant) locations and darkening the rest [5, 6]. The first neural correlate of that phenomenon has been discovered by Moran and Desimone [7] in V4 where neurons respond preferentially to a given feature in their receptive field. When a preferred and a non-preferred stimulus for a neuron are presented at the same time in its receptive field, the response becomes an average between the strong response to the preferred feature and the weak response to the non-preferred one. But when one of the two stimulus is at-tended, the response of the neuron represents the attended stimulus alone (strong or poor), as if the non-attended were ignored. The same kind of modulation of neural responses by attention has been found in each map of the ventral stream but also in the dorsal stream (area MT encoding for stimulus movement, LIP representing stimuli in a head-centered reference frame). All these findings are consistent with the "biased competition hypothesis" [8] which states that visual objects compete for neural representation under top-down modulation. This top-down modulation, perhaps via feedback connections, increases the importance of the desired features in the competition inside a map, but also between maps, to lead to a coherent representation of the target throughout the visual cortex. Importantly, when a subject is asked to search for a colored target before its ap-pearance, sustained elevation of the baseline activity of color-sensitive neurons in V4 has been noticed, although the target had not appeared yet [9].

Another question is the origin of attention, which can be viewed as a supra-modal cognitive mechanism, independent from perception and action [10], or on the contrary as a consequence of the activation of circuits mediating sensorimotor transformations. This "premotor theory of attention" [11, 12] implies that covert attention (attention to extra-foveal stimuli) is the preparation of a motor action to this stimulus, but finally inhibited. Several studies support that theory, especially in [13, 14, 15], showing that covert attention engage the same structures than overt orienting. These structures comprise the frontal eye field (FEF), the superior colliculus, the pulvinar nuclei of the thalamus, LIP (also called parietal eye field) among others. FEF appears as the main source of modulation of area LIP because of their anatomical reciprocal connections: a sub-threshold modulation of FEF increases the discrimination of a target [16], and although LIP encodes the position of visual stimuli in head-centered coordinates, this representation is shifted before a saccade is made to its estimated new position [17].

This strong link between action and attention has the advantage to account for the fact that attention can be either maintained or switched under volitional and behaviourally relevant control. In serial search, attention is sequentially attracted to different potentially interesting locations until the correct target is found. Which mechanism does ensure that attention can effectively move its focus when the enlightened object is not the expected one, but stick to it when it is found? In their seminal paper, Posner and Cohen [18] discovered that the processing of a stimulus displayed just after attention is attracted to its location is enhanced (what is coherent with the notion of attention), but is decreased a certain amount of time after (around 200-300ms depending of the task). This phenomenon called "inhibition of return" (IOR) can be interpreted as a mechanism ensuring that attention can not be attracted twice to the same location in a short period of time, therefore encouraging exploring new positions.

This quick overview of attention can be summarized by saying that attention is an integrated mechanism distributed over sensorimotor structures, whose purpose is to help them to focus on a small number of regions in the input space in order to achieve relevant motor behaviours. Therefore, virtually all structures involved in behaviour have to deal with attention: for example the link between working memory and attention has been established in [19] and [20]. Attention is a motivated and integrated process.

3   Continuum Neural Field Theory

Even if the whole neural networks domain often draws (more or less tightly) on biological inspiration, core mechanisms like the activation function or learning rules often deny the inner temporal nature of neurons. They are usually designed with no reference to time while it is perfectly known that a biological neuron is a complex dynamic system that evolves over time together with incoming information. If such artificial neurons can be easily manipulated and used in classical networks such as the Multi-Layer Perceptron (MLP), Kohonen networks or Hopfields maps, they can hardly pretend to take time into account, see [21] for a complete review.

In the same time, the Continuum Neural Field Theory (CNFT) has been extensively analyzed both for the one-dimensional case [22, 23, 24] and for the two-dimensional case [25] where much of the analysis is extendable to higher dimensions. These theories explain the dynamic of pattern formation for lateral-inhibition type homogeneous neural fields with general connections. They show specifically that, in some conditions, continuous attractor neural networks are able to maintain a localised bubble of activity in direct relation with the excitation provided by the stimulation.

3.1   A Dynamic Equation for a Dynamic Neuron

We will use the notations introduced in [25] where a neuronal position is labelled by the vector $\mathbf{x}$ which represents a two-component quantity designing a position on a manifold M in bijection with $[-0.5, 0.5]^2$. The membrane potential of a neuron at the point $\mathbf{x}$ and time t is denoted by $u(\mathbf{x}, \mathbf{t})$ and it is assumed that there is a lateral connection weight function $w(\mathbf{x} - \mathbf{x}')$ as a function of the distance $|\mathbf{x} - \mathbf{x}'|$. There exists also an afferent connection weight function $s(\mathbf{x}, \mathbf{y})$ from the position $\mathbf{y}$ in the manifold M' to the point $\mathbf{x}$ in M. The membrane potential $u(\mathbf{x}, t)$ satisfies the following equation (1):

$$
\tau \frac{\partial u(\mathbf{x}, t)}{\partial t} = -u(\mathbf{x}, t) + \int_M w_M(\mathbf{x} - \mathbf{x}') f[u(\mathbf{x}', t)] d\mathbf{x}'
$$
$$
+ \int_{M'} s(\mathbf{x}, \mathbf{y}) I(\mathbf{y}, t) d\mathbf{y} + h \ .
$$

(1)

where f is a transfer function from the membrane potential u to a mean firing rate (either linear or sigmoidal or hyperbolic), $I(\mathbf{y}, t)$ is the input to the position $\mathbf{y}$ at time t in M and h is the neuron threshold. w_M is given by the equation (2).

$$
w_M(\mathbf{x} - \mathbf{x}') = A e^{-\frac{|\mathbf{x} - \mathbf{x}'|^2}{a^2}} - B e^{-\frac{|\mathbf{x} - \mathbf{x}'|^2}{b^2}} \text{ with } A, B, a, b \in \Re^{*+} \ .
$$

(2)

3.2   Some Properties of the CNFT

There are several models using population codes focusing on noise clean-up such as in [26, 27] or more general types of computation such as sensorimotor transformations, feature extraction in sensory systems or multisensory integration [28, 29, 30]. Deneve et al. [27] were able to show through analysis and simulations that it is indeed possible to implement an ideal observer using biologically plausible models of cortical circuitry and it comes as no surprise that this model relies heavily on lateral interactions. We also designed a model [31] that uses lateral interactions, as proposed by the CNFT, and fall into the more general case of *recurrent network whose activity relaxes to a smooth curve peaking at a position that depends on the encoded variable* that was analyzed as being a good implementation of a Maximum Likelihood approximator [27]. This dynamic model of attention has been described using the Continuum Neural Field Theory that explains attention as being an emergent property of a neural population. Using distributed and iterative computation, this model has been proven very robust and able to track one static or moving target in the presence of noise with very high intensity or in the presence of a lot of distractors, possibly more salient than the target. The main hypothesis concerning target stimulus is that it possesses a spatio-temporal continuity that should be observable by the model, i.e. if the movement of the target stimulus is too fast, then the model can possibly loose its focus. Nonetheless, this hypothesis makes sense when considering *real world* robotic applications.

4   A Computational Model of Spatial Visual Attention

The first model that has been designed in [31] demonstrated why and how CNFT can be used to attend to one moving stimulus and this model has been proven to be extremely robust against both noise and distractors. But, what has been considered to be a nice feature in this previous model is now viewed as a drawback since it prevents the model from switching to another stimulus when this is required to achieve a relevant behaviour. The natural solution to this situation is then to actively inhibit this behaviour in order to allow the model to switch to another stimulus. But then, the difficulty is to somehow ensure that the model will not switch back and forth between two stimuli only. Since the ultimate goal of the model is the active exploration of the visual scene, it needs a working memory to be able to memorize what has been already seen and what has not. This is even more difficult when considering camera movements that result in having any stimulus moving on the retina image. A static working memory system would be useless in this situation because it is generally disconnected from perception, while for a visual exploration task the working memory system has to track down every attended stimuli in order to prevent attending them again. There are neurophysiological evidences [32] that inhibition of return (tightly linked with working memory) can follow moving targets. In the following paragraphs, we will describe the role and connectivity of each map in the model represented in Figure 1. In a few words, there are three sub-systems: the INPUT-VISUAL-FOCUS ensemble, whose role is to process the visual input and to generate a focus of attention; the FEF-WM ensemble, designed to remember the previously focused locations; the switching sub-architecture, used to dynamically change the current focus of attention. Even if some maps have biologically inspired names, discussing about this plausibility is out of the scope of this paper.

4.1   Architecture

Input Map. The INPUT map in the model (cf. Figure 1) is a pre-processed representation of the visual input. As our aim is not to focus on visual processing but on motor aspects of attention, we did not model any local filtering nor recognition. What we use as input in our model is a kind of "saliency map" (see [33]) which represents in retinotopic coordinates the relative salience of the objects present in the visual field. This may be the role of the area LIP in monkey as discovered by Gottlieb et al. [34], but this issue is still controversial. In the simulation, we will generate bubbles into that map of 40 × 40 units, but

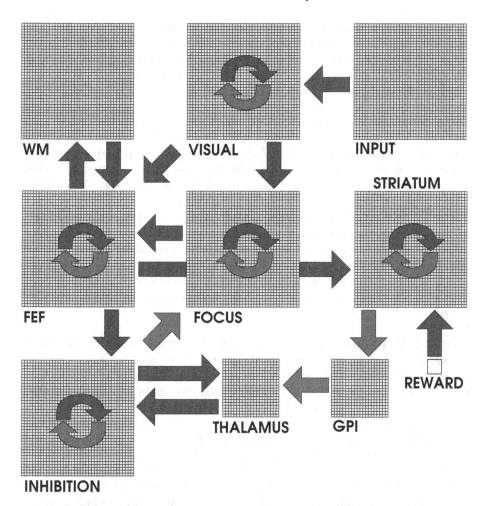

Fig. 1. The different maps of the model, with schematic connections. Red (dark) arrows represent excitatory connections, blue (light) arrows represent inhibitory connections, circular arrows represent lateral connections. See text for details

we will explain in Section 4.3 how it is implemented on the robot. This map has no dynamic behaviour, it just represents visual information. In contrast, all the following maps have dynamics like in equation 1 with mexican-hat shaped lateral connectivity like in equation 2. Parameters will be given in Appendix.

Visual Map. The VISUAL map receives excitatory inputs from the INPUT map with a "receptive-field"-like connection pattern that allows topology to be conserved since the two maps have the same size. The lateral connectivity in the VISUAL map ensures that only a limited number of bubbles of activity can emerge anytime. As a consequence, the activity of the VISUAL map is virtually noiseless and expresses only the most salient stimuli present within the input. If too many stimuli are presented in the same time, then the dynamic interactions within the map will reduce this number to the most salient stimuli only. Roughly, in the present architecture, this number is around seven stimuli which can be presented simultaneously (this is mainly due to the size of the map compared to the lateral extent of the inhibitory lateral connections).

Focus Map. The FOCUS map receives excitatory inputs from the VISUAL map and have the same size as the VISUAL map to ensure that topology is loosely conserved. The lateral connectivity is wider than in the VISUAL map so that only one bubble of activity can emerge anytime. When no stimulus is present within the input, no activity is present within the FOCUS map. With these three maps (INPUT, VISUAL and FOCUS), the system can track one stimulus in the input map which will represented by only one bubble of activation in FOCUS. In [31] we demonstrated that this simple system had interesting denoising and stability properties. Now, to implement a coherent attention-switching mechanism, we need to add a switching mechanism coupled with a working memory system. The switching mechanism will be done by adding an inhibitory connection pattern from a map later labelled INHIBITION. Let's first describe the working memory system.

FEF and WM Maps. FEF and WM maps implement a dynamic working memory system that is able to memorize stimuli that have already been focused in the past together with the currently focused stimulus. The basic idea to perform such a function is to reciprocally connect these two maps one with the other where the WM map is a kind of reverbatory device that reflects FEF map activity. Outside this coupled system, the FEF map receives excitatory connections (using gaussian receptive fields to conserve topology) from both the VISUAL and FOCUS maps. Activity in the VISUAL map alone is not sufficient to generate activity in FEF; it needs a consistent conjunction of activity of both VISUAL and FOCUS to trigger some activity in FEF map. Since there is only one bubble of activity in the focus map, the joint activation of VISUAL and FOCUS only happens at the location of the currently focused stimulus. So, when the system starts, several bubbles of activation appear in VISUAL map, only one emerges in FOCUS, what allows the appearance of the same bubble in FEF map. As soon as this bubble appears, it is transmitted to WM which starts to show activity at the location of that bubble which in turn excites the FEF map. This is a kind of reverbatory loop, where mutual excitation leads to sustained activity.

One critical property of this working memory system is that once this activity has been produced, WM and FEF map are able to maintain this activity even when the original activation from FOCUS disappears. For example, when the system focuses on another stimulus, previous activation originating from the FOCUS map vanishes to create a bubble of activity somewhere else. Nonetheless, the previous coupled activity still remains, and a new one can be generated at the location of the new focus of attention.

Importantly, the system is also sensitive to the visual input and thus allows memorized stimuli to have a very dynamic behaviour since a bubble of activity within FEF and WM tends to track the corresponding bubble of activity within the VISUAL map. In other words, once a stimulus has been focused, it starts reverberating through the working memory system which can keep track of this stimulus, even if another one is focused. However, if the corresponding bubble in VISUAL disappears (e.g. get out of the image), the activity in FEF and WM vanishes. Another mechanism should be involved to remember out-of-view targets.

Switching Sub-architecture. The mechanism for switching the focus in the FOCUS map is composed of several maps (REWARD, STRIATUM, GPI, THALAMUS and INHIBITION). The general idea is to actively inhibit locations within the focus map to prevent a bubble of activity from emerging at these locations. This can be performed in cooperation with the working memory system which is able to provide the information on which locations have already been visited.

The STRIATUM map receives weak excitatory connections from the FEF map, which means that in the normal case no activity appears on STRIATUM map. But when the REWARD neuron (which sends a connection to each neuron in the STRIATUM) fires, it allows bubbles to emerge at the location they are potentiated by FEF. The REWARD activity is a kind of "gating" signal which allows the STRIATUM to reproduce or not the FEF activity.

The STRIATUM map sends inhibitory connections to the GPI, which has the property to be tonically active: if the GPI neurons receive no input, they will show a great activity. They have to be inhibited by the STRIATUM to quiet down. In turn, the GPI map sends strong inhibitory connections to the THAL map, which means that when there is no reward activity, the THAL map is tonically inhibited and can not show any activity. It is only when the REWARD neuron allows the STRIATUM map to be active that the GPI map can be inhibited and therefore the THAL map can be "disinhibited". Note that this is not a reason for the THAL to show activity, but it allows it to respond to excitatory signals coming from somewhere else.

This disinhibition mechanism is very roughly inspired by the structure of the basal ganglia, which are known as mediating selection of action [35]. It allows more stability than direct excitation of the THAL map by FEF.

The INHIBITION map is reciprocally and excitatory connected with the THAL map, in the same way as FEF and WM are. But the reverberatory mechanism is gated by the tonic inhibition of GPI on THAL. It is only when the REWARD neuron fires that this reverbation can appear. INHIBITION receives weak excitatory connections from FEF (not enough to generate activity) and sends inhibitory connections to FOCUS. The result is that when there is no reward, the inhibitory influence of the INHIBITION map is not sufficient to change the focus of attention in FOCUS, but when the REWARD neuron fires, INHIBITION interacts with THAL and shows high activity where FEF stores previously focused locations, what prevents the competition in FOCUS to create a bubble at a previously focused location, but rather encourages it to focus on a new location.

4.2   Simulated Behaviour

Having described the architecture of the model and the role of the different maps, a switching sequence, where we want the model to change the focused stimulus in favor of another unfocused one, is quite straightforward. As detailed in Figure 2, the dynamic of the behavior is ruled both by the existing pathways between the different maps (either excitatory or inhibitory) and the dynamic of the neurons.

The INPUT map is here clamped to display three noisy bubbles at three different locations in the visual field, so that the network can sequentially focus these points. In Figure 2-a), the three noisy bubbles in map INPUT are denoisified in the VISUAL map, allowing only one bubble to emerge in the FOCUS map which is immediately stored in FEF and WM. In Figure 2-b), a switch signal is explicitly sent to the network via the REWARD unit, allowing the STRIATUM to be excited at the location corresponding to the unique memorized location in the working memory system. This striatum excitation inhibits in turn the corresponding location within the GPI map. In Figure 2-c), the localized destabilization of the GPI prevents it from inhibiting the thalamus at this same location and allow the inhibition map to activate itself, still at the same location. In Figure 2-d), the INHIBITION map is now actively inhibiting the FOCUS map at the currently focused location. In Figure 2-e), the inhibition is now complete and another bubble of activity starts to emerge within the FOCUS map (precise location of the next bubble is unknown, it is only ensured that it can not be the previously visited stimulus). In Figure 2-f), once the focus is fully activated, it triggers the memorization of the new location while the previous one is kept in memory.

4.3   Experimental Results on a Robotic Platform

This model is built to deal with switching and focusing spatial selective attention on salient locations. It is not meant to model the whole attention network. In particular, we did not implement the recognition pathway and feature-selective attention because we only wanted to figure out how attention can sequentially scan equivalent salient locations. When we wanted to test this model on our PeopleBot robot, we therefore chose to consider identical targets, for example green lemons, which are artificially made salient for the system.

The experimental environment is the following (see Figure 3): we put the PeopleBot in front of three green lemons lying on a table. At start, the camera is directed somewhere on the table with each fruit somewhere in its viewfield. The task for the system is to sequentially gaze (by moving its mobile camera) at the three targets while never looking twice at the same fruit, even if the fruits are moved during the experiment.

To make the fruits artificially salient, we applied a gaussian filter on the image centered on the average color of a green lemon (H=80 S=50 in HSV coordinates). This results in three noisy patches of activation (between 0 and 1) in the transformed image (see Figure 4). These activations then feed the INPUT map to be represented by a smaller set of neurons (here 40×40). As the

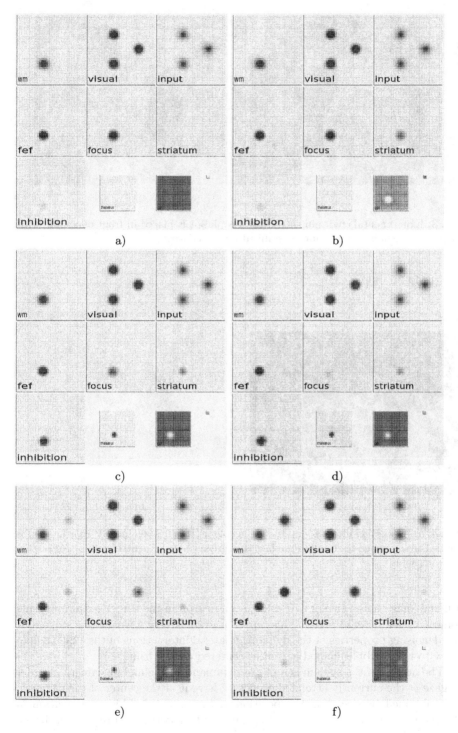

Fig. 2. A simulated sequence of focus switching. See text for details

a)   b)

Fig. 3. Experimental environment: a) the PeopleBot is placed in front of a table with three green lemons. b) The image grabbed by the camera

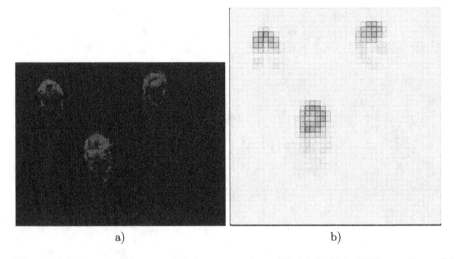

a)   b)

Fig. 4. a) A gaussian filter around the green colour (H=80 S=50 in HSV coordinates) is applied to the image to simulate the fact that green objects are attended. b) Activation in INPUT map

original image had a size of 640×480, each neuron in the INPUT map represents something like 12×12 pixels. This representation is very noisy at this stage, but the denoising properties of the dynamical lateral interactions in the VISUAL map allow to have bubble-shaped activities centered on the fruit.

The output of the system is a motor command to the mobile camera in order to gaze at the currently attended object (ie have it at the center of the camera). It is obtained by decoding the position of the unique bubble of activation in the FOCUS map in $[-0.5, 0.5]^2$ and by linearly transforming this position into a differential command to the effectors of the camera. This motor mapping is quite

Fig. 5. Some snapshots of the sequence executed by the robot when trying to sequentially gaze at three green lemons. First, the robot initially looks at somewhere on the table. Then it gazes successively at fruit 1 and fruit 2. While fixating fruit 2, even if someone exchanges fruit 1 and the third not previously focused fruit, the robot will fixate the third "novel" fruit

obvious but dependent on the type of mobile camera, so we will not describe it here. One important thing to notice here is that this command is differential, i.e. just a little percentage of the displacement needed to go to the target is actuated, then the network is updated with a new image and so on. We will discuss this limitation later.

The switching sequence there is the same as in Section 4.2, the only difference being the motor outputs. The user still has to send the switching signal by "clamping" the reward unit to its maximal value for one step, and leaving it decay with its own dynamic.

An example of behaviour of the model is given in Figure 5. The center of gaze of the camera is first directed somewhere on the table. The model randomly decides to focus its attention on the bottom-right fruit (let's understand "randomly" as "depending on the noise in the input image, the initial state of the network and so on") and step-by-step moves the camera to it. When the camera is on it, the user can decide whenever he wants to focus another fruit by clamping the reward neuron (in a biologically relevant scenario, the system would have to learn that he could obtain more reward by switching its focus and therefore make the reward neuron fire) which inhibits the currently focused object. The focus of attention then moves to one of the two remaining fruits (here the bottom-left one), what makes the camera gaze at it. At this point, the "working memory" system contains the current and the past focused fruits. If the user clamps again the reward unit, the new focused location will obligatorily be on the third fruit, even if one slowly exchanges the locations of the first and the third fruit, because the representations in the working memory are updated by perception.

5   Conclusion

Despite a massively distributed architecture, the model we presented is able to accurately switch between available stimuli in spite of noise present at several levels, fruit positions, distance of the robot from the table, lightening conditions, etc. The resulting serialization of the behavior is a direct consequence of both the dynamic of neurons and the existence of dedicated pathways between different maps. What is important to understand is that any neuron in any map at any time is always computing its activity from the available information it can perceive via both afferent and lateral connections. The point is that there is no such thing as a concept of layer, an order of evaluation nor a central executive (either at the level of maps or at the level of the whole model). This is quite a critical feature since it somehow demonstrates that the resulting and apparent serialization of behavior in this model is a simple emergent property of the whole architecture and consequently, there is no need of this famous central supervisor to temporally organize elementary actions.

As a part of the FET Mirrorbot project, this work does not aim to model visual attention as a whole, but rather to offer a mechanism allowing efficient visual search in a complex scene. As the Mirrobot scenario involves active exploration of a complex and natural environment [36], it is computationally untractable to consider the visual input as a raw data that would be finally transformed into the correct motor output: the strategy we propose relies on the fact that potentially interesting visual elements are often salient locations on the image, and that sequential scanning of these salient locations is sufficient to appropriately represent the environment. The different partners of the Mirrobot project are currently working at gathering their different models to create a global and coherent behaviour for the robot.

Nevertheless, the model presents a major drawback which is the speed at which the camera can change its gaze to a target: the network has to be updated after a camera displacement of approximately 5° so that the spatial proximity of a fruit on the image before and after a movement of the camera can be interpreted by the network as the representation of the same object. This is quite incoherent with the major mode of eye movement, namely saccades, as opposed to this "pursuit" mode which can not be achieved under voluntary control: pursuit eye movements are only reflexive. To implement saccades, we would need an anticipation mechanism that could foresee what would be the estimated position of a memorized object after a saccade is made. Such a mechanism has been discovered in area LIP of the monkey by [17] where head-centered visual representations are remapped before the execution of a saccade, perhaps via corollary motor plans from FEF. Ongoing work is addressing this difficult problem by implementing a mechanism whose aim is to predict (to some extents) the consequences of a saccade on the visual input.

Acknowledgement. The authors wish to thank the FET MirrorBot project and the Lorraine Region for their support.

References

1. A. Treisman and G. Gelade. A feature-integration theory of attention. *Cognitive Psychology*, 12:97–136, 1980.
2. L. G. Ungerleider and M. Mishkin. Two cortical visual systems. In D. J. Ingle, M. A. Goodale, and R. J. W. Mansfield, editors, *Analysis of Visual Behavior*, pages 549–586. The MIT Press, Cambridge, Mass., 1982.
3. N. Rougier. *Modèles de mémoires pour la navigation autonome*. PhD thesis, Université Henri Poincaré Nancy-I, 2000.
4. J. H. Reynolds and R. Desimone. The role of neural mechanisms of attention in solving the binding problem. *Neuron*, 14:19–29, 1999.
5. M. I. Posner. Orienting of attention. *Quarterly Journal of Experimental Psychology*, 32:3–25, 1980.
6. A. Treisman. Features and objects: The bartlett memorial lecture. *The Quarterly Journal of Experimental Psychology*, 40:201–237, 1988.
7. J. Moran and R. Desimone. Selective attention gates visual processing in the extrastriate cortex. *Science*, 229:782–784, 1985.
8. R. Desimone. Visual attention mediated by biased competition in extrastraite visual cortex. *Philosophical Transactions of the Royal Society London*, 353:1245–1255, 1998.
9. S. J. Luck, L. Chelazzi, S. A. Hillyard, and R. Desimone. Neural mechanisms of spatial attention in areas v1, v2 and v4 of macaque visual cortex. *Journal of Neurophysiology*, 77:24–42, 1997.
10. M. I. Posner and S. E. Petersen. The attentional system of the human brain. *Annual Review of Neurosciences*, 13:25–42, 1990.
11. G. Rizzolatti, L. Riggio, I. Dascola, and C. Ulmita. Reorienting attention across the horizontal and vertical meridians: Evidence in favor of a premotor theory of attention. *Neuropsychlogia*, 25:31–40, 1987.
12. G. Rizzolatti, L. Riggio, and B. M. Sheliga. Space and selective attention. In C. Ulmità and M. Moscovitch, editors, *Attention and Performance*, volume XV, pages 231–265. MIT Press, Cambridge, MA, 1994.
13. B. M. Sheliga, L. Riggio, L. Craighero, and G. Rizzolatti. Spatial attention-determined modifications in saccade trajectories. *Neuroreport*, 6(3):585–588, 1995.
14. A. C. Nobre, D. R. Gitelman, E. C. Dias, and M. M. Mesulam. Covert visual spatial orienting and saccades: overlapping neural systems. *NeuroImage*, 11:210–206, 2000.
15. L. Craighero, M. Nascimben, and L. Fadiga. Eye position affects orienting of visuospatial attention. *Current Biology*, 14:331–333, 2004.
16. T. Moore and M. Fallah. Control of eye movements and spatial attention. *Proceedings of the National Academy of Sciences*, 98(3):1273–1276, 2001.
17. C. L. Colby, J. R. Duhamel, and M. E. Goldberg. Visual, presaccadic, and cognitive activation of single neurons in monkey lateral intraparietal area. *Journal of Neurophysiology*, 76:2841–2852, 1996.
18. M. I. Posner and Y. Cohen. Components of visual orienting. In H. Bouma and D. Bouwhuis, editors, *Attention and Performance*, volume X, pages 531–556. Erlbaum, 1984.
19. J. W. DeFockert, G. Rees, C. D. Frith, and N. Lavie. The role of working memory in visual selective attention. *Science*, 291:1803–1806, 2001.
20. S. M. Courtney, L. Petit, J. M. Maisog, L. G. Ungerleider, and J. V. Haxby. An area specialized for spatial working memory in human frontal cortex. *Science*, 279:1347–1351, 1998.

21. H. Frezza-Buet, N. Rougier, and F. Alexandre. *Neural, Symbolic and Reinforcement Methods for Sequence Learning*, chapter Integration of Biologically Inspired Temporal Mechanisms into a Cortical Framework for Sequence Processing. Springer, 2000.
22. H. R. Wilson and J. D. Cowan. A mathematical theory of the functional dynamics of cortical and thalamic nervous tissue. *Kybernetic*, 13:55–80, 1973.
23. J. Feldman and J.D. Cowan. Large-scale activity in neural nets. i. theory with applications to motoneuron pool responses. *Biological Cybernetics*, 17:29–38, 1975.
24. S.-I. Amari. Dynamical study of formation of cortical maps. *Biological Cybernetics*, 27:77–87, 1977.
25. J. G. Taylor. Neural bubble dynamics in two dimensions: foundations. *Biological Cybernetics*, 80:5167–5174, 1999.
26. R. J. Douglas, C. Koch, M. Mahowald, K. A. Martin, and H. H. Suarez. Recurrent excitation in neocortical circuits. *Science*, 269:981–985, 1995.
27. S. Deneve, P. Latham, and A. Pouget. Reading populatiopn codes: a neural implementation of ideal observers. *Nature Neuroscience*, 2:740–745, 1999.
28. K. Zhang. Representation of spatial orientation by the intrinsic dynamics of the head-direction cell ensemble: A theory. *Journal of Neuroscience*, 16:2112–2126, 1996.
29. S. Deneve, P.E. Latham, and A. Pouget. Efficient computation and cue integration with noisy population codes. *Nature Neuroscience*, 4(8):826–831, 2001.
30. S. M. Stringer, E. T. Rolls, and T. P. Trappenberg. Self-organising continuous attractor networks with multiple activity packets, and the representation of space. *Neural Networks*, 17:5–27, 2004.
31. N. Rougier and J. Vitay. Emergence of attention within a neural population. *Submitted*, 2004.
32. S. P. Tipper, J. C. Brehaut, and J. Driver. Selection of moving and static objects for the control of spatially directed action. *Journal of Experimental Psychology: Human Perception and Performance*, 16:492–504, 1990.
33. L. Itti. Visual attention. In M. A. Arbib, editor, *The Handbook of Brain Theory and Neural Networks*, pages 1196–1201. MIT Press, 2nd edition, 2003.
34. J. P. Gottlieb, M. Kusunoki, and M. E. Goldberg. The representation of visual salience in monkey parietal cortex. *Nature*, 391:481–484, 1998.
35. O. Hikosaka, Y. Takikawa, and R. Kawagoe. Role of the basal ganglia in the control of purposive saccadic eye movements. *Physiological Reviews*, 80(3):953–978, 2000.
36. M. Elshaw, S. Wermter, C. Weber, C. Panchev, H. Erwin, and W. Schmidle. Mirrorbot scenario and grammar. Technical Report 2, Mirrorbot, 2002.

Appendix

Dynamic of the Neurons

Each neuron *loc* in a map computes a numerical differential equation given by equation 3, which is a numerized version of equation 1:

$$act_{loc}(t+1) = \sigma(act_{loc}(t) + \frac{1}{\tau} \cdot (-(act_{loc}(t) - baseline)$$

$$+ \frac{1}{\alpha} \cdot (\sum_{aff} w_{aff} \cdot act_{aff}(t) + \sum_{lat} w_{lat} \cdot act_{lat}(t)))) \,. \tag{3}$$

Table 1. Parameters for each map: number of units and baseline activities

Map	Size	Baseline
VISUAL	40*40	0.0
FOCUS	40*40	-0.05
FEF	40*40	-0.2
WM	40*40	0.0
INHIBITION	40*40	-0.1
THAL	20*20	0.0
GPI	20*20	0.8
STRIATUM	40*40	-0.5
REWARD	1*1	0.0

where:

$$\sigma(x) = \begin{cases} 0 & \text{if } x < 0, \\ 1 & \text{if } x > 1, \\ x & \text{else .} \end{cases} \tag{4}$$

and τ is the time constant of the equation, α is a weighting factor for external influences, aff is a neuron from another map and lat is a neuron from the same map.

All maps have the values $\tau = 1$ and $\alpha = 13$ except the REWARD map where $\tau = 15$. The size and baseline activities of the different maps are given in Table 1.

Connections Intra-map and Inter-map

The lateral weight from neuron lat to neuron loc is:

$$w_{lat} = Ae^{-\frac{\text{dist}(loc,lat)^2}{a^2}} - Be^{-\frac{\text{dist}(loc,lat)^2}{b^2}} \text{ with } A, B, a, b \in \Re^{*+} \text{ and } loc \neq lat . \tag{5}$$

where dist(loc, lat) is the distance between lat and loc in terms of neuronal distance on the map (1 for the nearest neighbour).

In the case of a "receptive field"-like connection between two maps, the afferent weight from neuron aff to neuron loc is:

$$w_{aff} = Ae^{-\frac{\text{dist}(loc,aff)^2}{a^2}} \text{ with } A, a \in \Re^{*+} . \tag{6}$$

The connections in the model are described in Table 2.

These parameters have been found experimentally to ensure that the global functioning of the network is correct. Nevertheless, they are only orders of magnitude because small variations in their value do not affect drastically the performance of the network. For example, the fact that all lateral connections have the same parameters (except for the FOCUS map which has a wider inhibitory extent – $b = 17.0$ – to allow the emergence of only one bubble throughout the map) is only for the sake of simplicity. The variations in the parameter a of the "receptive-field" connections are explained by the smaller size of the THAL and GPI maps (as they are supposed to be small subcortical nuclei) and by the fact

Table 2. Connections between maps: parameters refer to equations 5 and 6

Source Map	Destination Map	Type	A	a	B	b
INPUT	VISUAL	receptive-field	2.0	2.0	-	-
VISUAL	VISUAL	lateral	2.5	2.0	1.0	4.0
VISUAL	FOCUS	receptive-field	0.25	2.0	-	-
FOCUS	FOCUS	lateral	1.7	4.0	0.65	17.0
VISUAL	FEF	receptive-field	0.25	2.0	-	-
FOCUS	FEF	receptive-field	0.2	2.0	-	-
FEF	FEF	lateral	2.5	2.0	1.0	4.0
FEF	WM	receptive-field	2.35	1.5	-	-
WM	FEF	receptive-field	2.4	1.5	-	-
FEF	INHIBITION	receptive-field	0.25	2.5	-	-
INHIBITION	FOCUS	receptive-field	-0.2	3.5	-	-
INHIBITION	INHIBITION	lateral	2.5	2.0	1.0	4.0
INHIBITION	THAL	receptive-field	3.0	1.5	-	-
THAL	INHIBITION	receptive-field	3.0	1.5	-	-
FEF	STRIATUM	receptive-field	0.5	2.5	-	-
STRIATUM	STRIATUM	lateral	2.5	2.0	1.0	4.0
STRIATUM	GPI	receptive-field	-2.5	2.5	-	-
GPI	THAL	receptive-field	-1.5	1.0	-	-
REWARD	STRIATUM	one-to-all	8.0	-	-	-

that the inhibitory connection from INHIBITION to FOCUS has to be wider to achieve successful switching.

An interesting issue would be to incorporate a learning mechanism to set these parameters, but it is very difficult to define an error function as the desired behaviour of the network is strongly temporal. In order to do so, we would have to integrate this mechanism into a more general framework, where reinforcement learning could play a major role in weighting the connections through trial-and-error learning phases.

A Hybrid Architecture Using Cross-Correlation and Recurrent Neural Networks for Acoustic Tracking in Robots

John C. Murray, Harry Erwin, and Stefan Wermter

Hybrid Intelligent Systems,
School for Computing and Technology,
University of Sunderland, Tyne-and-Wear, SR6 0DD, UK
John.Murray@sunderland.ac.uk

Abstract. Audition is one of our most important modalities and is widely used to communicate and sense the environment around us. We present an auditory robotic system capable of computing the angle of incidence (azimuth) of a sound source on the horizontal plane. The system is based on some principles drawn from the mammalian auditory system and using a recurrent neural network (RNN) is able to dynamically track a sound source as it changes azimuthally within the environment. The RNN is used to enable fast tracking responses to the overall system. The development of a hybrid system incorporating cross-correlation and recurrent neural networks is shown to be an effective mechanism for the control of a robot tracking sound sources azimuthally.

1 Introduction

The way in which the human auditory system localizes external sounds has been of interest to neuroscientists for many years. Jeffress [1] defined several models of how auditory localization occurs within the Auditory Cortex (AC) of the mammalian brain. He developed a model for showing how one of the acoustic cues, namely that of the Interaural Time Difference (ITD) is calculated. This model describes the use of neurons within the auditory cortex as coincidence detectors [2]. Jeffress also describes the use of coincidence detectors for other auditory cues, namely Interaural Level Difference within the auditory cortex. These two cues (ITD and ILD) together enable the auditory system to localize a sound source within the external environment, calculating both the azimuth and distance from the observer.

Recently, robotics research has become interested in the ability to localize sound sources within the environment [3-4]. Audition is a vital sense for interpreting the world around us as audition enables us to perceive any object with an acoustic element. For localization and navigation purposes, the primary modality in robotics has been that of vision [5-6]. However, audition has some advantages over vision in that for us to visually see an object it must be within line of sight, i.e. not hidden by other objects. Acoustic objects however do not have to be within line of sight of the observer and can be detected around corners and when obscured by other objects.

S. Wermter et al. (Eds.): Biomimetic Neural Learning, LNAI 3575, pp. 73–87, 2005.
© Springer-Verlag Berlin Heidelberg 2005

This paper describes an acoustic tracking robotic system that is capable of sound source angle estimation and prediction along the horizontal plane. This system draws from some basic principles that exist in its biological equivalent, i.e. that of ITD, trajectory predictions, and the mammalian auditory cortex. Our system also has the ability to detect the angle of incidence of a sound source based on Interaural time difference.

2 Motivation for Research

How does the mammalian auditory system track sound sources so accurately within the environment? What mechanisms exist within the AC to enable acoustic localization and how can these be modeled? The AC of the mammalian brain works with excellent accuracy [7] and quick response times to the tracking and localization of dynamic sound sources.

With the increasing use of robots in areas such as service and danger scenarios [4], we are looking into the mechanisms that govern the tracking and azimuth estimation and predictions of sound sources within the mammalian AC to guide a model for sound source tracking within a robotic system. Our motivation comes from being able to create an acoustic sound source tracking robot capable of tracking azimuthally the angle of a dynamically moving stimulus.

With the scenario of interactive service robots, we can envisage the robot as a waiter in a restaurant serving drinks to the patrons. In order for this to be possible the customers would need to be able to attract the robot waiter's attention. The robot would need to detect the direction the sound comes from and attend to it. Fig. 1 shows an example of this scenario.

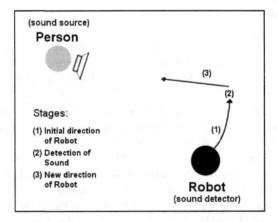

Fig. 1. Shows an example of a robot attending to sounds

3 The System Model

The model for our system has two main components; these are Azimuth Estimation and Neural Prediction. The first component in our model determines the azimuth of the sound source from the environment using Cross-Correlation and presents this angle to the Neural Predictor for estimation of the next predicted angle in the sequence using an RNN. The Neural Predictor receives a set of input angles passed from estimations of the angle of incidence from the azimuth estimation stage in the model and uses these to predict the angle for the robot to attend to next. The idea behind this is to enable the robot to move to the next position where it expects to hear the sound and then waits to see if it hears it. The system therefore has a faster response time to its tracking ability as the robot is not constantly calculating the position of the sound source, then attending and repeating this phase recursively, as this would mean the robot would be in constant lag of the actual position for the sound source. Instead our system has an active model for predicting the location of the sound source.

The system requires two valid sound inputs (as discussed in section 3.2). When the network receives its second input at time t_2 the network provided an output activation to attend to next. This is when the robot is informed to go at time t_2 as opposed to the passion the sound was detected at during time t_2 itself. Our system therefore provides a faster response in attending to the position of the dynamic sound source enabling more real-time tracking.

3.1 Azimuth Estimation

Azimuth estimation is the first stage of the overall system model and is used to determine the angle of incidence of the dynamic sound source from the environment. The azimuth estimation is performed by a signal processing variation of Cross-Correlation (Eq. 1). It has also been shown that the AC employs the use of Cross-Correlation as discussed by Licklider [8] for angle estimation. Therefore, we have employed Cross-Correlation to analyze the two signals $g(t)$ and $h(t)$ received at the left and right microphones in our system. Ultimately, Cross-Correlation as discussed in [9] is used for determining the ITD with the use of coincidence detectors [1].

Within our model the Cross-Correlation method is used to check $g(t)$ and $h(t)$ for the position of maximum similarity between the two signals, which results in the creation of a product vector C where each location represents the products of signals $g(t)$ and $h(t)$ at each time step. The robot records a 20ms sample of sound at each microphone resulting in an N x M matrix of 2 x 8820 where each row represents the signal received at each of the microphones. To correlate the two signals they are initially offset by their maximum length. At each time step signal $h(t)$ is 'slid' across signal $g(t)$ and the product of the signals is calculated and stored in the product vector C.

$$Corr(g,h)_j(t) \equiv \sum_{k=0}^{N-1} g_{j+k} h_k \qquad (1)$$

Fig. 2 below shows an example of how the two signals $g(t)$ and $h(t)$ are checked for similarity. As can be seen in the graph of Fig. 2a we can see that the function starts by offsetting the right channel $h(t)$ to the beginning of the left channel $g(t)$ and gradually 'slides' across until $h(t)$ leads $g(t)$ by the length of the matrix (graph in Fig. 2c). When the signals are in phase (shown by the shaded area in the graph of Fig. 2b) the resultant correlation vector will produce a maximum value at this time step position in the product vector C.

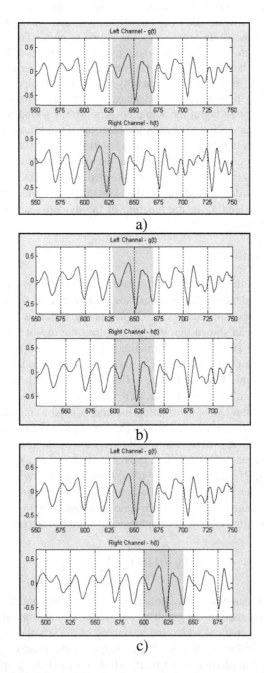

Fig. 2. Shows the 'sliding' of the signals presented to the cross-correlation function

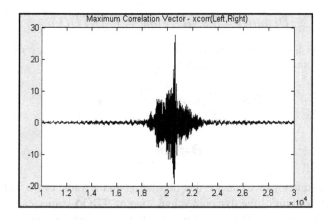

Fig. 3. Shows the product vector *C* of the Cross-Correlation of signals *g(t)* and *h(t)*

The maximum position within the resultant correlation vector represents the point of maximum similarity between *g(t)* and *h(t)*. If the angle of incidence was at $0°$ then this would result in *g(t)* and *h(t)* being in phase with each other therefore resulting in the maximum value in *C* being in the middle location of the vector. See Fig. 4 for an example of the creation of a correlation vector from two slightly offset signals.

Therefore, to determine the amount of delay offsets for the ITD we subtract the location of the maximum point in *C* from the size of *C/2*. We divide the correlation vector *C* by 2 as the method of cross-correlation creates a vector that is twice the size of the original signal (due to the sliding window) and therefore to find the mid point of the vector (i.e. zero degrees) we divide by 2. This result is used to help determine the angle of incidence of the sound source. The sounds within the system are recorded with a sample rate of 44.1 KHz resulting in a time increment Δ of 22.67µs. Therefore, each position within the correlation vector is equal to Δ and represents 2.267^{-5} seconds of time. Knowing the amount of time per delay increment and knowing the number of delay increments (from the correlation vector *C*) then using equation 2 we can calculate the ITD, or more precisely in terms of our model, the time delay of arrival (TDOA) of the sound source between the two microphones.

$$\text{TODA} = ((\text{length}(C) / 2) - C_{\text{MAX}})*\Delta \qquad (2)$$

This result gives us the time difference of the reception of the signal at the left and right microphones; this in turn is used in conjunction with trigometric functions to provide us with Θ the angle of incidence of the sound source in question. Looking at Fig. 5 we can determine which trigometric function we need to use to determine the angle of incidence.

We have a constant value for side 'c' set at 0.30 meters (the distance on the robot between the two microphones, see Fig. 10) and 'a' can be determined from the ITD or TDOA from Eq. 2 substituted into Eq. 3. Therefore, in order to determine the azimuth of the sound source we can use the inverse sine rule as shown in Eq 5.

$$Distance = Speed \times Time = 384 \times TDOA \tag{3}$$

where, Speed is the speed of sound = 384m/s at room temperature of 24°C at sea level. TDOA is the value returned from Eq. 2.

$$Sin\Theta = \frac{a}{c}, Cos\Theta = \frac{b}{c}, Tan\Theta = \frac{c}{b} \tag{4}$$

$$\Theta = Sin^{-1}\frac{a}{c} \tag{5}$$

From these equations we can see that depending on the value of the TDOA we can determine the azimuth of the dynamic sound source. TDOA values can range from -90° to +90°. The values of Θ returned are used to provide input to the recurrent neural network of the second stage within the system for prediction of azimuth positions. This initial part of the model has shown that it is indeed possible to create a robotic system capable of modeling ITD to emulate similar robotic tasks. That is, we have looked at functional mechanisms within the AC and represented the output of these mechanisms within our robot model.

Signals:	g(t) 111112222111	h(t) 111111112222	
		Product Vector C	
		Location	Value
111112222211100000000000		1	1
000000000001111111112222			
111112222211110000000000		2	2
000000000011111111112222			
111112222211100000000 0		3	3
000000000111111111 2222			
111112222211100000000		4	5
00000000111111112222			
111112222211110000000		5	7
000000011111111 2222			
111112222211100000 0		6	9
000000111111112222			
111112222211100000		7	11
000001111111112222			
111112222211100000		8	12
00001111111112222			

Fig. 4. Shows how the sliding-window of the Cross-Correlation method builds the Correlation vector C. As the signals get more in phase the value in C increases

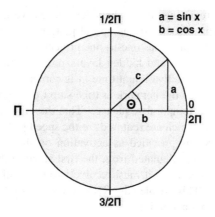

Fig. 5. Geometric diagram of sin(x) and cos(x)

3.2   Speed Estimation and Prediction

Speed estimation and prediction is the second stage in our model and is used to estimate the speed of the sound source and predict the next expected location. It has been shown that within the brain there exists a type of short term memory that is used for such prediction tasks and in order to predict the trajectory of an object it is required that previous positions are remembered [10] to create temporal sequences.

Within this stage of our model, we create a recurrent neural network (RNN) with the aim to train this network to detect the speed of a sound source and provide estimated prediction positions for the robot to attend to. This stage in the model receives its input from the previous azimuth estimation stage as activation on the relevant neuron within the input layer of the network. Each neuron within the input and output layers represent $2°$ of azimuth, therefore an angle of $1°$ will cause activation on the first input neuron whilst an angle of $3°$ will cause activation on the second input neuron. As can be seen in Fig. 6 the input and out layers of the network have 45 units each with each unit representing $2°$ of azimuth. Therefore, the layers only represent a maximum of $90°$ azimuth, however as the sign (i.e. + or – angle recorded by the cross-correlation function) is used to determine if the source is left or right of the robots center then the network can be used to represent $+90°$ and $-90°$ thus covering the front hemisphere of the robot.

The RNN consists of four separate layers with weight projections connecting neurons between layers. The architecture of the network is as follows:

Layer 1 – Input – 45 Units
Layer 2 – Hidden – 30 Units
Layer 3 – Context – 30 Units
Layer 4 – Output – 45 Units

Fig. 6 shows the layout of the architecture within the network along with the fully connected layers. The network developed is based on the Elman network [11] which provides a method for retaining context between successive input patterns. In order for the network to adapt to sequential temporal patterns a context layer is used. To provide context the hidden layer has one-to-one projections to the context layer within the network (both the context and hidden layers must contain the same amount of neurons). The hidden layer activation at time t_{-1} is copied to the context layer so that the activation is available to the network at time step t during the presentation of the second pattern within the temporal sequence. This therefore enables the network to learn temporal sequences which are required for the speed prediction task.

The input to the RNN is provided as activation on the input neuron that corresponds to the current angle calculated from the first stage in the model. In order for the network to make a prediction it must receive two sequential input activation patterns at times t_{-1} and t_0. This enables the RNN to recognize the temporal pattern and provide the relevant output activation.

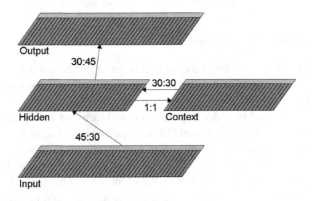

Fig. 6. Recurrent Neural Network architecture used in model

Due to the RNN output activation only being set for the predicted angle to attend to, it is necessary to still provide the system with an angle of incidence to move to before the prediction is made. The angle recorded by the initial stage of the system is referred to as the perceived angle due to this being the angle of the sound source relative to the robot. The system also maintains a variable containing the current angle of the robot from its initial starting position of $0°$. Therefore the input activation for the RNN is calculated from Current angle + Perceived angle with the output activation being the angle from the starting position. Therefore, the angle to move to is RNN output angle − current angle.

The weight update algorithm for the network is based on that of the normal backpropagation as shown in Eq 7, with the stopping criterion for the network being set to 0.04 for the Sum Squared Error (SSE). This value was chosen as it was the highest value of the SSE that classified all patterns within the least number of epochs. The value of the SSE is checked after each epoch. If the change in the SSE is below 0.04 between epochs then training stops and the network is said to have converged, otherwise the weights are adapted and presentation of training patterns continues.

$$a_i^C(t+1) = a_i^H(t) \tag{6}$$

$$\Delta w_{ij}(n+1) = \eta_i {}_j o_i + a\Delta w_{ij}(n) \tag{7}$$

where, Δw_{ij} = the weight change between neurons i and j, n = current pattern presented to the network, η = learning rate = 0.25, δ_j = error of neuron j, o_i = output of neuron i, α = momentum term used to prevent the weight change entering oscillation by adding a small amount of the previous weight change to the current weight change.

The RNN architecture provides the system with the ability to recognize a predetermined number of speeds (provided during training) from a sequential temporal pattern and therefore introducing a form of short-term memory into the model. After the RNN receives two time steps a prediction of the next location of the sound source is provided for the robot to attend to. This enables a faster response from the system and therefore enables a more real-time implementation of the sound source tracking system. This is due to the fact that the system does not have to wait for a subsequent third sound sample in order to determine the location in azimuth of the sound source.

3.3 Training the RNN

In order to train the RNN to recognize the various speeds, a separate training sub-group was created within the training environment for each individual speed. Each sub-group within the training environment contains the events required to train the network to individual speeds.

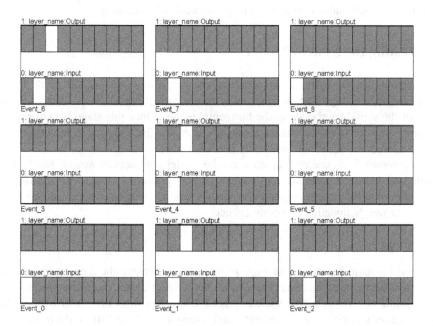

Fig. 7. Sub-group training environment for speed = 1 showing the required input activations in order to create an output activation remembering that the patterns are temporal

The events (input sequences i.e. angle representations) are activations of '1' on the neuron they represent and are presented for training to the network sequentially in the order expected from a sound source (and shown in Table 2). This is to ensure the network learns the correct temporal sequence for the speeds it needs to recognize and provide prediction for.

The environment shown in Fig. 7 presents the nine events in a sequential manner, that is, every time the pattern is presented to the network the events are given in the same order Event_0 → Event_8. However, the last two events (Event_7 and Event_8) within the training sub-group deviate from the temporal order and output activation of the first 7 events. These two events are provided to ensure the network does not only learn to provide activation output on the presentation of input activation on neuron 2 in Fig. 7 but also ensures that past context is taken into account and output activation is only set if a valid temporal pattern is provided to the network, in this case, at time t_{-1} activation on input neuron 1 and at time t_0 activation on input neuron 2 resulting in an output activation at time t_0 on the third output neuron in the network.

Within the training environment 20 sub-groups were created with each sub-group representing a set speed, as each sub-group contains 9 training events this gives us a total of 180 events to present to the network. As previously mentioned, we trained the network by presenting the events within the sub-groups in a sequential order. However, each sub-group was presented in a random fashion to the network so as to prevent the network learning the presentation sequence of the sub-groups themselves. The network took on average 35000 epochs to converge, this varied slightly however due to varying initialization weights in the network when training.

4 Testing the System Model

The testing is carried out in two separate phases for the system model, with the azimuth estimation stage first being tested to ensure correct operation, i.e. correct estimation of the sound source along the horizontal plane. Once this stage is confirmed to be operating correctly, the output results are stored to file and used as the input to the second stage in that model, the Neural Predictor stage.

The results from the second stage on the model are checked against results from both predetermined input activation and randomly generated output activations to ensure the system does not respond erroneously due to unexpected input sequences, or incorrect weight updating and convergence.

4.1 Stage 1

We test the azimuth estimation stage of the model to ensure the correct azimuth values were being calculated and presented to the RNN. For this the robot was placed in the middle of a room with a speaker placed 1.5 meters from the center of the robot. The speaker was placed at 10 separate angles around the front 180° of the robot. Each angle was tested five times with the results shown in table 1.

Table 1. Tests of Azimuth Estimation stage of model

Test	Actual Angle	Robot Position (Average)	Accuracy %
Test 1	-90	±4	95.5
Test 2	-50	±2	96
Test 3	-40	±1	97.5
Test 4	-30	±0	100
Test 5	0	±2	98
Test 6	+10	±2	80
Test 7	+20	±1	95
Test 8	+35	±2	94.3
Test 9	+45	±2	95.6
Test 10	+70	±3	95.7

As can be seen from table 1 the maximum average error was ±1.9°. That is, the averages in column 3 summed and divided by number of test cases to give the average system error (±19/10 = 1.9). As shown in [7] the human auditory system can achieve an accuracy of ±1.5° azimuth. Therefore, the results from the initial tests for this stage in our model show that the use of cross-correlation for calculating TDOA and ultimately the angle of incidence is an effective system for determining the azimuth position of a sound source.

Furthermore, the results passed into the second stage of our model are also accurately representative of the actual position of the sound source within the environment and therefore a useful input into the RNN for predicting the next angle.

4.2 Stage 2

Testing of the RNN after training was done with the aid of azimuth results, i.e. a data file was created with the angles returned of the initial stage of the model as the sound source was moved around the robot at various speed levels. This data was then presented to the network in order to check the response of the system to actual external data as opposed to simulated environments.

Fig. 8 shows the response of the network when the test data presented activation in an accepted sequential order. The first angle presented in Fig. 8a was within the range 0° → 2° and therefore provided input activation to the first neuron. Next, in Fig. 8b the angle presented was within the range 2.01° → 4° and therefore activated input neuron 2; this resulted in a recognizable temporal pattern therefore providing output activation for the next predicted position as shown in the output layer of the network in Fig. 8b.

Fig. 9 shows the response of the RNN to a different sequence (speed) to that presented in Fig. 8. Fig 9a shows the first pattern at t_{-1} with an activation on the first input neuron, representing an azimuth estimation of 0° → 2°. The second pattern presented at time t_0 (Fig. 9b) is on the 11 input neuron and so represents an azimuth angle of 20° → 21.9°. Output activation is also provided in Fig. 9b on the 21st output neuron representing an angle of azimuth of 40° → 41.9° for the robot to attend to.

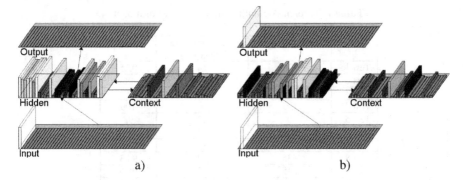

Fig. 8. Shows the response of the RNN after input activations at t_{-1} and t_0 for speed 1

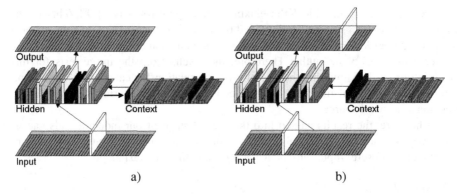

Fig. 9. Shows the response of the RNN after input activations at t_{-1} and t_0 for speed 8

There are always going to be cases when a set of angles presented to the network do not match any predetermined speed representations. To ensure the system provides correct output activation for unforeseen input sequences a testing environment of 10000 randomly generated events was created.

Once the environment was created it was first passed through an algorithm to analyze the input layer activations within the training environment to determine the correct output activation that should be seen; these 'desired' input (with calculated) output activation patterns are then stored to file to later compare with the actual output activation received from the network once the randomly generated test environment has been passed to the system.

The output activations of the network were recorded and compared with the 'desired' stored results to ensure they matched. The comparison showed that from the randomly created test environment only one pair of unforeseen sequences caused a misclassification. Fig. 10 shows the particular misclassification found within the RNN during the specific temporal pair of input sequence patterns. Fig. 10a shows at t_{-1} input activation falls on neuron 28 and at time t_0 Fig. 10b shows that input activation falls on neuron 18. Clearly this is not one of out trained speeds (as the sequence goes backwards) however output activation is set at time t_0 to neuron 39.

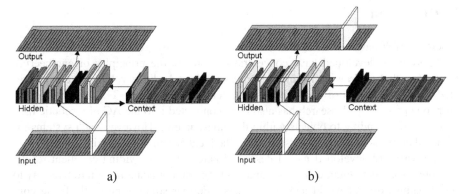

a)   b)

Fig. 10. Shows an misclassification within the trained RNN providing undesired output activation to the system

Table 2. Representation of input activations for the first 4 speeds

Speed	Input Representation
1_{t1}	1000000000................
1_{t2}	0100000000................
2_{t1}	1000000000................
2_{t2}	0010000000................
3_{t1}	1000000000................
3_{t2}	0001000000................
4_{t1}	1000000000................
4_{t2}	0000100000................

Table 2 gives an example of the input sequences for the first four trained speeds. The binary pattern represents activation on the input neurons of the RNN where '1' shows activation on the relevant neuron. As can be seen from the table, each speed is determined by the increment in neuron activation between t_1 and t_2 in the temporal sequence.

The results from the testing of the RNN shows that it is possible to encode several temporal patterns within a RNN using a context layer to act as a form of short-term memory within the system to provide history information for the network to use in classifying temporally encoded patterns. With the exception of the temporal sequence shown in Fig. 10 which shows a misclassification in the system for a temporal sequence pair, all other combinations of sequences within the testing environments 10000 events provided correct desired output activation i.e. either no output at all for undesired temporal pairs or single neuron output activation for desired temporal sequence pairs.

With the introduction of new sub-groups within the training environment it is possible to remove anomalies from the network. This would be accomplished by including the temporal sequence shown in Fig. 10 but having no output activation set. However misclassifications will not be detected until the sequence that generates them is presented to the network.

5 Discussion

Much research has been conducted in the field of acoustic robotics. However, many of the systems developed have concentrated more on the principles of engineering rather than that of drawing inspiration from biology. Auditory robots have been created which use arrays of microphones to calculate independent TDOA between microphone pairs [12]. Research has also been conducted into the AC of a robotic barn owl [13] for attending to objects within the environment. However, such systems include other modalities such as vision to aid the localization of the spatial object.

The currently developed model described here, whilst providing excellent results for sound source azimuth estimation and tracking can not adapt in an intuitive way to the dynamics of real world acoustic objects. Further study is currently being conducted into creating an architecture that can learn to recognize the temporal sequences of new speeds the system may encounter but does not have in its training set. Using this adaptive network to recognize new temporal patterns, it may also be possible for the network to learn how to recognize acceleration and deceleration patterns through this adaptive model. This adaptive network would provide the system with the ability to more accurately track sound sources whose dynamic motion is not a fixed constant but rather varies its speed randomly.

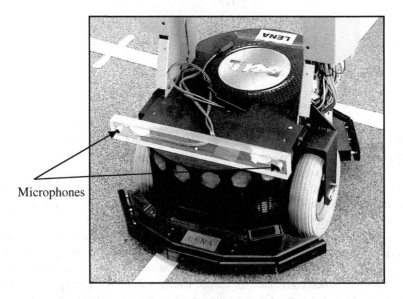

Fig. 11. The robot used for sound source tracking with the two microphones as ears

6 Conclusion

A hybrid architecture has been presented with inspiration drawn from the mechanisms that have been shown to exist within the AC [1, 7, 14] of the mammalian. By using biological inspiration we can take advantage of the cues and mechanisms that already exist to build our model. The model has been shown to utilize the ITD cue to determine the angle of incidence of the sound source and present this to a RNN for temporal processing to determine the current speed and predict the next location for the robot to attend to. The hybrid architecture shown has proven to have excellent potential for developing robotic sound source tracking system which draws inspiration from their biological counterpart. The results of our model have shown comparable with the capabilities of the human AC with the azimuth localization differing by an average of ±0.4°.

As more information on the workings of the AC becomes known it would be possible to further adapt and create neural network architectures that emulate the functionality of the various components of the AC giving rise to robotic system which operate in the acoustic modality in much the same manner as the mammalian.

References

[1] Joris, P.X., Smith, P.H., and Yin, T.C.T., Coincidence detection in the Auditory System: 50 years after Jeffress. Neuron, 1998. 21(6): p. 1235-1238.

[2] Hawkins, H.L., et al., *Models of Binaural Psychophysics, in Auditory Computation*, 1995, Springer. p. 366-368.

[3] Wang, Q.H., Ivanov, T., and Aarabi, P., *Acoustic robot navigation using distributed microphone arrays.* Information Fusion, 2004. 5(2): p. 131-140.

[4] Macera, J.C., et al., *Remote-neocortex control of robotic search and threat identification.* Robotics and Autonomous Systems. 2004. 46(2): p. 97-110.

[5] Bohme, H.J., et al., *An approach to multi-modal human-machine interaction for intelligent service robots.* Robotics and Autonomous Systems, 2003. 44(1): p. 83-96.

[6] Lima, P., et al., *Omni-directional catadioptric vision for soccer robots.* Robots and Autonomous Systems, 2001. 36(2-3): p. 87-102

[7] Blauert, J., *Table 2.1, in Spatial Hearing – The Psychophysics of Human Sound Localization.* 1997, p. 39.

[8] Licklider, J.C.R., *Three auditory theories, in Koch ES (ed) Psychology: A Study of a Science.* Study 1, Vol. 1, New York: McGraw-Hill: p. 41-144.

[9] Hawkins, H.L., et al., *Cross-Correlation Models, in Auditory Computation.* 1995, Springer. p. 371-377.

[10] Kolen, J.F., Kremer, S.C., The search for context, in A Field Guide to Dynamical Recurrent Networks, 2001. IEEE Press. p. 15-17.

[11] Elman, J.L., *Finding structure in time.* Cognitive Science, 1990. 14(2): p. 179-211.

[12] Valian, J-M., et al. *Robust Sound Source Localization Using a Microphone Array on a Mobile Robot.* In Proceedings of the 2003 IEEE/RSJ International Conference on Intelligent Robotics and Systems, p. 1228 – 1233

[13] Rucci, M., Wray, J., Edelman, G.M., *Robust localization of auditory and visual targets in a robotic barn owl.* Robotics and Autonomous Systems 30 (2000) 181-193.

[14] Griffiths, T.D., Warren, J.D., *The planum temporale as a computational hub.* TRENDS in Neurosciences. 2002, 25(7): p. 348-353.

Image Invariant Robot Navigation Based on Self Organising Neural Place Codes

Kaustubh Chokshi, Stefan Wermter, Christo Panchev, and Kevin Burn

Centre for Hybrid Intelligent Systems,
School of Computing and Technology, University of Sunderland St. Peter's Campus,
Sunderland SR6 0DD. United Kingdom
kaustubh.chokshi@sunderland.ac.uk
http://www.his.sunderland.ac.uk

Abstract. For a robot to be autonomous it must be able to navigate independently within an environment. The overall aim of this paper is to show that localisation can be performed even without having a pre-defined map given to the robot by humans. In nature place cells are brain cells that respond to the environment the animal is in. In this paper we present a model of place cells based on Self Organising Maps. We also show how image invariance can improve the performance of the place cells and make the model more robust to noise. The incoming visual stimuli are interpreted by means of neural networks and they respond only to a specific combination of visual landmarks. The activities of these neural networks implicitly represent environmental properties like distance and orientation to the visual cues. Unsupervised learning is used to build the computational model of hippocampal place cells. After training, a robot can localise itself within a learned environment.

1 Introduction

Despite progress made in the fields of AI and Robotics, robots today still remain vastly inferior to humans or animals in terms of performance [1]. One reason for this is that robots do not possess the neural capabilities of the brain. Human and animal brains adapt well to diverse environments, whereas artificial neural networks are usually limited to a controlled environment, and also lack the advantage of having millions of neurons working in true parallelism.

In an mammal's brain place cells fire when the animal occupies a familiar portion of its environment, known as its place field. However, the activity of cells, or even a collection of such cells, simply indicates different locations; it informs the animal where it is, but it cannot directly inform the animal where it should go [2, 3, 4]. One role of the place cells is to associate a path integrator and local view so that when an animal enters a familiar environment, it can reset its path integrator to use the same coordinate system as during previous experiences in the environment [5].

To navigate in familiar environments, an animal must use a consistent representation of its positions in the environments. In other words, the animal must localise in order to navigate within the environment. Visual clues that support a local view to inform the animal of its initial position may be ambiguous or incomplete and there must be

S. Wermter et al. (Eds.): Biomimetic Neural Learning, LNAI 3575, pp. 88–106, 2005.
© Springer-Verlag Berlin Heidelberg 2005

a way to settle on a consistent representation of localisation [4]. The evidence from neurophysiology suggests that place cells are well suited for this role. Spikes fired by dentate granule cells, CA1 and CA3 pyramidal cells, are strongly correlated with the location.

In a experimental environment, place cells have clearly shown a firing rate that relates to that environment. From the experimental evidence [5] we can summarise their properties as follows:

1. When distinct landmarks move, place fields also move proportionately.
2. Place cells continue to show clean place fields when landmarks are removed.
3. The firing rate correlates to more than the location of the animal.
4. Place cells show different place fields in the environment.
5. Place cells are directional when the animal takes a limited path, but non-directional when wandering around randomly in open fields.
6. Place cells are multi-modal. They can integrate various input sensors to localise with vision being the primary one. In the case of no vision or restricted vision, they localise using other sensors such as odour, or whiskers.

In this paper we evaluate our computational place code model in a realistic context, using a Khepera robot. Visual information is provided by a linear vision system. Eight infra-red sensors are used to provide reactive behaviour. This paper is structured as follows: we describe the basis of the model in section 2, outline of the model in section 3, following with the experiments and results in section 4.

2   Self Organising Map for Localisation

In the brain, hippocampal pyramidal cells called place cells have been identified that fire when an animal is at a certain location within its environment. In our model, we show that place cells based on SOMs have potential to provide locations to the path integrator and place cells can localise the robot in a familiar environment. Self-localisation in animals or humans often refers to the internal model of the world outside. As seen in a white water maze experiment [4], even though a rodent was not given any landmarks, it could still reach its goal by forming its own internal representation of landmarks of the world outside. It is seen in humans and animals that they can create their own landmarks, depending on the firing of place cells [6]. These cells change their firing patterns in an environment when prominent landmarks are removed. With this evidence from computational neuroscience, it is reasonable to assume that a model of place cells might prove to be an efficient way of robot localisation using vision.

One possibility is to build a place code model that is based on Self Organising Maps (SOM). SOM [7] networks learn to categorise input patterns and to associate them with different output neurons, or a set of output neurons. Each neuron, j, is connected to the input through a synaptic weight vector $\mathbf{w}_j = [w_{j1}....w_{jm}]^T$. At each iteration, the SOM finds a winning neuron $\mathbf{v}$ by minimising the following equation:

$$v(x) = arg\ min_j \|x(t) - w_j\|, \qquad j = 1, 2, ...n \qquad (1)$$

x belongs to an *m*-dimensional input space, $\|.\|$ is the Euclidean distance, while the update of the synaptic weight vector is done in the following way:

$$w_j(t+1) = w_j(t) + \alpha(t)h_{j,v(x)}(t)[x(t) - w_j(t)], \qquad j = 1, 2, \ldots\ldots n, \qquad (2)$$

This activation and classification are based on features extracted from the environment by the network. Feature detectors are neurons that respond to correlated combinations of their inputs. These are the neurons that give us symbolic representations of the world outside. In our experiments, once we get symbolic representations of the features in the environment we use these to localise the robot in that environment.

The sparsification performed by competitive networks is very useful for preparing signals for presentation to pattern associators and auto associators, since this representation increases the number of patterns that can be associated or stored in such networks [8, 9]. Although the algorithm is simple, its convergence and accuracy depend on the selection of the neighbourhood function, the topology of the output space, a scheme for decreasing the learning rate parameter and the total number of neuronal units [10].

The removal of redundancy by competition is thought to be a key aspect of how the visual system operates [8, 9]. Competitive networks also reduce the dimensions of the input vector as a set of input patterns, in our case pixels of the input image vector. The representation of a location is achieved by activation of a neuron.

An important property of SOMs is feature discovery. Each neuron in a SOM becomes activated by a set of consistently active input stimuli and gradually learns to respond to that cluster of coactive inputs. We can think of SOMs as feature discovery in the input space. The features in the input stimuli can thus be defined as consistently coactive inputs and SOMs thus show that feature analysers can be built in without any external teachers [8]. This is a very important aspect of place cells, as they have to respond to unique features or landmarks in the input space in order to localise the robot.

3   Scenario and Architecture

Our approach is to try to model aspects of neural visual localisation present in human and animal brains. The main emphasis of this research is to build a robot that uses robust localisation, with the objective that it has learning and autonomy. The central objective is on natural vision for navigation based on neural place codes. This section summarises the scenario and architecture of our approach.

3.1   Scenario

In our experiments the overall goal for the robot (a Khepera robot, figure 1(a)) was to to localise itself between two desired locations. In order to facilitate natural vision experiments, we provided random colour-coded squares on the wall, along with some distinguishable features like cubes, cylinders and pyramids randomly kept in the environment as shown in figure 1**(b)**. During experimentation, the robot should be able to create its own internal representation of the world model based on unsupervised learning for neural place codes.

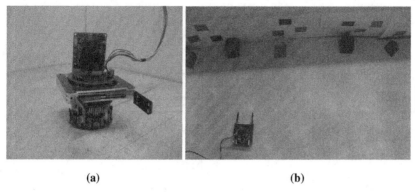

(a)   (b)

Fig. 1. (a) A Khepera robot used during experimentation. **(b)** A birds eye view of the overall experiment setup and the cage in which the robot was allowed to move in

3.2   Overall Architecture of the Model

Humans and animals use various sensors to navigate [11, 12]. In our robot model, we are primarily using vision as a global navigation strategy and for our local navigation strategy we have employed the use of infra red sensors.

Our approach is based upon functional units each of which uses a neural network. An overview of the different functional units can be seen in figure 2. SOMs are used for the visual landmarks, which enable the robot to generate its internal representation of the world based on the most salient features in its visual field. A primitive visual landmark allows us to implement simple, visually-based behaviour. The transform invariance and pattern completion modules are based on MLPs, the output of which forms the input to the SOM. Furthermore, self-localisation and target representation are based on SOMs.

In figure 2, 'visual information derivation' is a module which is responsible for getting the images from the robot's camera. The Visual information derivation module is responsible for image pre-processing and normalising the images for the network. Transform invariance, a part of our localisation module (figure 5) makes use of associative memory and pattern completion for noise reduction. The localisation module is responsible for the localisation of the robot in the environment.

Effective navigation depends upon the representation of the world the robot is using [11]. In our architecture the world representation is called 'spatial representation'. This provides the path planning module with necessary information from the localisation module and visual target module. It maps both the current location and the location of the target into the same map and enables the path planning module to compute the most appropriate path. Once we can map both the visual target and the current location of the robot into the same spatial representation, the 'path-planning module' directs the robot to its goal. The path planning can derive a path which is the shortest and quickest way towards the goal.

There are various ways in which the robot can be instructed as to where its target for navigation is. We are exploring how to translate the place code output and target representation into a spatial representation. The path planning module provides output to the 'motors control'. This forms the global navigation strategy.

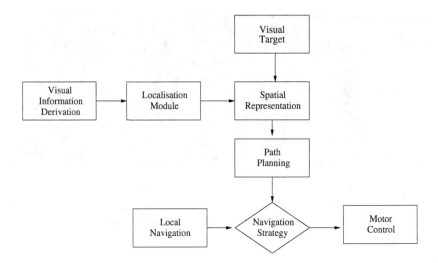

Fig. 2. Overall architecture used in the visual navigation strategy of the robot. This shows the flow of the model

We have implemented the local navigation strategy using reactive behaviour. Both the global and local navigation strategies meet each other in the navigation strategy module, which is mostly responsible for choosing motor commands either from local or global behaviours. Accordingly, it chooses the output from either the global navigation strategy or local navigation strategy to generate motor control commands.

Real-time processing of artificial neural networks requires vast amounts of computational power, especially those algorithms that require real-time vision. Therefore, we make use of distributed and decentralised processing. The robot onboard computer is primarily responsible for robot control. At the same time we are making use of various off-board computers in order to achieve the real time navigation. Each module in figure 2 can run on a different computer/CPU as a part of distributed architecture.

3.3   Overview of the Model Implementation

The focus of this paper is on robot localisation and therefore in this section we will describe in detail the implemented models. It consists of a hierarchical series of five layers of hybrid neural networks, corresponding to the transform invariance layers and place code layer. Figure 4 shows the forward connections to individual layers derived from the modular arrangement of the layers.

Local Navigation: Reactive Behaviour. A lot of recent research in intelligent robotics involves reactive behaviour [13, 14]. In a reactive robot system, all sensors are wired to the motor controls. This enables the motors to react on the sensory state. In these systems internal representations play a limited role or no role at all in determining the motor control output for the robot. Even though reactive behaviour robots do not have an internal representation of the outside world, they are able to solve many complex tasks, since the robot can react to different sensory states in a different manner based upon coordination of perception and action [15, 16].

As in biological systems, reactive behaviours have a direct mapping of sensory inputs to motors actions [11]. The reactive behaviour emerges as a result of **SENSE** and **ACT** strongly coupled together. Sensing in reactive behaviour is local to each behaviour, or in other words it is behaviour-specific. One behaviour is unaware of what the other behaviour is doing, i.e. the behaviours are independent of each other and they do not interact with each other. This is the fundamental difference between local and global navigation strategies.

Our neural network design for reactive behaviour (figure 3) is based on Braitenberg's Vehicle [17, 16], with eight infrared sensors forming the input layer. The inputs were pre-processed to toggle the actual input between 0 and 1. The output layer had two nodes, one connected to the left wheel, another to the right wheel and direction was determined by the value of activation between -1 and 1: positive activation for the forward direction and negative activation for backwards.

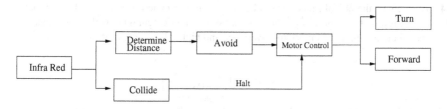

Fig. 3. Control system for robots reactive behaviour

Constructing a local navigation system by behaviours is often referred to as programming by behaviour [11], since the fundamental component of any implementation is a behaviour. Behaviours are inherently modular and easy to test in isolation i.e. they can be tested independently of the global navigation. Behaviours also support incremental expansion of the capabilities of the robot and a robot becomes more "intelligent" with more behaviours in it. The reactive behaviour decomposition results in an implementation that works in real time and is computationally inexpensive. If the behaviours are implemented poorly, then the reactive implementation can be slow. But generally, the reaction speed of a reactive behaviour is equivalent to the stimulus-response time in animals [11].

Global Navigation and Self Localisation. The global navigation strategy is crucial for how systems behave. Global navigation requires perception and motor skills in order to provide complex sensor-motor integration enabling the system to reach its goal. The global navigation strategy is the strategy which uses an internal representation, or map, of the environment while local navigation does not make use of such representations or maps. Many of these global planning methods are based on paths without obstacles [18] and their main advantage is to prove the existence of a solution that will permit the robot to reach its destination. Thus both reactive and deliberate planning are needed, not only bottom-up reactive, but also top-down predictive behaviour.

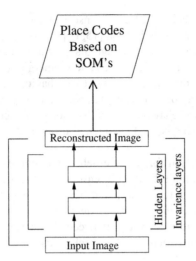

Fig. 4. Overview of the neural model that is being used in our experiments. Hybrid neural networks for localisation based on vision. The first part of the hybrid neural networks is associative memory based on associative memory for invariance and higher layers are SOM for place codes

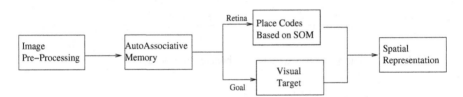

Fig. 5. An overview of the localisation in our model. This is the overall implementation of the localisation module as been before in figure 2. The image pre-processing is responsible for getting the images from the robot camera and resizing. Then the associative memory is responsible for image-invariant processing. Localisation of the robot is done by place codes. Visual Target represents the main goal for the navigation. Spatial representation will take activations from both the neural network regions and represent it in the environmental space

Although localisation has been investigated for more then a decade, there is still no universally accepted solution [19]. Some general methods have been employed for encoding prior knowledge of the environment and matching it with local sensor information. Some of the previous methods of localisation are *(i) Topological Maps*: the environment is mapped into a number of distinct locations, usually connected with each other [20]. Typically these maps are learned during the exploration stage. *(ii) Evidence grids* [21]: in this method each location in the environment is represented by a grid point in the global map. For localisation, the system constructs local grip maps with occupancy probability for each grid point which are matched to the global map. *(iii) Markov Models*: in this method of place code localisation the probability distribution is computed for all possible locations in the environment [22]. *(iv) Landmarking*: in this method the robot encodes a number of distinctive locations [23, 20, 24, 25, 2].

Our method of localisation is based on landmarks. We use landmarks for localisation, mainly because this enables us to make internal representation of the environment and does not involve human interference for determining the landmarks in the environment. As the robot generates its own landmarks depending on the features in the environment, we call it "Self-Localisation". Our method of localisation is distinct from other methods described above because there are not maps given to the robot and the neural network creates an internal representation of the world based on the visual stimuli. This network is inspired from neural place codes, making the model more robust and efficient.

4   Experiments and Results

Once the environment was learned by the robot, and when the landmarks were presented to it, it could be seen clearly that the activation in the SOM's map would represent a place code. There were some interesting observations made, including that the landmarks were not only self-generated, but also that when the robot starts to approach the landmark there was activation in neighbouring fields before it reached it, as discussed in section 4.5. Another observation was that the neuron responsible for the landmark would have increasing levels of activation when it was approached by the robot.

4.1   Experimental Setup

The experiments were conducted on a Khepera robot. The robot was introduced in a closed environment, as seen in figure 6 of about 2m x 1.5m, which was divided into four parts: north, south, east and west. The environment was further divided into a grid of 10 cm x 10 cm squares. This grid was only used for the purpose of calculation of the error by the place cells. All the landmarks were placed against the wall of the cage. There were cubes and pyramids of different colour codes spread across the walls of the cage randomly. The walls also had randomly colour-coded figures on it.

Each square represents a place code. Each cell was given a name, depending on where it was located, for example a cell in the southern part within the eastern part was given name "$se10$". The naming convection was simple; the first letter represents which hemisphere, the second letter which block and the numbers indicate the x and y coordinates. This information was purely for our use in order to test the results and set up the experiments. This information was not provided to the robot. For training purposes there were 4 images taken from each of the place codes. For testing there were 10 new images from each place code in the environment.

4.2   Representation of Visual Input

It is more convenient to perform neural analysis on smaller versions of an image while retaining all the essential information of interest, in our case to detect a landmark in an image using neural networks. If the landmark is just as evident in the smaller image, it

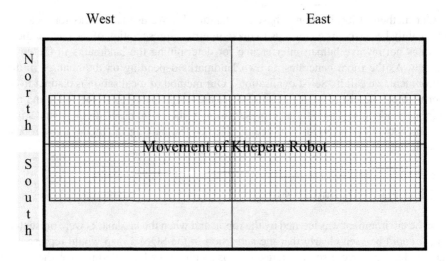

Fig. 6. The arena of the robot. The robot was allowed to moved in a limited space, as we did not want it to be too close to, nor too far away from, the landmarks. The area of movement of the robot was divided into smaller grids of 10cm x 10cm, giving us an approximate position of the robot

is more efficient to reduce the image size before applying neural methods. Thus computational time saving occurs as the smaller images contain fewer pixels and the recall time for associative memory and self organising map is reduced, since the number of the neurons is reduced.

There are number of techniques that can be used to enlarge or reduce images [26, 27, 28]. These generally have a tradeoff between speed and the degree to which they reduce salient visual features. The simplest methods to reduce the image keep every n^{th} pixel. However, this results in aliasing of high frequency components. Therefore, a more general case of changing the size of an image by a given factor requires interpolation of colours. The simplest method is called "nearest neighbourhood", which is currently used by us. Using this method one finds the closest corresponding pixel in the original image (i, j) for each pixel in the reduced image (i', j'). If the original image has the dimensions width (w) and height (h), and the reduced image would be of w' and h', then the point in the destination is given by

$$i' = iw'/w \qquad (3)$$
$$j' = jh'/h \qquad (4)$$

where the division (equation 4) is a integer, the remainder being ignored. In other words, in the nearest neighbour method of resizing, the output pixel is assigned the value of the pixel that the point falls within. The number of pixels considered affects the complexity of the computation.

Once the images are resized, they are arranged to a single dimension vector from a two dimensional vector. All images were in 24 bit colour RGB (Red Green Blue)format (equations 8), N represents the whole image.

$$N = (a_{ijk}) \in \mathbb{A}^{3mn}, \quad i = 1, 2, \ldots, m \quad j = 1, 2, \ldots, n \quad k = 1, 2, 3 \tag{5}$$

$$R = (r_{ij}) \in \mathbb{A}^{mn}, \quad i = 1, 2, \ldots, m \quad j = 1, 2, \ldots, n \quad r_{ij} := a_{ij1} \tag{6}$$

$$G = (g_{ij}) \in \mathbb{A}^{mn}, \quad i = 1, 2, \ldots, m \quad j = 1, 2, \ldots, n \quad g_{ij} := a_{ij2} \tag{7}$$

$$B = (b_{ij}) \in \mathbb{A}^{mn}, \quad i = 1, 2, \ldots, m \quad j = 1, 2, \ldots, n \quad b_{ij} := a_{ij3} \tag{8}$$

$\mathbb{A}$ is the set of possible pixel values. The values are between 0 and 255, and can be represented as shown in 9.

$$\mathbb{A} = \{0, 1, 2, \ldots, 255\}, \qquad m = 17, \qquad n = 27 \tag{9}$$

Each image was reduced to a size of 17 x 27. This image in turn was converted into a single vector to be presented to the network. It was done as explained from equations 10 to 12.

$$A = \begin{pmatrix} a_{11} & \cdots & a_{1n} \\ a_{21} & \cdots & a_{2n} \\ \vdots & \ddots & \vdots \\ a_{m1} & \cdots & a_{mn} \end{pmatrix} \qquad A_i = (a_{i1}, \ldots, a_{in}) \qquad i = 1, 2, \ldots, m \tag{10}$$

$$V = (v_l) := (A_1, \ldots, A_m) \in \mathbb{A}^{mn} \qquad l = 1, 2, \ldots, mn \tag{11}$$

Equation 11 is a concatenation of A_i of A. In other words,

$$v_{(i-1)n+j} := a_{ij} \qquad i = 1, 2, \ldots, m \qquad j = 1, 2, \ldots, n \tag{12}$$

4.3   Training and Testing Procedure

The stimuli used for training and testing our model are specially constructed to investigate the performance of localisation using the self organising maps. To train the network, a sequence of 200 images was presented to represent over 20 landmarks. At each representation the winning neuron was selected and the weight vector of the winning neuron was updated along with the distance vector. The presentation of all the stimuli across all the landmarks consists of one epoch of training. In this manner the networks were trained using backpropagation in Multi-Layered Perceptrons. Invariance and the place code networks were trained separately.

Table 1. Training and Testing procedure

	Training	Testing
No. of Images	200	485
No. of Epoch (Invariance)	1800	-
No. of Epoch (Place Code)	3000	-

4.4   Results for Transform Invariance

This method of representation shows promise, although it is not ideal for invariance such as size, view etc. It is observed that, compared to the traditional template matching methods, it is computationally and memory efficient. After the neural networks memorised the various landmarks, and after a new image has been given at the retina, our method finds the image nearest to the image previously memorised.

The main purpose of transform invariance was to reconstruct the image that was on the retina for the SOM. It was seen that due to this process of reconstructed images, we could also achieve a certain degree for independence from light conditions. The independence of this was achieved due to the generalisation feature of neural networks, which would generalise the effect of light over various colours in the reconstructed image. Transform invariance has improved the performance of place codes based on SOMs in various way, and the results will be described in sections 4.5. and 4.6.

4.5   Discussion of Results for SOMs Place Codes

Activation Activity of a Neuron. When an animal approaches a desired landmark, the place cells representing the landmark increase activation and, when the animal is at that the desired landmark, the activation is maximum. This property is observed in biological place cells and has been discussed in 1.1.

To show that our model also follows the same principles of biological place cells, we have taken readings of activation of various neurons and we are presenting here activation levels of a neuron responsible for different place codes. In figure (7) we can

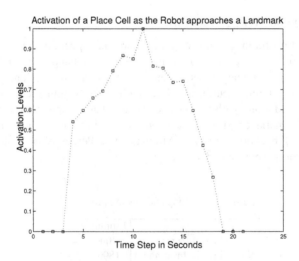

Fig. 7. This figure shows activation levels and activity for a single place cell. As the robot approaches the place code, the activation rises and when the robot is at the location represented by the place cell the activation is maximum

see that when the robot starts to approach the landmark, there is a sudden steep rise in the activation of the neuron responsible. As the robot gets closer, the activation keeps on rising, until it is at the desired landmark. Once there, the activation is 1. As the robot moves away from the desired landmark, there is a gradual fall in the activation of the neuron, and as soon as the landmark is out of sight of the robot, the activation is set to 0. This is shown in figure 7.

Activation Activity of a Place Code Cluster. Another property of place cells is that when the animal is approaching the desired landmark, the neighbouring neurons would also be active as described in section 1. These results are shown in figure 8. It is seen that when the robot is within a region there is a cluster of place codes responding to the robot's location. In section 4.6, we will see the advantages of having clusters of place cells for noise handling. The activation in the cluster provides us the robot's grid location. Each neuron within the cluster provides us with a more precise location of the robot, within 2cm of precision.

Clustering of Place Codes. The basic property of a SOM network is to form clusters of information relating to each other, in our case landmarks. A cluster is a collection of neurons which are next to each other representing the same landmark. Figure 8 shows that when the robot was approaching the desired landmark, there were activations in the neighbouring neurons. This is due to clustering of similar images around the landmark. There are multiple similar images that are being represented by a single neuron, making the cluster smaller and richer in information. This is achieved with the invariance module.

On the other hand, figure 8(c) shows the landmarks which were at a distance to the location represented in figure 8(d). Two landmarks that were given to the robot at a distance would be mapped not only into different clusters, but also distant from each other. By their very definition, landmarks are features in the environment. This was the reason behind a formation of these clusters by SOMs. The landmarks that were chosen by the SOM were quite significant in the image and distinguished features from the rest of the environment, and other landmarks.

Distinction Between North and South. We have observed that there is a clear distinction between the north and the south on the place code map. Figure 9(a) shows all the neurons that are responsible for landmarks in the north and figure 9 (b) represents the neurons responsible for the south. The reason for this distinction is that the field of view is quite different in both hemispheres. It has been observed that in the northern hemisphere the object of attention for the landmark selection was very much limited to an object in the field of view. In contrast, in the southern hemisphere, the robot had a much larger field of view, therefore the object of attention was not focused only on a single object, but on various objects.

Overlap of East and West in the South Section. In the north it was observed that there was a clear distinction between the east and west, whereas in the south, there is overlap. The overlap was caused by the field of view of the robot retina. From the south section of the cage, the robot is able to see more landmarks than in the north section.

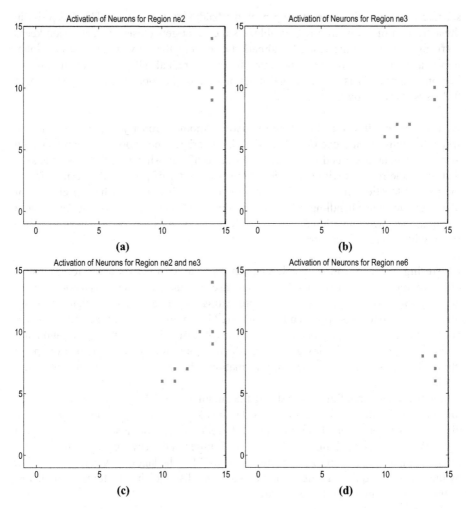

Fig. 8. Each of the graph shows activation of winning neurons. It is seen in the images that neighbouring regions in environment are neighbours to each other on the place code map. There is also a clear overlap of a neuron in both regions. The reason for the overlap is because in the field of view of the robot between both locations, both prominent landmarks can be seen

After small movements in the north, the object of attention changes, whereas in the southern hemisphere there are various objects of attention that lie in the field of view. Therefore, minor movements in the southern hemisphere, do not lead to drastic changes in the visual field. The overlap is caused by landmarks which are nearer the borders of east and west.

Directional Independence. It was clearly observed that the model was directionally independent. It was seen that in whichever direction the robot travelled within the environment, as it came across the landmark, it would activate the place cells responsible

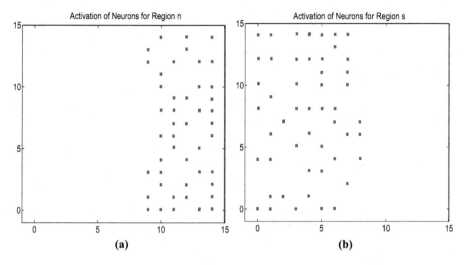

Fig. 9. The various activations of neurons that represent landmarks in the northern and southern hemisphere. We can also see that they are not topological

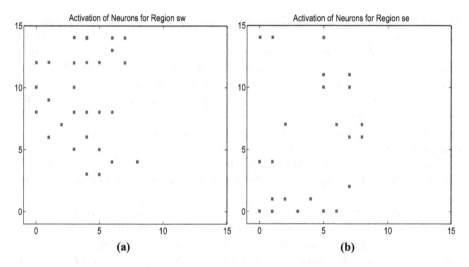

Fig. 10. There is a clear overlap of the regions in the (a) south-eastern and (b) south western. The overlap is due to the sharing of landmarks between both the regions

for the landmark. Therefore it did not matter in which direction the robot travelled; it could localise itself as soon as it came across a significant landmark.

For testing purposes, the model was presented with random images with different landmarks. It was even seen that once the landmark was seen, the place cell responsible for it would be activated.

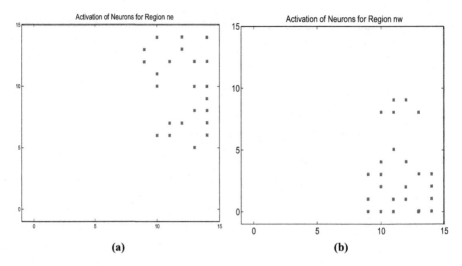

Fig. 11. In the images above, there is a clear distinction between the (a) north eastern and (b) northwestern. The reason for this is that the field of view here is restricted to a particular landmark

Manipulating the Landmarks. There were two experiments that were conducted for testing robustness with the robot. In the first experiment, we removed two blocks from eastern part and blocks from western part randomly. It was observed that when the robot was in the northern hemisphere and came across the missing blocks, it would activate an unknown cluster or completely wrong neuron. This happened because the field of view within the northern hemisphere is very limited to one or two objects. So, when the blocks were removed, the environment was not rich enough to provide the robot with enough clues as to where it was, but as soon as the robot came across a known landmark, it localised itself again. This was not observed in the southern hemisphere. The southern hemisphere visual field was large, hence the removal of blocks did not affect the activations of neurons. The visual field was rich enough to provide the necessary clues to localise.

In the second experiment, the blocks that were removed were now replaced, but not in their original positions. It was observed that the activations in the southern hemisphere were still representing the right location. In the northern hemisphere, the activations were not in the unknown cluster, but for the neurons representing those landmarks.

Reduction in Cluster Size. It was observed in [23], that the main cause for large clusters of place codes was due to the SOM trying to handle transform invariance by having the neighbouring neurons responding to the invariance. With the use of associative memory for transform invariance, the size of the clusters was reduced. In the present model, the SOM does not represent the invariance, rather it represents the place codes. Images were collected at every 10th frame i.e. approximately half a second between images. This causes large amounts of overlap and large amounts of transform invariance. The associative memory clustered the similar images and reduced the transform invariance. The number of neurons per location reduced, since there were fewer neurons

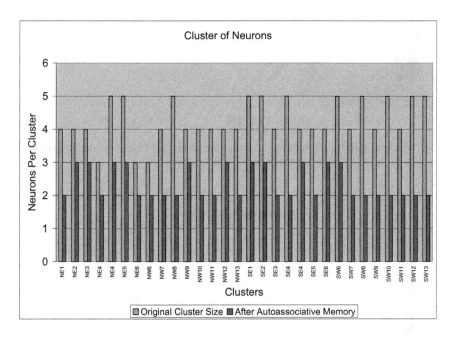

Fig. 12. The cluster sizes i.e. number of neurons per cluster representing a particular location of the robot. It is seen that There is a drastic reduction in the size of the clusters after the use of autoassociative memory before the place code map. This makes it possible for us to have more place code on the same size of the map

required to represent the same location if there was a shift in the images. This also has additional benefits, mainly now SOM can represent more place codes without actually growing or increasing the size of the map.

4.6   Performance of Network

To test the performance of the network, we tested it with white noise with a mean noise ranging from 0.0 to 0.8 and variance of 0.01 to the image. The effects of the noise on the images can be seen in figure 13. The aim of the neural network is to localise the robot within its environment, based on the internal representations it has formed. As the place cells are based on SOMs, there is a cluster of neurons responsible for a place code in the environment. The neuron representing that place code would be more accurate then the neighbouring neurons. To have a more precise localisation, we need the neuron responding to the place to be active.

There are various reasons where we would need an approximate localisation. It was noted that with the increasing noise, it was more likely for the robot to be 'lost' (unable to localise itself) [23]. During these times, approximate coordinates would help the robot to localise itself. We consider two options for localisation with noise handling: one being that a neuron responsible for the robot response and another being another neuron in the cluster of neurons responding for the same place responding. In the later case, localisation may not be very accurate.

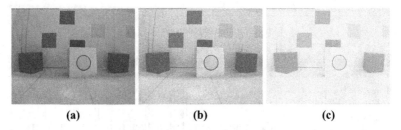

(a)   (b)   (c)

Fig. 13. Effects of different noise levels added to the image. **(a)** Image without any noise. **(b)** Image with 0.2 mean deviation **(c)** Image with 0.5 mean deviation

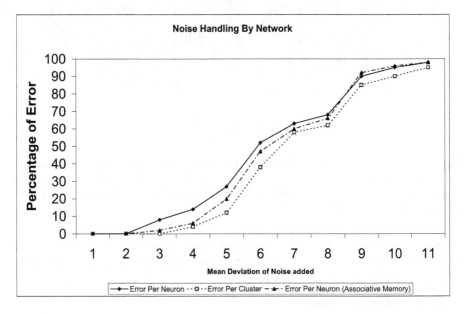

Fig. 14. This figure shows the noise handling by the network, with transform invariance, based on associative memory, and also without transform invariance. It shows that the network performs much better with transform invariance

As seen in figure 14, the clusters are more robust than the neurons with regard to noise handling. As the amount of noise increases, the neurons or cluster response to localisation becomes more random. However the network performance is quite impressive until 0.5 mean deviation of noise where the error for localisation is below 30%. The cluster for a place codes still performs much better and is still below 20%.

Noise handling by the neural networks was also improved by adding an additional layer of associative memory below the place codes. The associative memory reduces the noise before the outputs are given to the place codes layer in the architecture. It can be seen in figure 14 that associative memory helps place cells to perform better, giving less error per neuron.

It was also noted that as the noise level increased, the performance of the network decreased along with associative memory. This was mainly seen for localisations where higher levels of noise were present. With noise of 70%, the performance was not significantly different, with or without associative memory. At higher levels of noise of more than 80% the noise handling of the associative memory failed, making the place codes performance performs worse than without the layer. At can be expected, with 80% noise level, the performance of the SOMs is completely random to get the right localisation.

For our experiments we were not expecting more than 30% of noise levels. Since this would mostly be caused by interference in wireless signals. Even if the noise levels for a few frames is more then 30%, it would be possible for the robot to localise itself with the following input frames. With the use of the associative memory, the performance up to 30% noise level improved substantially.

5   Conclusion

The aim of this research, was to investigate whether the reliability of robot localisation can be improved using SOMs based on place codes. In this paper we have described a place cell model based on a SOM for localisation. The model was successful in learning the locations of landmarks even when tested with distorted images. Visual landmarks were associated with locations in a controlled environment. This model clusters neighbouring landmarks next to each other. The landmarks that are distant from each other are also relatively distant in the place code map. Rather than pre-programming localisation algorithms as internal modules, our place code based SOMs architecture demonstrates that localisation can be learnt in a robust model based on external hints from the environment. This model was developed to learn landmarks in an environment, by having maps divided into clusters of neurons for different parts of the environment. It is considered to have a lot of potential for learning the localisation of the robot within an environment.

References

1. J. Ng, R. Hirata, N. Mundhenk, E. Pichon, A. Tsui, T. Ventrice, P. Williams, and L. Itti, "Towards visually-guided neuromorphic robots: Beobots," in *Proc. 9th Joint Symposium on Neural Computation (JSNC'02), Pasadena, California*, 2002.
2. D. Foster, R. Morris, and P. Dayan, "A model of hippocampally dependent navigation, using the temporal difference learning rule," *Hippocampus*, vol. 10, pp. 1–16, 2000.
3. A. Arleo, F. Smeraldi, S. Hug, and W. Gerstner, "Place cells and spatial navigation based on 2d visual feature extraction, path integration, and reinforcement learning," in *NIPS*, pp. 89–95, 2000.
4. A. D. Redish, *Beyond Cognitive Map from Place Cells to Episodic Memory*. London: The MIT Press, 1999.
5. A. D. Redish, *Beyond Cognitive Map*. PhD thesis, Carnegie Mellon University, Pittsburgh PA, August 1997.
6. E. T. Rolls, S. M. Stringer, and T. P. Trappenberg, "A unified model of spatial and episodic memory," *Proceedings of the Royal Society B*, vol. 269, pp. 1087–1093, 2002.

7. T. Kohonen, *Self-organizing Maps*. Springer Series in Information Sciences, Berlin, Germany: Springer-Verlag, 3rd ed ed., 2001.

8. E. Rolls and G. Deco, *Computational Neuroscience of Vision*. New York: Oxford University Press, 2002.

9. S. Wermter, J. Austin, and D. Willshaw, *Emergent Computational Neural Architectures based on Neuroscience*. Heidelberg: Springer, 2001.

10. M. Haritoppulos, H. Yin, and N. M. Allinson, "Image denoising using self-organisising map-based non-linear independent component analysis," *Neural Networks: Special Issue, New Developments in Self Organising Maps*, vol. 15, pp. 1085–1098, October - November 2002.

11. R. R. Murphy, *Introduction to AI Robotics*. London, England: The MIT Press, 2000.

12. R. A. Brooks, *Cambrian Intelligence: The Early History of the New AI*. Cambridge, Massachusetts: The MIT Press, 1999.

13. N. E. Sharkey, "The new wave in robot learning," *Robotics and Automation system*, vol. 22, pp. 179–186, 1997.

14. S. Wermter and R. Sun, eds., *Hybrid Neural Systems*. Berlin: Springer, 2000.

15. S. Nolfi and D. Floreano, *Evolutionary Robotics: The Biology, Intelligence, and Technology of Self Organizing Machines*. The MIT Press, 2000.

16. R. Pfeifer and C. Scheier, *Understanding Intelligence*. The MIT Press, 1999.

17. V. Braitenberg, *Vehicles*. Cambridge - Massachusetts: MIT Press (A Bradford Book), 1984.

18. H. Maaref and C. Barret, "Sensor based navigation of mobile robot in an indoor environment," *Robotics and Automation Systems*, vol. 38, pp. 1–18, 2002.

19. U. Gerecke, N. E. Sharkey, and A. J. Sharkey, "Common evidence vectors for reliable localization with som ensembles," in *Proceedings of Engineering Application of Neural Networks*, EANN-2001, 2001.

20. C. Owen and U. Nehmzow, "Landmark-based navigation for a mobile robot," in *Simulation of Adaptive Behaviour*, (Zurich), The MIT Press, 1998.

21. H. Moravec and A. Elfes, "High resolution maps from wide angle sonar," in *In. proc. IEEE International Conference on Robotics and Automation*, pp. 116–121, 1985.

22. W. Burgard, A. Cremers, D. Fox, D. Hahnel, G. Lakemeyer, D. Schulz, W. Steiner, and S. Thrun, "Experiences with an interactive museum tour-guide robot," *Artificial Intelligence*, vol. 114, no. 1-2, pp. 3–55, 1999.

23. K. Chokshi, S. Wermter, and C. Weber, "Learning localisation based on landmarks using self organisation," in *ICANN 03*, 2003.

24. U. Nehmzow, *Mobile Robotics: A Practical Introduction*. London: Springer Verlag, 2000.

25. S. Marsland, *Online Novelty Detection Through Self Organisation, With Application to Inspection Robotics*. PhD thesis, University of Manchester, Manchester UK, December 2001.

26. K. R. Castleman, *Digital Image Processing*. New Jersey: Prentice Hall, 1996.

27. L. G. Shapiro and G. C. Stockman, *Computer Vision*. New Jersey, USA: Prentice Hall, 2001.

28. M. Seul, L. O'Gorman, and M. J. Sammon, *Practical Algorithms for Image Analysis: Description, Examples, and Code*. Cambridge, UK: Cambridge University Press, 2000.

Detecting Sequences and Understanding Language with Neural Associative Memories and Cell Assemblies

Heiner Markert[1], Andreas Knoblauch[1,2], and Günther Palm[1]

[1] Abteilung Neuroinformatik,
Fakultät für Informatik, Universität Ulm,
Oberer Eselsberg, D-89069 Ulm, Germany
Tel: (+49)-731-50-24151; Fax: (+49)-731-50-24156
{markert, knoblauch, palm}@neuro.informatik.uni-ulm.de
[2] Honda Research Institute Europe GmbH, Carl-Legien-Str. 30,
D-63073 Offenbach/Main, Germany
Tel: (+49)-69-89011-761; Fax: (+49)-69-89011-749
andreas.knoblauch@honda-ri.de

Abstract. Using associative memories and sparse distributed representations we have developed a system that can learn to associate words
with objects, properties like colors, and actions. This system is used in
a robotics context to enable a robot to respond to spoken commands
like "bot show plum" or "bot put apple to yellow cup". This involves
parsing and understanding of simple sentences and "symbol grounding",
for example, relating the nouns to concrete objects sensed by the camera
and recognized by a neural network from the visual input.

1 Introduction

When words referring to actions or visual scenes are presented to humans, distributed networks including areas of the motor and visual systems of the cortex
become active (e.g. [1]). The brain correlates of words and their referent actions
and objects appear to be strongly coupled neuronal assemblies in defined cortical areas. The theory of cell assemblies [2, 3, 4, 5, 6] provides one of the most
promising frameworks for modeling and understanding the brain in terms of
distributed neuronal activity. It is suggested that entities of the outside world
(and also internal states) are coded in groups of neurons rather than in single ("grandmother") cells, and that a neuronal cell assembly is generated by
Hebbian coincidence or correlation learning [7, 8] where the synaptic connections are strengthened between co-activated neurons. Thus models of neural
(auto-)associative memory have been developed as abstract models for cell assemblies [9].

One of our long-term goals is to build a multi-modal internal representation
for sentences and actions using cortical neuron maps, which will serve as a basis
for the emergence of action semantics and mirror neurons [10, 11, 12]. We have

S. Wermter et al. (Eds.): Biomimetic Neural Learning, LNAI 3575, pp. 107–117, 2005.
© Springer-Verlag Berlin Heidelberg 2005

developed a model of several visual, language, planning, and motor areas to enable a robot to understand and react to spoken commands in basic scenarios of the MirrorBot project [11, 12, 13, 14] that is described in the first part of this book. The essential idea is that different cortical areas represent different aspects (and correspondingly different notions of similarity) of the same entity (e.g., visual, auditory language, semantical, syntactical, grasping related aspects of an apple) and that the (mostly bidirectional) long-range cortico-cortical projections represent hetero-associative memories that translate between these aspects or representations. This involves anchoring symbols such as words in sensory and motor representations where invariant association processes are required, for example recognizing a visually perceived object independent of its position, color, or view direction. Since word symbols usually occur in the context of other words specifying their precise meaning in terms of action goals and sensory information, anchoring words is essentially equivalent to language understanding.

In this work we present a neurobiologically motivated model of language processing based on cell assemblies [2, 3, 4, 5]. We have developed a system that can learn to associate words with objects, properties like colors, and actions. This system is used in a robotics context to enable a robot to respond to spoken commands like "bot show plum" or "bot put apple to yellow cup". The scenario for this is a robot close to one or two tables on which there are certain kinds of fruit and/or other simple objects. We can demonstrate part of this scenario where the task is to find certain fruits in a complex visual scene according to spoken or typed commands. This involves parsing and understanding of simple sentences and relating the nouns to concrete objects sensed by the camera and recognized by a neural network from the visual input.

In the first section we outline the concept of cell assemblies as a model for sequential associative processing in cortical areas and how our model is related to discrete finite automates and sequence detector networks by Pulvermüller [1, 15], also explained in this book [16]. Then we briefly describe our robot architecture used for implementing simple scenarios of associating words to objects, and detail the language module. Finally, we summarize and discuss our results.

2 Language and Cell Assemblies

A large part of our model is based on associative memory and cell assemblies. Anchoring a symbol first requires understanding the context in which the symbol occurs. Thus, one requirement for our system is language processing and understanding.

2.1 Regular Grammars, Finite Automates, and Neural Assemblies

Noam Chomsky developed a hierarchy for grammar types [17, 18]. For example, a grammar is called *regular* if the grammar can be expressed by rules of the type

$$A \rightarrow a$$
$$B \rightarrow bC$$

DFA   Neural Network

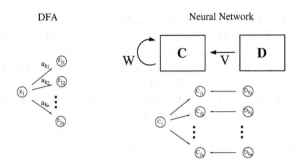

Fig. 1. Comparison of a deterministic finite automaton (DFA, left side) with a neural network (right side) implementing formal language. Each δ transition $\delta(z_i, a_k) = z_j$ corresponds to synaptic connections from neuron C_i to C_j and from input neuron D_k to C_j (see text for details)

where lower case letters are *terminal symbols* (i.e. elements of an alphabet Σ), and upper case letters are *variables*. Usually there is a starting variable S which can be expanded by applying the rules. A sentence $s \in \Sigma^*$ (which is a string of alphabet symbols of arbitrary length) is called *valid with respect to the grammar* if s can be derived from S by applying grammatical rules and resolving all variables by terminal symbols.

There are further grammar types in the Chomsky hierarchy which correspond to more complex rules, e.g. context-free and context-sensitive grammars, but here we will focus on regular grammars. It is easy to show that regular grammars are equivalent to deterministic finite automata (DFA). A DFA can be specified by $M = (Z, \Sigma, \delta, z_0, E)$ where $Z = \{z_0, z_1, ..., z_n\}$ is the set of states, Σ is the alphabet, $z_0 \in Z$ is the starting state, $E \subseteq Z$ contains the terminal states, and the function $\delta : (Z, \Sigma) \rightarrow Z$ defines the (deterministic) state transitions. A sentence $s = s_1 s_2 ... s_n \in \Sigma^*$ is valid with respect to the grammar if iterated application of δ on z_0 and the letters of s transfers the automaton's starting state to one of the terminal states, i.e., if $\delta(...\delta(\delta(z_0, s_1), s_2), ..., s_n) \in E$ (cf. left side of Fig. 1).

In the following we show that DFAs are equivalent to binary recurrent neural networks such as the model architecture described below (see Fig. 3). As an example, we first specify a simpler model of recurrent binary neurons by $N = (C, I, W, V, c_0)$, where $C = \{C_0, C_1, ..., C_n\}$ contains the local cells of the network, $D = \{D_1, D_2, ..., D_m\}$ is the set of external input cells, $W = (w_{ij})^{n \times n}$ is a binary matrix where $w_{ij} \in \{0, 1\}$ specifies the strength of the local synaptic connection from neuron C_i to C_j, and, similarly, $V = (v_{ij})^{m \times n}$ specifies the synaptic connections from input cell D_i to cell C_j. The temporal evolution of the network can be described by

$$c_i(t+1) = \begin{cases} 1, \text{if } \sum_j w_{ji} c_j(t) + \sum_j v_{ji} d_j(t) \geq \Theta_i \\ 0, \text{otherwise.} \end{cases}$$

where $c_i(t)$ is the output state of neuron C_i at time t, and Θ_i is the threshold of cell C_i. Figure 1 illustrates the architecture of this simple network.

The network architecture can easily be adapted to simulate a DFA. We identify the alphabet Σ with the input neurons, and the states Z with the local cells, i.e. each $a_i \in \Sigma$ corresponds to input cell D_i, and, similarly, each $z_i \in Z$ corresponds to a local cell C_i. Then we can specify the connectivity as follows: Synapses w_{ij} and v_{kj} are active if and only if $\delta(z_i, a_k) = z_j$ for the transition function δ of the DFA (see Figure 1). In order to decide if a sentence $s = a_{i(0)} a_{i(1)} a_{i(2)} \ldots$ is valid with respect to the language we can specify the activation of the input units by $d_i(t) = 1$ and $d_j = 0$ for $j \neq i(t)$. By choosing threshold $\Theta_i = 2$ and a starting activation where only cell c_0 is active, the network obviously simulates the DFA. That means, after processing of the last sentence symbol, one of the neurons corresponding to the end states of the DFA will be active if and only if s is valid.

To be precise, the above algorithm represents a slight oversimplification. In this implementation it might happen that the next state is not determined uniquely. To avoid this problem it would be sufficient to use states representing pairs (z_i, a_k) of a state z_i and an action a_k leading to it. Then the synapses from (z_i, a_l) to (z_j, a_k) and from a_k to (z_j, a_k) are activated for the transition $\delta(z_i, a_k) = z_j$.

The described neural network architecture for recognizing formal languages is quite simple and reflects perfectly the structure of a DFA even on the level of single neurons. In addition, this network is also very similar to the network of sequence detectors discussed in [15, 16]. Essentially an elementary sequence detector can be identified with a 3-state automaton, where the sequence ab is identified by two transitions from an initial state z_0 into a final state z_2 as depicted in the left part of figure 2. The resulting network (depicted in the right part of figure 2) resembles the sequence detector idea of Pulvermüller ([1, 15]) and they obey the same underlying principle: Pulvermüller's simplest linear sequence detector (e.g. for ab) needs a weak input for a followed by a strong input for b. This works because the strong input decays faster then the weak input (in absolute terms; in relative terms they both decay exponentially fast). Thus Pulvermüller's idea works whenever the second input b decays faster than the first input a. This makes the sequence ab more effective than ba, because more activity is left from a when b occurs, than vice versa. This simple principle also holds for an automaton (see figure 2) and the corresponding network, because a feeds back onto itself (it is an auto-associative assembly) and therefore is more

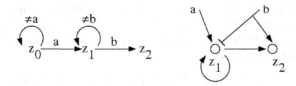

Fig. 2. Identification of an elementary sequence detector representing the sequence ab with a 3-state automaton. Left: schematic diagram, right: neural implementation, where $\rightarrow$ means excitation and $\dashv$ means inhibition

persistent than b. Ideally, this scheme works for arbitrary delays between a and b, in contrast to Pulvermüller's simple sequence detector that only works for a small range of delays ([16]). The idea of using auto-associative persistent patterns with different persistency in different areas is used extensively in the model derived in the next section. Also the sequencing or automaton idea is used at least in one area (A4, see figure 3), where both auto-associative memories stabilizing individual patterns and hetero-associative memories representing patterns sequences are stored in the same local network.

Coming back to the cell assembly perspective, clearly a network as in figure 1 or 2, where single neurons are used to code the different states, is biologically not very realistic, since, for example, such an architecture is not robust against partial destruction and it is not clear how such a delicate architecture could be learned. The model becomes more realistic if we interpret the nodes in figure 1 or figure 2 not as *single* neurons but as groups of nearby neurons which are strongly interconnected, i.e., as local cell assemblies. This architecture has two additional advantages: First, it enables *fault tolerance* since incomplete input can be completed to the whole assembly. Second, overlaps between different assemblies can be used to express similarity, hierarchical and other relations between represented entities. In the following subsection we briefly describe a model of associative memory which allows to implement the assembly network analogously to the network of single neurons in figure 1 or 2.

2.2   Cell Assemblies and Neural Associative Memory

We decided to use the *Willshaw associative memory* [19, 20, 4, 21, 22, 23] as a single framework for the implementation of cell assemblies in cortical areas. A *cortical area* consists of n binary neurons which are connected with each other by binary synapses. A *cell assembly* or *pattern* is a binary vector of length n where k one-entries in the vector correspond to the neurons belonging to the assembly. Usually k is much smaller than n. Assemblies are represented in the synaptic connectivity such that any two neurons of an assembly are bidirectionally connected. Thus, an assembly consisting of k neurons can be interpreted as a k-clique in the graph corresponding to the binary matrix A of synaptic connections. This model class has several advantages over alternative models of associative memory such as the most popular Hopfield model [24]. For example, it better reflects the cortical reality where it is well known that activation is sparse (most neurons are silent most of the time), and that any neuron can have only one type of synaptic connection (either excitatory or inhibitory).

Instead of classical one-step retrieval we used an improved architecture based on spiking associative memory [25, 13]. A cortical area is modeled as a local population of n neurons which receive input from other areas via Hebbian learned hetero-associative connections. In each time step this external input initiates pattern retrieval. The neurons receiving the strongest external input will fire first, and all emitted spikes are fed back immediately through the Hebbian learned auto-associative connections resulting in activation of single assemblies. In comparison to the classical model, this model has a number of additional advantages.

For example, assemblies of different size k can be stored, and input superpositions of several assemblies can more easily be separated.

In the following section we present the architecture of our cortical model which enables a robot to associate words to visually recognized objects, and thereby anchoring symbolic word information in sensory data. This model consists of a large number of interconnected cortical areas, each of them implemented by the described spike counter architecture.

3   Cell-Assembly Based Model of Cortical Areas

We have designed a cortical model consisting of visual, tactile, auditory, language, goal, and motor areas (see figure 3), and implemented parts of the model on a robot. Each cortical area is based on the spike counter architecture described in the previous section. The model is simulated synchronously in discrete time steps. That means, in each time step t each area computes its output vector $y(t)$ as a function of the output vectors of connected areas at time $t - 1$. In addition to the auto-associative internal connection within each area there are also hetero-associative connections between theses areas (see figure 4).

3.1   Overall Architecture

Figure 3 illustrates the overall architecture of our cortical model. The model consists of auditory areas to represent spoken or typed language, of grammar areas to interpret spoken or typed sentences, visual areas to process visual input, goal areas to represent action schemes, and motor areas to represent motor output. Additionally, we have auxiliary areas or fields to activate and deactivate the cortical areas (activation fields), to compare corresponding representations in different areas (evaluation fields), and to implement visual attention.

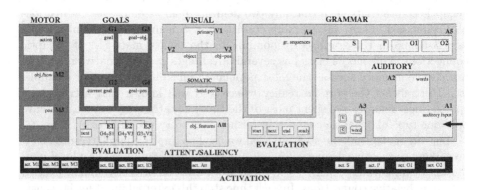

Fig. 3. Cortical architecture involving several inter-connected cortical areas corresponding to auditory, grammar, visual, goal, and motor processing. Additionally the model comprises evaluation fields and activation fields (see text)

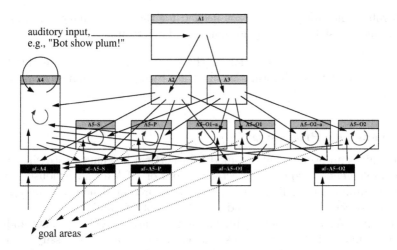

Fig. 4. The language system consisting of 10 cortical areas (large boxes) and 5 thalamic activation fields (small black boxes). Black arrows correspond to inter-areal connections, gray arrows within areas correspond to short-term memory

Each small white box corresponds to an associative memory as described in the previous section. The auditory areas comprise additional neural networks for processing of acoustic input, i.e. they perform basic speech recognition. The main purpose of the visual fields is to perform object recognition on the camera image.

Currently, we have implemented most parts of the model on a robot. The object recognition system basically consists of three components:

1. The *visual attention control system* localizes the objects of interest based on an attention control algorithm using top-down information from higher cortical areas.
2. The *feature extraction system* analyzes a window taken from the camera image corresponding to the region of interest. Scale and translation invariance is achieved by rescaling the window and using inherently invariant features as input for the classification system. The extracted features comprise local orientation and color information.
3. The *classification system* uses the extracted features as input to a hierarchical neural network which solves the classification task. The basic idea of using hierarchical neural networks is the division of a complex classification task into several less complex classification tasks by making coarse discrimination at higher levels of the hierarchy and refining the discrimination with decreasing depth of the hierarchy. Beneficial side effects of the hierarchical structure are the possibility to add additional classes quite easily at run-time, which means that the system will be able to learn previously untrained objects online, and the possibility to investigate the classification task at intermediate states. The latter can be useful if, for example, the full

classification information is not needed for the task at hand, or simply to gain more insight in the performance of the network.

For a more detailed description of the whole model, see [11] in this book. For additional detailed information on the object recognition system see for example [26].

3.2   Language Processing

Figure 4 gives an overview of our model for cortical language processing. It basically implements a sequence detector network or DFA for a previously defined regular grammar (as in figure 1). The system presented here can understand simple sentences like "bot show red plum" or "bot lift apple".

Areas A1, A2 and A3 are primary auditory areas, A1 represents auditory input by primary linguistic features, whereas area A2 and A3 classify the input with respect to function and content, respectively. Areas A2 and A3 serve as input areas to the sequence detector circuit.

Essentially, areas A4 and A5-X implement the DFA. The sequence detector network is split into several areas to enable the model to keep track of the path of states the automaton took to achieve the final state. This leads to a very useful representation of the parsed sentence in the end where each of the areas A5-X is filled with one word in one special grammatical context (e.g. A5S holds the subject of the sentence). Further interpretation of the semantics of the sentence then becomes relatively easy ([11] shows how our model for action planning uses this information).

Areas A5-X stores the state (z_i) the automaton has reached, together with the input (a_k) leading to that state. In our example, the inputs a_k are words and the states basically reflect the grammatical role of the input word in the sentence. The A5-X areas are used to explicitly store the input words a_k that lead to their activation. They implicitly store the grammatical context, because subjects will be stored in area A5S, predicates in A5P and so on. This corresponds to storing the pairs (z_i, a_k) for each state of the automaton.

The possible state transitions are stored in area A4. Hetero-associative feedback connections with longer delay from area A4 onto itself represent the possible state transitions, the current state is kept persistent by auto-associative feedback with short delays. To perform a state transition, the whole area A4 is inhibited for a short while, eliminating the effect of the auto-associative feedback. Because of the longer delay, the hetero-associative connection will still be effective if the inhibition is released. Biased by the new input pattern, it will then switch to the next state according to the next input symbol.

Area A4 only represents the state transition matrix; the usage of pairs in A4 is not necessary in this model because the required information is represented in areas A5-X, which all project onto area A4 making the state transitions unique.

Figure 5 shows the state of the system after processing the input sequence "bot put plum (to) green apple", where "to" does not belong to the actual input sequence due to simplification of the grammatical rules. The input sentence is

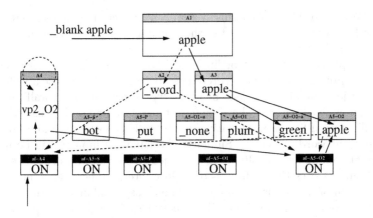

Fig. 5. System state of the language model after 30 simulation steps when processing the sentence "Bot put plum to green apple". (Processing of a word requires about 5-6 steps on average; during each simulation step the state of the associative network is synchronously updated)

segmented into subject, predicate and the two objects, and the automaton will reach its final state when the input "apple" disappears and a symbol for the end of the input sentence will be activated in area A1 and A2.

4 Conclusion

We have presented a cell assembly based model for cortical language processing that can be used for associating words with objects, properties like colors, and actions. This system is used in a robotics context to enable a robot to respond to spoken commands like "bot put plum to green apple". The model shows how sensory data from different modalities (e.g., vision and speech) can be integrated to allow performance of adequate actions. This also illustrates how symbol grounding could be implemented in the brain involving association of symbolic representations to invariant object representations.

Although we have currently stored only a limited number of objects and sentence types, it is well known for our model of associative memory that the number of storable items scales with $(n/\log n)^2$ for n neurons [19, 20, 21]. However, this is true only if the representations are sparse and distributed which is a design principle of our model. As any finite system, our language model can implement only regular languages, whereas human languages seem to involve context-sensitive grammars. On the other hand, also humans cannot "recognize" formally correct sentences beyond a certain level of complexity.

The neural implementation of this language understanding system not only shows that this comparatively intricate logical task can be mastered by a neural network architecture in real time, it also gives some additional advantages in terms of robustness and context-awareness. Indeed, we have shown at the NeuroBotics Workshop [11, 12] that this system can correct ambiguous input on the single word level due to the context of the whole sentence and even the complete sensory-motor situation.

For example the sentence "bot lift bwall" with an ambiguous word between "ball" and "wall" is correctly interpreted as "bot lift ball", because a wall is not a liftable object. Similarly, a sentence like "bot show/lift green wall" with an artificial ambiguity between "show" and "lift", can be understood as "bot show green wall", even if the disambiguating word "wall" comes later and even across an intermittent word ("green"). Similarly the language input could be used to disambiguate ambiguous results of visual object recognition, and vice versa.

This demonstrates the usefulness of a close interplay between symbolic and subsymbolic information processing (also known as "symbol grounding") in autonomous robots, which can be easily achieved by biologically inspired neural networks.

References

1. Pulvermüller, F.: Words in the brain's language. Behavioral and Brain Sciences **22** (1999) 253–336
2. Hebb, D.: The organization of behavior. A neuropsychological theory. Wiley, New York (1949)
3. Braitenberg, V.: Cell assemblies in the cerebral cortex. In Heim, R., Palm, G., eds.: Lecture notes in biomathematics (21). Theoretical approaches to complex systems. Springer-Verlag, Berlin Heidelberg New York (1978) 171–188
4. Palm, G.: Neural Assemblies. An Alternative Approach to Artificial Intelligence. Springer, Berlin (1982)
5. Palm, G.: Cell assemblies as a guideline for brain research. Concepts in Neuroscience **1** (1990) 133–148
6. Palm, G.: On the internal structure of cell assemblies. In Aertsen, A., ed.: Brain Theory. Elsevier, Amsterdam (1993)
7. Palm, G.: Local rules for synaptic modification in neural networks. Journal of Computational Neuroscience (1990)
8. Palm, G.: Rules for synaptic changes and their relevance for the storage of information in the brain. Cybernetics and Systems Research (1982)
9. Palm, G.: Associative memory and threshold control in neural networks. In Casti, J., Karlqvist, A., eds.: Real Brains - Artificial Minds. North-Holland, New York, Amsterdam, London (1987)
10. Rizzolatti, G., Fadiga, L., Fogassi, L., Gallese, V.: Resonance behaviors and mirror neurons. Archives Italiennes de Biologie **137** (1999) 85–100
11. Fay, R., Kaufmann, U., Knoblauch, A., Markert, H., Palm, G.: Combining visual attention, object recognition and associative information processing in a neurobotic system. [27]
12. Fay, R., Kaufmann, U., Knoblauch, A., Markert, H., Palm, G.: Integrating object recognition, visual attention, language and action processing on a robot in a neurobiologically plausible associative architecture. accepted at Ulm NeuroRobotics workshop (2004)
13. Knoblauch, A.: Synchronization and pattern separation in spiking associative memory and visual cortical areas. PhD thesis, Department of Neural Information Processing, University of Ulm, Germany (2003)

14. Knoblauch, A., Fay, R., Kaufmann, U., Markert, H., Palm, G.: Associating words to visually recognized objects. In Coradeschi, S., Saffiotti, A., eds.: Anchoring symbols to sensor data. Papers from the AAAI Workshop. Technical Report WS-04-03. AAAI Press, Menlo Park, California (2004) 10–16

15. Pulvermüller, F.: The neuroscience of language: on brain circuits of words and serial order. Cambridge University Press, Cambridge, UK (2003)

16. Knoblauch, A., Pulvermüller, F.: Sequence detector networks and associative learning of grammatical categories. [27]

17. Hopcroft, J., Ullman, J.: Formal languages and their relation to automata. Addison-Wesley (1969)

18. Chomsky, N.: Syntactic structures. Mouton, The Hague (1957)

19. Willshaw, D., Buneman, O., Longuet-Higgins, H.: Non-holographic associative memory. Nature **222** (1969) 960–962

20. Palm, G.: On associative memories. Biological Cybernetics **36** (1980) 19–31

21. Palm, G.: Memory capacities of local rules for synaptic modification. A comparative review. Concepts in Neuroscience **2** (1991) 97–128

22. Schwenker, F., Sommer, F., Palm, G.: Iterative retrieval of sparsely coded associative memory patterns. Neural Networks **9** (1996) 445–455

23. Sommer, F., Palm, G.: Improved bidirectional retrieval of sparse patterns stored by hebbian learning. Neural Networks **12** (1999) 281–297

24. Hopfield, J.: Neural networks and physical systems with emergent collective computational abilities. Proceedings of the National Academy of Science, USA **79** (1982) 2554–2558

25. Knoblauch, A., Palm, G.: Pattern separation and synchronization in spiking associative memories and visual areas. Neural Networks **14** (2001) 763–780

26. Fay, R., Kaufmann, U., Schwenker, F., Palm, G.: Learning object recognition in an neurobotic system. acctepted at 3rd workshop SOAVE2004 - SelfOrganization of AdaptiVE behavior, Illmenau, Germany (2004)

27. Wermter, S., Palm, G., Elshaw, M., eds.: Biomimetic Neural Learning for Intelligent Robots. Springer, Heidelberg, New York (2005)

Combining Visual Attention, Object Recognition and Associative Information Processing in a NeuroBotic System

Rebecca Fay, Ulrich Kaufmann, Andreas Knoblauch,
Heiner Markert, and Günther Palm

University of Ulm,
Department of Neural Information Processing,
D-89069 Ulm, Germany

Abstract. We have implemented a neurobiologically plausible system on a robot that integrates visual attention, object recognition, language and action processing using a coherent cortex-like architecture based on neural associative memories. This system enables the robot to respond to spoken commands like "bot show plum" or "bot put apple to yellow cup". The scenario for this is a robot close to one or two tables carrying certain kinds of fruit and other simple objects. Tasks such as finding and pointing to certain fruits in a complex visual scene according to spoken or typed commands can be demonstrated. This involves parsing and understanding of simple sentences, relating the nouns to concrete objects sensed by the camera, and coordinating motor output with planning and sensory processing.

1 Introduction

Detecting and identifying objects as well as processing language and planning actions are essential skills for robots performing non-trivial tasks in real world environments. The combination of object recognition, language understanding and information processing therefore plays an important role when developing service robots. We have implemented a neurobiologically inspired system on a robot that integrates visual attention, object recognition, language and action processing using a coherent cortex-like architecture based on neural associative memories (cf. [1]). Neural networks and associative memories which are both neurobiologically plausible and fault tolerant form the basic components of the model. The model is able to handle a scenario where a robot is located next to one or two tables with different kinds of fruit and other simple objects on them (cf. figure 4). The robot ("bot") is able to respond to spoken commands such as "bot show plum" or "bot put apple to yellow cup" and to perform tasks like finding and pointing to certain fruits in a complex visual scene according to a spoken or typed command. This involves parsing and understanding of simple sentences, relating the nouns to concrete objects sensed by the camera, and coordinating motor output with planning and sensory processing.

S. Wermter et al. (Eds.): Biomimetic Neural Learning, LNAI 3575, pp. 118–143, 2005.
© Springer-Verlag Berlin Heidelberg 2005

The underlying cortical architecture is motivated by the idea of distributed cell assemblies in the brain [2][3]. For visual preprocessing we use hierarchically organized radial-basis-function networks to classify objects selected by attention, where hidden states in this hierarchical network are used to generate sparse distributed cortical representations. Similarly, auditory input pre-processed by standard Hidden-Markov-Model architectures can be transformed into a sparse binary code for cortical word representations. In further cortical areas for language and action the sensory input is syntactically and semantically interpreted and finally translated into motor programs. The essential idea behind the cortical architecture is that different cortical areas represent different aspects (and correspondingly different notions of similarity) of the same entity (e.g., visual, auditory language, semantical, syntactical or grasp-related aspects of an apple) and that the (mostly bidirectional) long-range cortico-cortical projections represent hetero-associative memories that translate between these aspects or representations. These different notions of similarity can synergistically be used, for example, to resolve ambiguities within or across sensory modalities.

2   Architecture

Our architecture can roughly be divided into two large parts: (1) Sensory preprocessing and (2) cortical model. In the preprocessing part features are extracted from auditory and visual input selected by attention control. For the cortical model several different neural architectures are used. For object recognition and speech processing more artificial neural networks such as radial-basis-function networks are utilized, while a biologically more plausible architecture of many interconnected neural associative memories is used to model cortical information processing. Figure 1 shows the different components of our model, their interactions and the division into two parts.

For the implementation of cortical cell assemblies [4][5][6][7][8][9], we decided to use the *Willshaw model* of associative memory as an elementary architectural framework. A *cortical area* consists of n binary neurons which are connected with each other by binary synapses. A *cell assembly* or *pattern* is a sparse binary vector of length n where k one-entries in the vector correspond to the neurons belonging to the assembly. Usually k is much smaller than n. Assemblies are represented in the synaptic connectivity such that any two neurons of an assembly are bidirectionally connected. Thus, an assembly consisting of k neurons can be interpreted as a k-clique in the graph corresponding to the binary matrix A of synaptic connections. This model class has several advantages over alternative models of associative memory such as the most popular Hopfield model [10]. For example, it better reflects the cortical reality where it is well known that activation is sparse (most neurons are silent most of the time), and that any neuron can have only one type of synaptic connection (either excitatory or inhibitory).

Instead of classical one-step retrieval we used an extended algorithm based on spiking associative memory [11][12]. A cortical area is modeled as a local neuron population which receives input from other areas via hetero-associative Hebbian

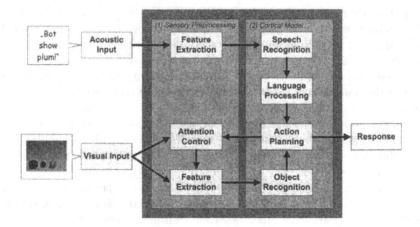

Fig. 1. The architecture is roughly divided into two parts: Sensory preprocessing and cortical model

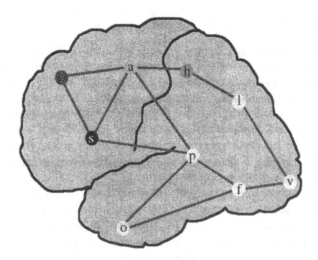

Fig. 2. Interaction of the different areas of the cortical model (v : visual, l : location, f : contour features, o: visual objects, h: haptic / proprioceptive, p: phonetics, s: syntactic, a: action / premotoric, g: goals / planning) and their rough localization in the human brain

synaptic connections. In each time step this external input initiates pattern retrieval. The neurons receiving the strongest external input will fire first, and all emitted spikes are fed back immediately through the auto-associative Hebbian synaptic connections which allows both the activation of single assemblies and the representation of superpositions. In comparison to the classical model, this model has a number of additional advantages. For example, assemblies of dif-

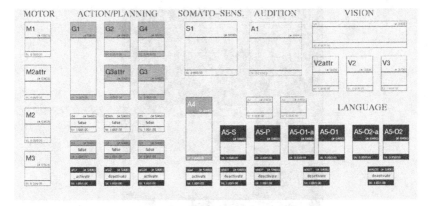

Fig. 3. Cortical architecture involving several inter-connected cortical areas corresponding to auditory, grammar, visual, goal, and motor processing. Additionally the model comprises evaluation fields and activation fields (see text)

ferent size k can be stored, input superpositions of several assemblies can more easily be separated, and it is possible to transmit more subtle activation patterns (e.g., about ambiguities or superpositions) in the spike timing.

As already shown in figure 1, our cortical model divides into four main parts. Some of those again are quite complex tasks that may, considering the situation in the human brain, involve some different cortical areas. In figure 2, a rough overview of the cortical areas that are somehow reflected in our system is given. The interconnections shown there correspond to the flow of information we realized in our model.

Figure 3 illustrates the overall architecture of our cortical model. Each box corresponds to a local neuron population implemented as spiking associative memory. The model consists of phonetic auditory areas to represent spoken language, of grammar related areas to interpret spoken or typed sentences, visual areas to process visual input, goal areas to represent action schemes, and motor areas to represent motor output. Additionally, we have auxiliary areas or fields to activate and deactivate the cortical areas (activation fields), to compare corresponding representations in different areas (evaluation fields), and to direct visual attention. The primary visual and auditory areas are part of sensory preprocessing and comprise additional (artificial) neural networks for processing of camera images and acoustic input.

The suggested approach is implemented on the PeopleBot base by ActivMedia. To integrate the implemented functionality on the robot we used MIRO [13], a robot middleware framework that allows control of the robot's hardware and facilitates communication with other programs by using Corba. MIRO supports distributed computing, i.e. time consuming calculations with low i/o-rates can be outsourced to other computers. MIRO also facilitates the usage of the same application on different robot platforms. Hence the software developed here runs on the PeopleBot as well as on other robot bases.

Fig. 4. In the test scenario the robot is situated in front of a table. Different objects are laying on this table. The robot has to grasp or point to specified objects

3   Sensory Preprocessing

The object classification is performed in three stages, i.e. the system consists of three components which are activated consecutively. First objects of interests are to be localized within the robot's camera image. Since neither the camera image contains solely the object of interest but also background as well as possible other objects nor is it guaranteed that the object is located in the center of the image, it is necessary to perform a pre-processing of the image. In the course of this pre-processing a demarcation of the objects from the background and from each other as well as a localization of the objects takes place. For this a color-based visual attention control algorithm is used to find regions of interests within the camera images. These regions contain objects to be classified. Once the objects are localized, characteristic features like orientation histograms or color information are extracted, which should be invariant to lighting conditions. In order to save computing time the features are not extracted from the entire image but only from the region of interest containing the object. These features are used to classify the localized objects one after the other. Due to the fact that the objects are centered within the regions of interest, the extracted features are made shift and scaling invariant. This yields improved classification results. The fact that in a first step meaningful objects are separated from meaningless background further improves the classification. Moreover, this accounts for invariance of the classification regarding background conditions. An overview of the whole process is shown in figure 5.

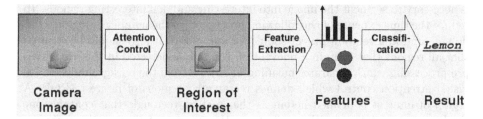

Fig. 5. The classification system consists of three components arranged in successive order: object localization, feature extraction and classification. The figure depicts the interconnections as well as the inputs and outputs of the miscellaneous components. Starting with the robot's camera image the classification process is shown. The attentional system currently works on visual sensory input. In the first step, the camera image is searched for regions containing interesting low level features (e.g., blobs of desired color). In the second step, additional features are extracted from the region of interest to be used by the object recognition system

3.1 Attentional System

The attentional system [14] receives input from some higher goal and motor areas which specify the current interest of the robot (e.g., searching for red-colored objects; areas M2 or M2attr in Fig. 3). Subsequently, the camera image is processed by standard computer vision algorithms in order to find regions of interest (see Fig. 5). If an interesting region is found, this part of the image is analyzed in more detail. Features (e.g., orientation histograms) are extracted and transmitted to the object recognition system which is explained in detail in the next section. The object recognition system classifies the object in the region of interest and transmits the information to the corresponding visual cortical area (areas V2 and V2attr). After some time, attention shifts to the next region of interest, until this process is interrupted by a cortical area controlling attention (e.g., areas G1 and M1).

When locating relevant objects in complex visual scenes it would be too time-consuming to simply look at all possible regions. Therefore pre-processing

Fig. 6. Meaningful objects are separated from the background and are marked as regions of interest. The object colour is used to identify objects of interest. This also allows to detect partially occluded objects

is necessary to segment the image into interesting and non interesting regions. To reduce the time expense of the following process steps the number of regions that do not contain a relevant object should be minimized. It also should be ensured that all regions that contain a relevant object will be detected in this step. This pre-processing which separates meaningful objects from the background is called visual attention control which defines rectangular regions of interest (ROI). A region of interest is defined herein as the smallest rectangle that contains one single object of interest. Figure 6 shows the process of placing the region of interest.

Especially in the field of robotics real-time requirements have to be met, i.e. a high frame processing rate is of great importance. Therefore the usage of simple image processing algorithms is essential here.

The attention control algorithm consists of six consecutive steps. Figure 7 shows the intermediate steps of the attention control algorithm. Camera images in the RGB format constitute the starting point of the attention control algorithm. At first the original image is smoothed using a 5×5 Gaussian filter. This also reduces the variance of colors within the image. The smoothed image is then converted to HSV color space, where color information can be observed independent of the influence of brightness and intensity. In the following only the hue components are considered. The next step is a color clustering. For this purpose the color of the object of interest and a tolerance are specified. The HSV image is searched for colors falling into this predefined range. Thereafter the mean hue value of the so identified pixels is calculated. This mean value together with a predefined tolerance is now used to once again search for pixels with colors in this new range. The result is a binary image where all pixels that fall into this color range are set to black and all others are set to white. Owing to this adaptivity object detection is more robust against varying lighting conditions. The results are further improved using the morphological closing operation [15]. The closing operation has the effect of filling-in holes and closing gaps. In a last

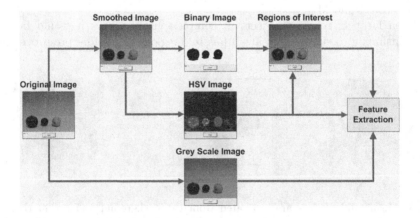

Fig. 7. The intermediate steps of the attention control algorithm

step a floodfill algorithm [16] is applied to identify coherent regions. The region of interest is then determined by the smallest rectangle that completely encloses the region found. Simple heuristics are used to decide whether such a detected region of interest contains an object or not. For the problem at hand the width-to-height ratio of the region of interest as well as the ratio of pixels belonging to the object to pixels belonging to the background in the region of interest are determined. This method allows for detecting several objects in one scene and can handle partial occlusion to a certain degree.

3.2   Feature Extraction

Simply using the raw image data for classification would be too intricate. Furthermore dispensable and irrelevant information would be obtained. Therefore it is necessary to extract characteristic features from the image which are suitable for classification. These features are more informative and meaningful for they contain less redundant information and are of a lower dimensionality than the original image data. To reduce calculation time, features will be extracted from the detected regions of interest only. Thus even expensive image processing methods can be used to calculate the features. If the object was localized precisely enough this yields translation and partial scaling invariance.

The selection of the features depends, inter alia, on the objects to be classified. For the distinction of different fruits, appropriate features are those representing color and form of the object present. Among others we use the mean color values of the HSV representation of the detected region of interest as well as orientation histograms [17] summing up all orientations (directions of edges represented by the gradient) within the region of interest.

To determine the mean color values the camera image is converted from RGB color space to HSV color space [18]. For each color channel the mean value of the localized object within the region of interest is calculated. Color information is helpful to distinguish e.g. between green and red apples. Advantages of color information are its scale and rotation invariance as well as its robustness to partial occlusion. Furthermore it can be effectively calculated.

To calculate the orientation histograms which represent the form of the object the gradient in x and y direction of the grey value image is calculated using the Sobel edge detector [19]. The gradient angles are discretized. Here we divided the gradient angle range into eight sections. The discrete gradient directions are weighted with the gradient value and summed to form the orientation histogram. The orientation histogram provides information about the directions of the edges and their intensity.

It has been found that the results achieved could be improved if not only one orientation histogram per region of interest is used to represent the form of the object but several orientation histograms are calculated from different parts of the region of interest as information about the location of the orientations is not completely disregarded. The region of interest is therefore split into $m \times m$ parts of the same size. For each part a separate orientation histogram is calculated. The concatenated orientation histograms then form the feature vector. If the

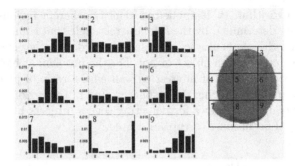

Fig. 8. The image is split into sub-images. For each sub-image an orientation histogram is calculated by summing up the orientations that occur in this sub-image. For reasons of simplicity non-overlapping sub-images are depicted

parts overlap by about 20% the result improves further since it becomes less sensitive to translation and rotation. We chose to divide the image into 3×3 parts with approximately 20% overlap. Figure 8 illustrates how an orientation histogram is generated.

The dimension of the feature vector depends only on the number of subimages and the number of sections used to discretise the gradient angle. The orientation histogram is thus largely independent of the resolution of the image used to extract the features.

4   Cortical Model

Our cortical model consists of four parts, namely speech recognition, language processing, action planning and object recognition (see also figure 1).

For speech recognition standard Hidden-Markov-Models on single word level are used, but for simplicity it is also possible to type the input directly via a computer terminal.

The resulting word stream serves as input to the language processing system, which analyzes the grammar of the stream. The model is capable of identifying regular grammars. If a sentence is processed which is incorrect with respect to the learned grammar, the systems enters an error state, otherwise, the grammatical role of the words is identified.

If the sentence is grammatically interpreted, it becomes possible to formulate what the robot has to do, i.e. the goal of the robot. This happens in the action planning part, which receives the corresponding input from the language processing system. The robot's goal (e.g. "bot lift plum") is then divided into a sequence of simple subgoals (e.g. "search plum", "lift plum") necessary to archive the goal. The action planning part initiates and controls the required actions to archive each subgoal, e.g. for "search red plum", the attentional system will be told to look for red objects. The detected areas of interest serve as input for the object recognition system, the fourth part of our cortical model.

We use hierarchically organized RBF (radial basis function) networks for object recognition, which are more accurate compared to a single RBF network, while still being very fast. The output of the object recognition again serves as input to the action planning system. If in our example a plum is detected, the action planning areas will recognize that and switch to the next subgoal, which here would be to lift the plum.

The language as well as the action planning part are using the theory of cell assemblies which is implemented using Willshaw's model of associative memory [4][5][6][7][8][9]. This results in an efficient, fault tolerant and still biological realistic system.

4.1   Visual Object Recognition

The visual object recognition system is currently implemented using a hierarchical arrangement of radial-basis-function (RBF) networks. The basic idea of hierarchical neural networks is the division of a complex classification task into several less complex tasks by making coarse discrimination at higher levels of the hierarchy and refining the discrimination with increasing depth of the hierarchy. The original classification problem is decomposed into a number of less extensive classification problems organized in a hierarchical scheme. Figure 9 shows a hierarchy for recognition of fruits and gestures which has been generated by unsupervised k-means clustering.

From the activation of the RBF networks (the nodes in Fig. 9) we have designed a binary code in order to express the hierarchy into the domain of cell assemblies. This code should preserve similarity of the entities as expressed by the hierarchy. A straightforward approach is to use binary vectors of length corresponding to the total number of neurons in all RBF networks. Then in a representation of a camera image those components are activated that correspond to the l strongest activated RBF cells on each level of the hierarchy. This results in sparse and translation invariant visual representations of objects.

That way, the result of the object recognition is transformed into the binary code, and using additional information about space and location from the attentional system, the corresponding neurons are activated in areas V1,V2,V2attr, and V3.

Hierarchical Neural Networks. Neural networks are used for a wide variety of object classification tasks [20]. An object is represented by a number of features, which form a d dimensional feature vector x within the feature space $X \subseteq \mathbb{R}^d$. A classifier therefore realizes a mapping from feature space X to a finite set of classes $C = \{1, 2, ..., l\}$. A neural network is trained to perform a classification task using supervised learning algorithms. A set of training examples $S := \{(x^\mu, t^\mu), \mu = 1, ..., M\}$ is presented to the network. The training set consists of M feature vectors $x^\mu \in \mathbb{R}^d$ each labeled with a class membership $t^\mu \in C$. During the training phase the network parameters are adapted to approximate this mapping as accurately as possible. In the classification phase unlabeled data

$x^\mu \in \mathbb{R}^d$ are presented to the trained network. The network output $c \in C$ is an estimation of the class corresponding to the input vector x.

The basic idea of hierarchical approaches to object recognition is the division of a complex classification task into several smaller and less complex ones [21]. The approach presented here hierarchically decomposes the original classification problem into a number of less extensive classification problems. Starting with coarse discriminations between few (but large) subsets of classes at higher levels of the hierarchy the discriminations are stepwise refined. At the lowest levels of the hierarchy there are discriminations between single classes. Thus the hierarchy emerges from successive partitioning of sets of classes into disjoint subsets, i.e the original set of classes is recurrently decomposed into several disjoint subsets until subsets consisting of single elements result. This leads to a decrease in the number of class labels processed in a decision node with increasing depth of this node.

The hierarchy consists of several simple neural networks that are stratified as a tree or more generally as a rooted directed acyclic graph, i.e. each node within the hierarchy represents a neural network that works as a classifier. The division of the complex classification problem into several less complex classification problems entails that instead of one extensive classifier several simple classifiers are used which are more easily manageable. This has not only a positive effect on the training effort, but also simplifies modifications of the design since the individual classifiers can be amended much more easily to the decomposed simple classification problems than one classifier could be adapted to a complex classification task. The use of different feature types additionally facilitates the classification tasks at different nodes, since for each classification task the feature type that allows for the best discrimination can be chosen. Hence for each data point within the training set a feature vector for each prototype is available. Moreover the hierarchical decomposition of the classification result provides additional intermediate information. In order to solve a task it might be sufficient to know whether the object to be recognized belongs to a set of classes and the knowledge of the specific category of the object might not add any value.

The hierarchy also facilitates a link between symbolic information and sub-symbolic information: The classification itself is performed using feature vectors which represent sub-symbolic information, whereas symbolic knowledge can be provided concomitantly via the information about the affiliation to certain subsets of classes. The usage of neural networks allows the representation of uncertainty of the membership to these classes since the original output of the neurons is not discrete but continuous. Moreover a distributed representation can easily be generated from the neural hierarchy. Since the hierarchy is generated using features which are based on the appearance of the objects such as orientation or color information it primarily reflects visual similarity. Thus it allows the generation of a sparse similarity preserving distributed representation of the objects. A straight-forward approach is the usage of binary vectors of length corresponding to the total number of neurons in the output layer of all networks in the hierarchy. The representation is created identifying the strongest activated output neurons for each node. The corresponding elements of the code vector are then set to 1, the remaining elements are set to 0. These properties are extremely useful in the field of neuro-symbolic integration [22] [23] [24]. For separate object localization and recognition a distributed representation may not be relevant, but in the overall system in the MirrorBot project this is an important aspect [25].

Hierarchy Generation. The hierarchy is generated by unsupervised k-means clustering [26]. In order to decompose the set of classes assigned to one node into disjoint subsets a k-means clustering is performed with all data points belonging to these classes. Depending on the distribution of the classes across the k-means clusters disjoint subsets are formed. One successor node corresponds to each subset. For each successor node again a k-means clustering is performed to further decompose the corresponding subset. The k-means clustering is performed for each feature type. Since the k-means algorithm depends on the initialization of the clusters, k-means clustering is performed several times per feature type. The number of clusters k must be at least the number of successor nodes or the number of subsets respectively but can also exceed this number. If the number of clusters is higher than the number of successor nodes, several clusters are grouped together so that the number of groups equals the number of successor nodes. All possible groupings are evaluated. In the following all equations only refer to clusterings for reasons of simplicity, i.e. the number of clusters k equals the number of successor nodes. A valuation function is used to rate the clusterings or groupings respectively. The valuation function prefers clusterings that group data according to their class labels. Clusterings where data are uniformly distributed across clusters notwithstanding their class labels receive low ratings. Furthermore clusterings are preferred which evenly divide the classes. Thus the valuation function rewards unambiguity regarding the class affiliation of the data assigned to a prototype as well as uniform distribution regarding the number of data points assigned to each prototype.

The valuation function $V(p)$ consists of two terms regulated by a scaling parameter $\lambda > 0$. The first term $E(p)$ calculates the entropy of the distribution of each class across the different clusters. This accounts for unambiguous distribution of the data considering the corresponding classes. The term $E(p)$ becomes minimal if it is ensured for all classes that all data belonging to one class is assigned to one cluster. It becomes maximal if all data belonging to one class is uniformly distributed across all clusters. The second term $D(p)$ computes the deviation from the uniform distribution. This term becomes minimal if each cluster is assigned the same number of data points. This allows for the even division of the classes into subsets. During the hierarchy generation phase we are looking for clusterings that minimize the valuation function $V(p)$. The influence of the respective term is regulated by the scaling parameter λ. Both terms are normalized so that they return values of the interval $[0, 1]$. The valuation function $V(p)$ is given by

$$V(p) = \frac{1}{l \log_2(k)} E(p) + \lambda \frac{1}{l(k-1)} D(p) \rightarrow \min$$

where $E(p) = \sum_{i=1}^{l}(-\sum_{j=1}^{k}(p_i^j \log_2(p_i^j)))$, $D(p) = \sum_{j=1}^{k}|\sum_{i=1}^{l}p_i^j - \frac{l}{k}|$ and $p_i^j = \frac{|X_i \cap Z_j|}{|X_i|}$ denotes the rate of patterns from class i, that belong to cluster j. Here $X_i = \{x_\mu | \mu = 1, ..., M; t^\mu = i\} \subseteq X$ is the set of data points that belong to class i, $R_j = \{x \in \mathbb{R}^d | j = \text{argmin}_{i=1,...,k}\|x - z_i\|\}$ denotes the Voronoi cell [27] defined by cluster j and $Z_j = R_j \cap X$ is the set of data points that were assigned to cluster j. z_i is the center of cluster i. The best clustering, i.e. the one that minimizes the valuation function $V(p)$, is chosen and is used for determining the division of the set of classes into subsets. Moreover this also determines which feature type will be used to train the corresponding classifier. So each classifier within the hierarchy can potentially use a different feature type and thereby operates in a different feature space. To identify which classes will be added to which subset the distribution of the data across the clusters is considered. The division in subsets C_j is carried out by maximum detection. The set of classes belonging to subset C_j is defined as $C_j = \{i \in C | j = \text{argmax}\{q_{i,1}, ..., q_{i,k}\}\}$ where $q_{i,j} = \frac{|X_i \cap Z_j|}{|Z_j|}$ denotes the rate of class i in cluster j. For each class it is determined to which cluster the majority of data points belonging to this class were associated. The class label will then be added to the corresponding subset.

To generate the hierarchy at first the set of all classes is assigned to the root node. Starting with a clustering on the complete data set the set off classes is divided into subsets. Each subset is assigned to a successor node of the root node. Now the decomposition of the subsets is continued until no further decomposition is possible or until the decomposition does not lead to a new division. An example of a classification hierarchy is shown in figure 9.

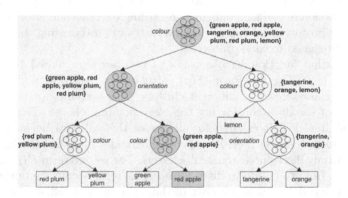

Fig. 9. Classifier hierarchy generated for the classification of different kind of fruits using two feature types: orientation histograms and color information. Each node within the hierarchy represents a neural network or a classifier respectively. The end nodes represent classes. To each node a feature type and a set of classes is assigned. The corresponding neural network uses the assigned feature type to discriminate between the assigned classes

Training and Classification. Within the hierarchy different types of classifiers can be used. Examples of classifiers would be radial basis function (RBF) networks, linear vector quantization classifiers or support vector machines. We chose RBF networks [28] as classifiers. They were trained with a three phase learning algorithm [29].

The hierarchy is trained by separately training each neural network within the hierarchy. The classifiers are trained using supervised learning algorithms. Each classifier is trained only with data points belonging to the classes assigned to the corresponding node hence the complete training set is only used to train the classifier that represents the hierarchy's root node. The classifiers are trained with the respective feature type identified during the hierarchy generation phase. To train the classifiers the data will be relabeled so that all data points of the classes that belong to one subset have the same label. The classifiers within the hierarchy can be trained independently, i.e. all classifiers can be trained in parallel.

The classification result is retrieved similar to the retrieval in a decision tree [27]. Starting with the root node the respective feature vector of the object to be classified is presented to the trained classifier. By means of the classification result the next classifier to categorize the data point is determined. Thus a path through the hierarchy from the root node to an end node is obtained which not only represents the class of the object but also the subsets of classes to which the object belongs. Hence the data point is not presented to all classifier within the hierarchy. If only intermediate results are of interest it is not necessary to evaluate the complete path.

4.2 Language Processing System

Our language system consists of a standard Hidden-Markov-based speech recognition system for isolated words and a cortical language processing system which can analyze streams of words detected with respect to simple (regular) grammars. For simplicity, the speech recognition system can also be replaced by direct text input via a computer terminal and a wireless connection to the robot.

Regular Grammars, Finite Automata and Neural Assemblies. Regular grammars can be expressed by generative rules of the type $A \mapsto a$ or $A \mapsto bC$, where capital letters are variables and lower case letters are terminal symbols, i.e. elements in of an alphabet Σ. There is usually a special starting variable S which can be expanded by applying the rules. Let Σ^* be the set of all possible strings of symbols from the alphabet Σ with arbitrary length. A sentence $s \in \Sigma^*$ is then called valid with respect to the grammar, if it can be derived from S by applying the grammatical rules and resolving all variables by terminal symbols. It is easy to show that regular grammars are equivalent to deterministic finite automata.

A deterministic finite automate (DFA) can be specified by the set $M = (Z, \Sigma, \delta, z_0, E)$, where $Z = \{z_0, z_1, \ldots, z_n\}$ is a finite set of states, Σ is the alphabet, $z_0 \in Z$ is the starting state and $E \subseteq Z$ contains the terminal states.

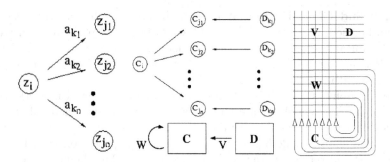

Fig. 10. Comparison of deterministic finite automate (DFA, left side) with a neural network (middle) and cell assemblies (right).
Each δ-transition $\delta(z_i, a_k) = z_j$ corresponds to synaptic connections from neuron C_i to C_j and from input neuron D_k to C_j in the neural network. For cell assemblies, the same basic architecture as for the neural net (lower part in the middle) is used, but instead of using single neurons for the states, assemblies of several cells are used

Finally, the function $\delta : (Z, \Sigma) \rightarrow Z$ defines the state transitions. A sentence $s = s_1 s_2 \ldots s_n \in \Sigma^*$ is valid with respect to the DFA, if iterated application of δ on z_0 and the letters of s transfers the automates starting state z_0 into one of the terminal states in E, i.e. if $\delta(\ldots \delta(\delta(z_0, s_1), s_2), \ldots, s_n) \in E$.

We now show that DFAs can easily be modelled by recurrent neural networks. For an overview, see also figure 10. This recurrent binary network is specified by $N = (C, D, W, V, C_1)$, where $C = (C_1, C_2, \ldots, C_n)$ contains the local cells of the network, $D = (D_1, D_1, \ldots D_m)$ is the set of external input cells, C_1 is the starting neuron and $W = (w_{ij})^{n \times n}$ as well as $V = (v_{ij})^{m \times n}$ are binary matrices. Here, w_{ij} describes the strength of the local synaptic connection between neuron C_i and neuron C_j, where v_{ij} is the synaptic strength between neuron D_i and neuron C_j. The network evolves in discrete time steps, where a neuron C_i is activated if the sum over its inputs exceeds threshold Θ_i, otherwise, it is inactive, i.e.

$$C_i(t+1) = \begin{cases} 0 \text{ , if } \sum_j w_{ji} C_j(t) + \sum_j v_{ji} d_j(t) \geq \Theta_i \\ 1 \text{ , otherwise} \end{cases}.$$

To simulate a DFA, we need to identify the alphabet Σ with the external input neurons D and the states Z with the local cells C. The synapses w_{ij} and v_{kj} are active if and only if $\delta(z_i, a_k) = z_j$ for the transition function δ of the DFA. Finally, set all thresholds $\Theta_i = 1, 5$ and activate only the starting neuron D_0 at time 0. A sentence $s = s_1 s_2 \ldots s_n$ is then presented by activating neuron D_{s_t} at the discrete time $t \in \{1, \ldots, n\}$ and to deactivate all other neurons D_j with $j \neq s_t$. Then, the network simulates the DFA, i.e. after presenting the last symbol of the sentence, one of the neurons corresponding to the terminal states of the DFA will be active if and only if the sentence was valid with respect to the simulated DFA.

Biologically, it would be more realistic to have cell assemblies (i.e. strongly interconnected group of nearby neurons) representing the different states. This

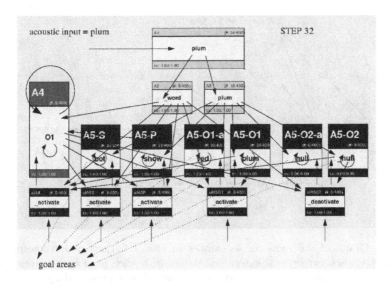

Fig. 11. The language system consisting of 10 cortical areas (large boxes) and 5 thalamic activation fields (small black boxes). Black arrows correspond to inter-areal connections, gray arrows within areas correspond to short-term memory

enables fault tolerance, since incomplete input can be completed to the whole assembly, furthermore, it becomes possible to express similarities between the represented objects via overlaps of the corresponding assemblies.

Cell Assembly Based Model. Figure 11 shows 15 areas of our model for cortical language processing. Each of the areas is modeled as a spiking associative memory of 400 neurons. Similar as described for visual object recognition, we a priori defined for each area a set of binary patterns constituting the neural assemblies stored auto-associatively in the local synaptic connections. The model can roughly be divided into three parts. (1) Primary cortical auditory areas A1, A2, and A3: First, auditory input is represented in area A1 by primary linguistic features (such as phonemes), and subsequently classified with respect to function (area A2) and content (area A3). (2) Grammatical areas A4, A5-S, A5-O1-a, A5-O1, A5-O2-a, and A5-O2: Area A4 contains information about previously learned sentence structures, for example that a sentence starts with the subject followed by a predicate (see Fig. 12). In addition to the auto-associative connections, area A4 has also a delayed feedback-connection where the state transitions are stored hetero-associatively. The other grammar areas contain representations of the different sentence constituents such as subject (A5-S), predicate (A5-P), or object (A5-O1, O1-a, O2, O2-a). (3) Activation fields af-A4, af-A5-S, af-A5-O1, and af-A5-O2: The activation fields are relatively primitive areas that are connected to the corresponding grammar areas. They serve to activate or deactivate the grammar areas in a rather unspecific way. Although establishing a concrete relation to real cortical language areas of the brain is beyond the scope

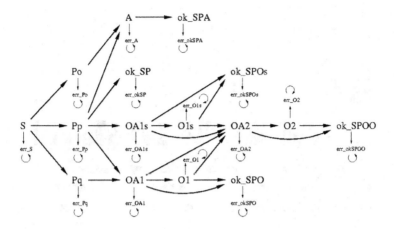

Fig. 12. Graph of the sequence assemblies in area A4. Each node corresponds to an assembly, each arrow to a hetero-associative link, each path to a sentence type. For example, a sentence "Bot show red plum" would be represented by the sequence (S,Pp,OA1,O1,ok_SPO)

of this work [30][31], we suggest that areas A1,A2,A3 can roughly be interpreted as parts of Wernicke's area, and area A4 as a part of Broca's area. The complex of the grammatical role areas A5 might be interpreted as parts of Broca's or Wernicke's area, and the activation fields as thalamic nuclei.

4.3   Planning, Action and Motor System

Our system for cortical planning, action, and motor processing can be divided into three parts (see Fig. 13). (1) The action/planning/goal areas represent the robot's goal after processing a spoken command. Linked by hetero-associative connections to area A5-P, area G1 contains sequence assemblies (similar to area A4) that represent a list of actions that are necessary to complete a task. For example, responding to a spoken command "bot show plum" is represented by a sequence (seek,show), since first the robot has to seek the plum, and then the robot has to point to the plum. Area G2 represents the current subgoal, and areas G3, G3attr, and G4 represent the object involved in the action, its attributes (e.g., color), and its location, respectively. (2) The "motor" areas represent the motor command necessary to perform the current goal (area G2), and also control the low level attentional system. Area M1 represents the current motor action, and areas M2, M2attr, and M3 represent again the object involved in that action, its attributes, and its location. (3) Similar to the activation fields of the language areas, there are also activation fields for the goal and motor areas, and there are additional "evaluation fields" that can compare the representations of two different areas. For example, if the current subgoal is "search plum", it is needed to compare between the visual input and the goals object "plum" in order to tell whether the subgoal is achieved or not.

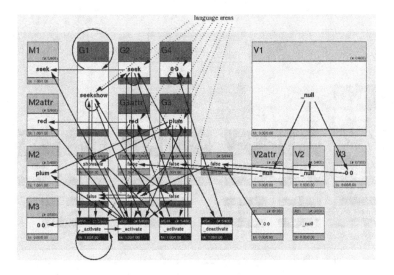

Fig. 13. The cortical goal and motor areas. Conventions are the same as for Fig. 11

5   Integrative Scenario

To illustrate how the different subsystems of our architecture work together, we describe a scenario where an instructor gives the command "Bot show red plum!", and the robot has to respond by pointing onto a red plum located in the vicinity.

To complete this task, the robot first has to understand the command. Fig. 14 illustrates the language processing involved in that task. One word after the other enters areas A1 and A3, and is then transferred to one of the A5-fields. The target field is determined by the sequence area A4, which represents the next sentence part to be parsed, and which controls the activation fields which in turn control areas A5-S/P/O1/O2. Fig. 14 shows the network state when "bot", "show", and "red" have already been processed and the corresponding representations in areas A5-S, A5-P, and A5-O1attr have been activated. Activation in area A4 has followed the corresponding sequence path (see Fig. 12) and the activation of assembly O1 indicates that the next processed word is expected to be the object of the sentence. Actually, the currently processed word is the object "plum" which is about to activate the corresponding representation in area A5-O1.

Immediately after activation of the A5-representations the corresponding information is routed further to the goal areas where the first part of the sequence assembly (seekshow,pointshow) gets activated in area G1 (see Fig. 15). Similarly, the information about the object is routed to areas G2, G3, and G3attr. Since the location of the plum is unknown, there is no activation in area G4. In area G2 the "seek" assembly is activated which in turn activates corresponding representations in the motor areas M1, M2, M3. This also ac-

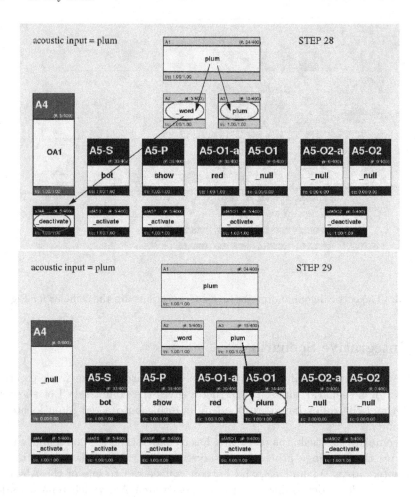

Fig. 14. The language areas when processing the sentence "Bot show red plum!". The first three words have already been processed successfully

tivates the attentional system which initiates the robot to seek for the plum as described in section 3.1. Fig. 15 shows the network state when the visual object recognition system has detected the red plum and the corresponding representations have been activated in areas V2, V2attr, and V3. The control fields detect a match between the representations in areas V2 and G3, which initiates area G1 to switch to the next part of the action sequence. Figure 16 shows the network state when already the "point" assembly in areas G1 and G2 has activated the corresponding representations in the motor areas, and the robot tries to adjust its "finger position" represented in area S1 (also visualized as the crosshairs in area V1). As soon as the control areas detect a match between the representations of areas S1 and G4, the robot has finished his task.

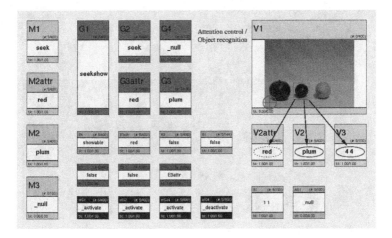

Fig. 15. System state of the action/motor model while seeking the plum

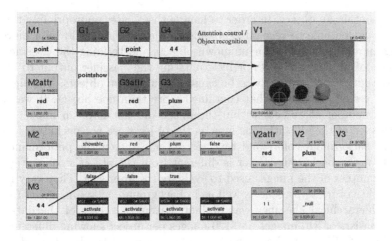

Fig. 16. System state of the action/motor model while pointing to the plum

6   Results

6.1   Evaluation of the Sensory Preprocessing

We evaluated the object localization approach using a test data set which consisted of 1044 images of seven different fruit objects. The objects were recorded under varying lighting conditions. The images contained a single object in front of a unicolored background. On this data set all 1044 objects were correctly localized by the attention control algorithm. No false-negative decisions were made, i.e. if there was an object in the scene it has been localized, and only 23 decisions were false-positive, i.e. regions of interest were marked although they

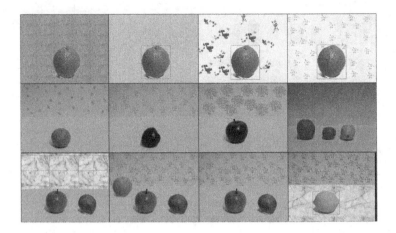

Fig. 17. Examples of the images used. They contain different fruits in front of various backgrounds

did not contain an object. In order to handle these cases appropriately the classifier should be able to recognize unknown objects. This could be achieved by adding an additional class for unknown objects and training the classifier with examples of unknown objects or by evaluating the classifier output to identify weak outputs which are likely to be caused by unknown objects. The algorithm was also tested on images that contained more than one object. It was found that the number of objects within the image did not have an impact on the performance of the algorithm as long as the objects are not occluded. Likewise different background colors and textures did not impact the performance provided that the background color is different from the object's color. Figure 17 shows examples of the images used to evaluate the approach. They vary in type and number of fruits present as well as in background color and structure.

The hierarchical classifier was evaluated using a subset of the data. This subset consists of 840 images, i.e. 120 images per object showing different views of the object. Figure 9 shows the hierarchy generated by the above described algorithm for the classification of seven fruit objects. Using 10-times 5-fold cross-validation experiments have been conducted on a set of recorded camera images in order to evaluate our approach.

We compared the performance of non-hierarchical RBF networks utilizing orientation histograms or color information respectively as feature type to the hierarchical RBF classifier architecture presented above. The classification rates are displayed in the box and whiskers plot in figure 19. Compared to simple neural networks that only used one feature type the performance of a hierarchical neural network is significantly better. The average accuracy rate of the hierarchical classifier was $94.05 \pm 1.57\%$ on the test data set and $97.92 \pm 0.68\%$ on the train data set. The confusion matrix in figure 18 shows that principally green apples were confused with red apples and yellow plums. Further confusions could be observed between red plums and red apples.

	green apple	red apple	tange-rine	orange	yellow plum	red plum	lemon
green apple	79.39	12.5	0	0	5.33	2.78	0
red apple	2.9	91.54	0	0	0	5.56	0
tange-rine	1.39	0	98.61	0	0	0	0
orange	0	0	0	100	0	0	0
yellow plum	2.78	0	0	0	97.22	0	0
red plum	0	6.95	0	0	0	93.05	0
lemon	0	0	0	1.39	0	0	98.61

Fig. 18. The confusion matrix shows row by row to which percentage objects of one class were classified as which class. If no confusions occurred the corresponding field on the matrix diagonal is marked in light grey. Confusions are marked in dark grey

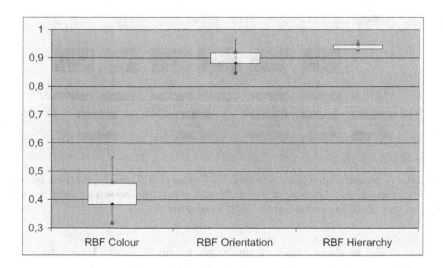

Fig. 19. The box and whisker plot illustrating the classification accuracy of non-hierarchical and hierarchical classifiers. The non-hierarchical classifiers were used to evaluate the different feature types used

The performance of the approach was tested on a Linux PC (Intel Pentium Mobile 1,6 GHz, 256 MB RAM). The average time for one recognition cycle including object localization, feature extraction and object classification was 41 ms.

Furthermore the object recognition system was tested on the robot. Thereto we defined a simple test scenario in which the robot has to recognize fruits placed on a table in front of it. This proved that the robot is able to localize and identify fruits not only on recorded test data sets but also in a real-life situation. A video of this can be found in [32].

6.2   Evaluation of the Cortical Information Processing

The cortical model is able to use context for disambiguation. For example, an ambiguous phonetic input such as "bwall", which is between "ball" and "wall", is interpreted as "ball" in the context of the verb "lift", since "lift" requires a small object, even if without this context information the input would have been resolved to "wall". Thus the sentence "bot lift bwall" is correctly interpreted. As the robot first hears the word "lift" and then immediately uses this information to resolve the ambiguous input "bwall", we call that "forward disambiguation". This is shown in figure 20.

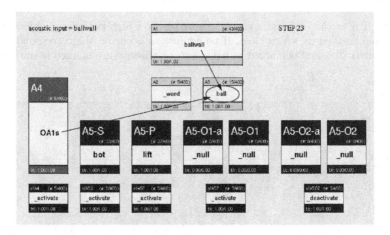

Fig. 20. A phonetic ambiguity between "ball" and "wall" can be resolved by using context information. The context "bot lift" implies that the following object has to be of small size. Thus the correct word "ball" is selected

Our model is also capable of the more difficult task of "backward disambiguation", where the ambiguity cannot immediately be resolved because the required information is still missing. Consider for example the artificial ambiguity "bot show/lift wall", where we assume that the robot could not decide between "show" and "lift". This ambiguity cannot be resolved until the word "wall" is recognized and assigned to its correct grammatical position, i.e. the verb of the sentence has to stay in an ambiguous state until enough information is gained to resolve the ambiguity. This is achieved by activating superpositions of the different assemblies representing "show" and "lift" in area A5P, which stores the verb of the current sentence. More subtle information can be represented in the spike times, which allows for example to remember which of the alternatives was the more probable one.

7   Conclusion

We have presented a cell assembly based model for visual object recognition and cortical language processing that can be used for associating words with objects, properties like colors, and actions. This system is used in a robotics context to enable a robot to respond to spoken commands like "bot put plum to green apple". The model shows how sensory data from different modalities (e.g., vision and speech) can be integrated to allow performance of adequate actions. This also illustrates how symbol grounding could be implemented in the brain involving association of symbolic representations to invariant object representations. The implementation in terms of Hebbian assemblies and auto-associative memories generates a distributed representation of the complex situational context of actions, which is essential for human-like performance in dealing with ambiguities.

Although we have currently stored only a limited number of objects and sentence types, our approach is scalable to more complex situations. It is well known for our model of associative memory that the number of storable items scales with $(n/\log n)^2$ for n neurons [4][5]. This requires the representations to be sparse and distributed, which is a design principle of our model. As any finite system, our language model can implement only regular languages, whereas human languages seem to involve more complex grammars from a linguistic point of view. On the other hand, also humans cannot "recognize" formally correct sentences beyond a certain level of complexity suggesting that in practical speech we use language rather "regularly".

Acknowledgement

This research was supported in part by the European Union award #IST-2001-35282 of the MirrorBot project.

References

1. Knoblauch, A., Fay, R., Kaufmann, U., Markert, H., Palm, G.: Associating words to visually recognized objects. In Coradeschi, S., Saffiotti, A., eds.: Anchoring symbols to sensor data. Papers from the AAAI Workshop. Technical Report WS-04-03. AAAI Press, Menlo Park, California (2004) 10–16
2. Hebb, D.: The organization of behavior. A neuropsychological theory. Wiley, New York (1949)
3. Palm, G.: Cell assemblies as a guideline for brain research. Concepts in Neuroscience 1 (1990) 133–148

4. Willshaw, D., Buneman, O., Longuet-Higgins, H.: Non-holographic associative memory. Nature **222** (1969) 960–962
5. Palm, G.: On associative memories. Biological Cybernetics **36** (1980) 19–31
6. Palm, G.: Neural Assemblies. An Alternative Approach to Artificial Intelligence. Springer, Berlin (1982)
7. Palm, G.: Memory capacities of local rules for synaptic modification. A comparative review. Concepts in Neuroscience **2** (1991) 97–128
8. Schwenker, F., Sommer, F., Palm, G.: Iterative retrieval of sparsely coded associative memory patterns. Neural Networks **9** (1996) 445–455
9. Sommer, F., Palm, G.: Improved bidirectional retrieval of sparse patterns stored by hebbian learning. Neural Networks **12** (1999) 281–297
10. Hopfield, J.: Neural networks and physical systems with emergent collective computational abilities. Proceedings of the National Academy of Science, USA **79** (1982) 2554–2558
11. Knoblauch, A., Palm, G.: Pattern separation and synchronization in spiking associative memories and visual areas. Neural Networks **14** (2001) 763–780
12. Knoblauch, A.: Synchronization and pattern separation in spiking associative memory and visual cortical areas. PhD thesis, Department of Neural Information Processing, University of Ulm, Germany (2003)
13. Utz, H., Sablatnög, S., Enderle, S., Kraetzschmar, G.K.: Miro – middleware for mobile robot applications. IEEE Transaction on Robotics and Automation, Special Issue on Object-Oriented Distributed Control Architectures **18** (2002) 493–497
14. Fay, R., Kaufmann, U., Schwenker, F., Palm, G.: Learning object recognition in an neurobotic system. 3rd IWK Workshop SOAVE2004 - SelfOrganization of AdaptiVE behavior, Illmenau, Germany. In press. (2004)
15. JShne, B.: Digitale Bildverarbeitung. Springer-Verlag (1997)
16. Foley, J., van Dam, A., Feiner, S., Hughes, J.: Computer Graphics. 2nd edn. Addisom-Wesley (2003)
17. Freeman, W.T., Roth, M.: Orientation histograms for hand gesture recognition. In: International Workshop on Automatic Face- and Gesture-Recognition, Znrich (1995)
18. Smith, A.R.: Color gamut transform pairs. Computer Graphics **12** (1978) 12–19
19. Gonzales, R.C., Woods, R.E.: Digital Image Processing. 2nd edn. Addison-Wesley (1992)
20. Bishop, C.M.: Neural Networks for Pattern Recognition. Oxford University Press (2000)
21. Schnrmann, J.: Pattern Classification. (1996)
22. Browne, A., Sun, R.: Connectionist inference models. Neural Networks **14** (2001) 1331–1355
23. Palm, G., Kraetzschmar, G.K.: Sfb 527: Integration symbolischer und subsymbolischer informationsverarbeitng in adaptiven sensomotorischen systemen. In Jarke, M., Pasedach, K., Pohl, K., eds.: Informatik '97 - Informatik als Innovationsmotor, 27. Jahrestagung der Gesellschaft fnr Informatik, Aachen, Springer (1997)
24. McGarry, K., Wermter, S., MacIntyre, J.: Hybrid neural systems: From simple coupling to fully integrated neural networks. Neural Computing Surveys **2** (1999) 62–93
25. Knoblauch, A., Fay, R., Kaufmann, U., Markert, H., Palm, G.: Associating words to visually recognized objects. In: AAAI Workshop on Anchoring Symbols to Sensor Data, San Jose, California. (2004)
26. Tou, J.T., Gonzales, R.C.: Pattern Recognition Principles. Addison-Wesley (1979)

27. Duda, R.O., Hart, P.E., Stork, D.G.: Pattern Classification. 2nd edn., New York (2001)
28. Broomhead, D., Lowe, D.: Multivariable functional interpolation and adaptive networks. Complex Systems **2** (1988) 321–355
29. Schwenker, F., Kestler, H.A., Palm, G.: Three learning phases for radial-basis-function netwoks. Neural Networks **14** (2001) 439–458
30. Knoblauch, A., Palm, G.: Cortical assemblies of language areas: Development of cell assembly model for Broca/Wernicke areas. Technical Report 5 (WP 5.1), MirrorBot project of the European Union IST-2001-35282, Department of Neural Information Processing, University of Ulm (2003)
31. Pulvermüller, F.: The neuroscience of language: on brain circuits of words and serial order. Cambridge University Press, Cambridge, UK (2003)
32. Fay, R., Kaufmann, U., Knoblauch, A., Markert, H., Palm, G.: Video of mirrorbot demonstration (2004)

Towards Word Semantics from Multi-modal Acoustico-Motor Integration: Application of the Bijama Model to the Setting of Action-Dependant Phonetic Representations

Olivier Ménard[1,2], Frédéric Alexandre[2], and Hervé Frezza-Buet[1]

[1] Supelec, 2 rue Edouard Belin, 57070 Metz, France
[2] Loria, Campus Scientifique, BP 239,
54506 Vandoeuvre-ls-Nancy, France

Abstract. This paper presents a computational self-organizing model of multi-modal information, inspired from cortical maps. It shows how the organization in a map can be influenced by the same process occurring in other maps. We illustrate this approach on a phonetic - motor association, that shows that the organization of words can integrate motor constraints, as observed in humans.

1   Introduction

In the evolutionary process, the appearance of the cerebral cortex has had dramatic consequences on the abilities of mammals, which reach their maximum in humans. Whereas it can be said that the limbic system has added an emotional dimension on purely reactive schemes [1], the cerebral cortex has offered a new substratum devoted to multimodal information representation [2]. When one considers the associated cost of this neuronal structure in terms of energy needs and size in the limited skull, it can be thought that the corresponding functions might be complex but highly interesting from an adaptive point of view.

Basically, the cerebral cortex is often described as a set of topological maps representing sensory or motor information, but also merging various representations in so-called associative maps. Concerning afferent connections toward cortical maps, the topological principle explains that information coming from sensors is represented along important dimensions, like retinotopy for the visual case. Moreover, at a lower level of description, some kind of filtering process allows to extract and represent onto the mapping other functional information [3], like orientation selectivity or color contrast in the visual case.

Concerning cortico-cortical connections, the important role of these internal links must be underlined. For example, the cerebral cortex and the cerebellum are reported as having approximately the same number of synapses (10^{12}) [4] and the big difference of volume between these structures can be explained by the fact that internal connections are much more numerous in the cerebral cortex (more than 75%). These internal connections inside the cerebral cortex are observed

S. Wermter et al. (Eds.): Biomimetic Neural Learning, LNAI 3575, pp. 144–161, 2005.
© Springer-Verlag Berlin Heidelberg 2005

as belonging to a map, to achieve the topological representation, but also as connecting maps, which is fundamental to create associative maps [5, 2].

From a functional point of view, the role of the cerebral cortex has often been described as unsupervised learning [6]. In the statistical domain, the goal of unsupervised models like the K-means, hierarchical classification, Principal Component Analysis (PCA), Independent Component Analysis (ICA) is to categorize information from the regularities observed in its distribution (as opposed to an external signal, seen as a teacher or a supervisor) or to select in a high dimensional space the most significant axes on which to project information. It must be underlined that such information processing is very consistent with the cortical organizational principles of topological representation and filtering.

From a modeling point of view, neuronal models are among the most well-known unsupervised techniques. The central one is certainly Kohonen's Self-Organizing Map [7], which has been proposed from its origin as a model of a cortical map (see also [8]) and has been applied in various sensory domains (see for example [9] for the visual case, [10] for the auditory case, etc.). Later, from this simple but powerful scheme, other more complicated models have been elaborated to fit more closely to the biological reality (cf. for example [11] for the visual case), but they all rely on the same fundamental principle of competitive learning, as observed in the cerebral cortex.

Interestingly, it must be noticed that most of these neuronal models lay emphasis on the representation of one sensory or motor information and not on the joint organization of several interacting flows of information (a notable exception being [12]). Nevertheless, evidence from neurosciences indicates that this function is also present in cortical processing. To tell it differently, the cortex is not only several self-organizing maps, each one representing its own modality (or set of modalities in the associative case) and communicating one with the other, but rather a set of maps acting all together to represent information of the external world from different but cooperating points of view in a global way.

Of course, such a holistic view cannot be obtained if, as it is often the case, one unique map is considered in the modeling process. The fact is that several biological data indicates that the cortical processing cannot be only summarized by independent self-organizations.

From a connectivity point of view, we have indicated above the important role which is given to recurrent cortico-cortical connections. This might be consistent with asking for a global consistency in representations on top of simple local competitions. Several electrophysiological studies have shown that a cortical region can change the kind of information it represents in case of a lesion (e.g. changing representation of the finger in a lesioned monkey [13]) or in case of sensory substitution (e.g. tactile stimulation for blind people [14]).

From a representational point of view, several brain imaging studies [15] have shown that word encoding within the brain is not only organized around phonetic codes but is also organized around action.

How this is done within the brain has not yet been fully explained but we would like to present how these action based representations naturally emerge in our model by virtue of solving constraints coming from motor maps. This model, called Bijama, is a general-purpose cortically-inspired computational framework that has also been used for rewarded arm control [16]. It is described in general terms in section 2, with particular attention to the effect of learning rules in section 3. Then, the actual use of the model for acoustico-motor integration in presented in section 4.

2   Bijama Model Features

The features of the model are presented briefly in the following sections. A more detailed presentation can be found in [17, 16], where the current model is applied to a simplified version of a target reaching problem with an artificial arm. This model is referred as the Bijama model in related papers, which stands for Biologically-Inspired Joint Associative MAps.

2.1   Maps, Units and Competition

The main computational block of the model is a set of computational units called a map. A map is a sheet made of a tiling of identical units. This sheet has been implemented as a disk, for architectural reasons described further. When input information is given to the map, each unit shows a level of activity, depending on the similarity of the information it receives with the information it specifically detects, as will be detailed in section 2.2. That activity, noted A^t, follows a Gaussian *tuning curve* in the model: A^t is a matching activity, that is maximal if input information exactly corresponds to the *prototype* of the unit, and gets weaker as input gets different from this prototype.

When an input is given to the map, the distribution of matching activities among units is a scattered pattern, because tuning curves are not sharp, which allows many units to have non null activities, even if prototypes don't perfectly match the input. From this activity distribution over the map, a small compact set of units that contains the most active units has to be selected. Unlike in SOMs where this decision is made by a centralized "winner-take-all" process, decision is made here by a numerical distributed process, emerging from a local competitive mechanism, as in [8].

In order to decide which units are locally the best matching ones inside a map, a local competition mechanism is implemented. It is inspired from theoretical results of the continuum neural field theory (CNFT) [18, 19], but it is adapted to become independent of the number of connections, thus avoiding disastrous border effects: The CNFT algorithm tends to choose more often units that have more connections. Thus, the local connection pattern within the maps must be the same for all units, which is the case for torus-like lateral connection pattern, with units in one border of the map connected to the opposite border, for example. Here, the field of units in the map computes a distribution of global activities A^*, resulting from the competition among current matching activity A^t. This competition process has been made insensitive to the actual position

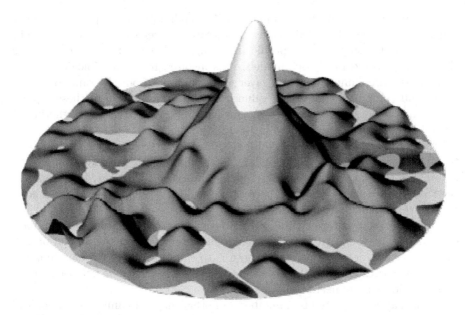

Fig. 1. Result of competition in a map among the A^t (dark gray) and resulting $A^\star$ (light gray). A bubble of $A^\star$ appears where A^t is the strongest in its neighborhood

of units within the map, in spite of heterogeneous local connection patterns at the level of border units in the Bijama model, as detailed further.

The result of this competition is the rising of a bubble of $A^\star$ activity in the map at places where A^t activities are the most significant (cf. figure 1). The purpose of the resulting $A^\star$ activity is twofold. First, this activity defines the main activity of the unit: This activity is the one that is viewed by other connected units in all activation rules detailed further. Second, all learning processes are modulated by this activity. That means that only units in $A^\star$ activity bubbles learn in the map.

The global behavior of the map, involving adaptive matching processes, and learning rules dependent on a competition, reminds the Kohonen SOM. However, the local competition algorithm used here allows the units to be feed with different inputs. The source of information received by a unit differs from one unit to its neighbors, because of the stripe connectivity described below in section 2.3. Another difference with SOM not previously detailed is that, in our model, competition and learning are not separated stages. Learning is dependent on $A^\star$, and also occurs during the $A^\star$ bubble setting.

2.2   Matching Activity Computation

It has been mentioned previously that competition is computed on the basis of a matching activity A^t. As detailed below, this activity is actually the merging of several matching results, and it may be considered as a global matching activity. Inside the units in the model, each matching result is performed by a computational module called a layer. Therefore, a layer in our model is a sub-part of a unit, computing a specific matching, and not a layer of neurons as classically reported in various models. It is inspired from the biological model of the cortical column by [20]. A layer gathers inputs from the same origin (a map), and computes a matching value from the configuration of these inputs. As a consequence, the behavior of a unit can be described as the gathering of several layers. These are detailed in the following.

First of all, some maps receive input from the external world. Each unit in the map reacts according to the fitting of this input to a preferred input. In the cortex, the thalamus plays a role in sending inputs to the cortex. In our model, the layer which tunes a preferred perception is called a thalamic layer. This layer provides a thalamic matching activity.

One other kind of layer is the cortical layer. It receives information from another map. The connectivity of this layer will be further discussed in section 2.3. Let us just say for now that its purpose is to compute a cortical matching activity that corresponds to the detection of some $A^\star$ activity distribution in the remote units it is connected to.

If the map is connected to a number n of other maps, its units have n cortical layers, thus computing n cortical matching results (one per cortical layer). These matchings are merged to form a global cortical matching. If the map has a thalamic layer, the thalamic matching result is then merged to the global cortical matching, to form the global matching A^t the competition is performed from.

To sum up, our model stresses the two kinds of cortico-cortical connections mentioned in section 1: Local connections and inter-map connections. The maps compute activity bubbles, that are a decision enhancing the most relevant units from local connections belonging to the map. This decision depends on external input, computed by the thalamic layer, but also on the state of other maps through cortical layers, that implement long range cortico-cortical connections. This computation is a multi-criteria decision, that has complex dynamics, since it performs a competition from input, but also from the competition that is performed in the same way in other maps. One consequence of this dynamics, central to the model, is that the self-organization in a map is modulated by the organization in the other maps, as illustrated in section 4.

2.3   Inter-map Stripe Connectivity and Disk-Shaped Maps

A cortical layer, that receives information from another map, doesn't receive inputs from all the units of the remote map, but only from one stripe of units (cf. fig. 3). For instance, a map may be connected row-to-row to another map: Each unit in any row of the first map is connected to every remote units in the corresponding row of the other map. These connections are always reciprocal in the model.

This limited connectivity is biologically grounded, as cortical zones are connected to other zones by stripes [2, 3]. Moreover, it has a computational purpose: If inter-map connectivity were total (if each unit in a map were connected to

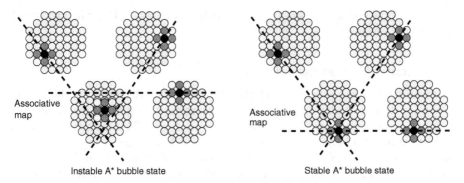

Fig. 2. Conditions for stable A^* activity states. On the left, instable state where bubbles of A^* don't stand at intersecting stripes. On the right, stable state: Any other bubble position in the same stripes on the three non-associative maps would also be stable

every unit in a connected remote map), the number of connections would rise too quickly as the size and the number of the maps increase and would lead to a combinatorial explosion. Since this model has been designed to handle multiple sensori-motor connections, the risk is real and map-to-map connectivity has to be limited.

A stripe has an orientation that is specific to a map-to-map connection: A map that is connected to many other ones has a different direction of connection for each kind of connection. In order to keep the model homogeneous, the shape of the map must not favor any direction. This is the reason why the maps are disk-shaped in our model.

Two analogous cortical layers of two neighboring units are connected to parallel, adjacent and overlapping stripes in the remote map. Neighboring units receive close but not identical inputs. That is why a winner-takes-all algorithm over the whole map isn't suitable, as already explained.

Through the inter-map connectivity, our model produces *resonance* between connected maps: Activity patches in connected maps can only stabilize within connected modular stripes. The role of reciprocally connected stripes is crucial for this resonance. Activity A^t is the basis for inner-map lateral competition (computation of A^*). As this A^t depends on some cortical activity(ies), computed from other cortical inputs that are fed with remote A^*, bubbles of A^* activities raise in the maps at some particular places: The bubble of activity that appears in an associative map is at the intersection of the stripes where activity bubbles coming from the connected maps stand (see figure 2).

In our model, this matching of activity can be compared with a phenomenon of resonance, as described in the ART paradigm by Grossberg [21], that produces stable and coherent states across the different maps. It ensures consistency of the activity bubbles across two connected cortical maps. Since units learning rate is modulated by their A^*, units whose A^* are activated simultaneously in the different maps learn together. We call this *coherent learning*. Learning strengthens the connection between these coherent units, so that they will tend to activate together again in the future.

2.4   Activation and Learning Rules

As mentioned before, cortical and thalamic layers of the units in the model have to perform a tuning from the input they receive, so that all matchings are merged to constitute the global matching activity A^t. This merging concerns all cortical and thalamic layers, and is computed from a geometric mean. This must be seen as a tricky way to compute some kind of numerical AND operator. Knowing these merging principles, let the computation of each elementary matching, and their associated learning rule, be detailed for both thalamic and cortical layers.

The thalamic layer in the model behaves similarly to neurons in Kohonen maps. This is a custom defined point in the model, depending on the actual entry format received by the map. For example, thalamic tuned activation can be a decreasing function of a well suited distance between the input and a prototype. Then learning consists of making the thalamic prototype be closer to the current input. This learning process has to be modulated by $A^\star$ activity for thalamic layer to be coherent with the remaining of the model. This is also what is done in Kohonen maps, where learning rate depends on a decreasing function of the proximity of a neuron with the winning one. This decreasing function in Kohonen algorithm is analogous to the $A^\star$ bubble of activity in the model.

The cortical layers all use the same matching and learning rules. Each cortical activity is computed from a cortical prototype pattern and the cortical input pattern, which is actually the $A^\star$ activity distribution in the connected stripe of remote units. The layer matching activity has to be high only when both the $A^\star$ of a remote unit and the corresponding value in the prototype are high: The cortical layer detects that a remote unit to which it is highly connected is active, and thus performs a computational "AND". The learning is, as for the thalamic layer, modulated by $A^\star$.

A unit learns only when it actively participates in the recognition process, i.e. when it is at a place where a $A^\star$ bubble stands. It learns both its thalamic and cortical prototypes, which creates and then maintains coherence between the different layers. The full unit model is summarized in figure 3.

2.5   Joint Organization

To conclude on the model behavior, the combination of self-organization and coherent learning produces what we call *joint organization*: Competition, although locally computed, occurs not only inside any given map, but across all maps. Moreover, the use of connection stripes limits the connectivity, which avoids the combinatorial explosion that would occur if the model were to employ full connectivity between the maps. Thus, coherent learning leads to both efficient data representation in each map and coordination between all connected maps.

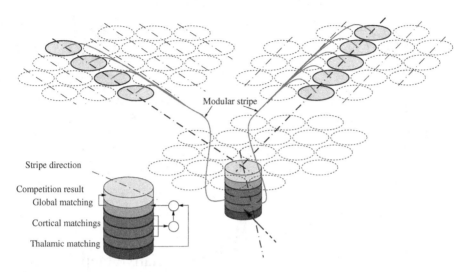

Stripe direction

Competition result
Global matching
Cortical matchings
Thalamic matching

Fig. 3. Full functional scheme: The cortical matching activities, obtained from the modular stripe inter-map connections, are merged together. The thalamic matching is merged with the result to form a global matching activity. This activity is then used in the competition process described in section 2.1

3   Multi-association and Cortical Learning Rules

A multi-associative model is intended to associate multiple modalities, regardless of however they are related. It must then handle the case where the associations between two modalities are not one-to-one, but rather one-to-many, or even many-to-many. This *multi-association* problem will now be presented on a simple example.

3.1   Associative Units and Multi-association

Let us consider an association between certain objects and the sounds they produce. A car, for example, could be associated with a motor noise. Certain objects produce the same noise. As a result, a single noise will be associated with multiple objects. For instance, a firing gun and some exploding dynamite produce basically both an explosion sound.

In our model, let us represent the sounds and the objects as two thalamic modalities on two different cortical maps. Let us now link both of these maps to another one, that we call an *associative* map. The sound representations and the object representations are now be *bound* together through the associative map (see fig. 4).

If we want a single unit to represent the "BANG" sound in the sound map, a single unit in the associative map has to bind together the "BANG" unit with both the gun and the dynamite units. This associative unit must then have the ability to perform *multi-associations*: It must have a strong cortical connection

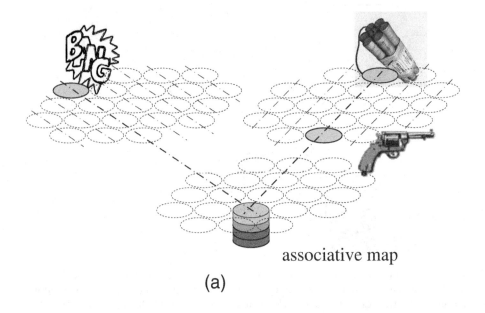

(a)

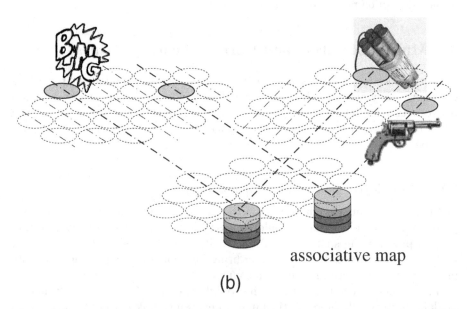

(b)

Fig. 4. A multi-association: A sound is associated with two different object, both of which may produce this actual sound. (a) A single sound unit is connected with the two object units by a unit in the associative map, that stands at the intersection of two sound and object stripes: This is possible because the unit's global connection strength is not distributed among all connections, but only among *active* connections (Widrow-Hoff rule). (b) When a unit's global connection strength is distributed among all connections, the sound unit must be duplicated (Hebb rule). See text for detail

to two different units (the gun and the dynamite units) in the same cortical stripe (see fig. 4a).

If associative units cannot perform multi-associations, the resulting self-organization process among all maps will duplicate the "BANG" representative unit. The reason is that, in this very case, an associative unit is able to listen to only one unit in each connected module. Each instance of that unit will then be bound, through its own associative unit, either to the dynamite or the gun unit (see fig. 4b). Moreover, the two object units cannot be in the same connectivity stripe, or else the model would try to perform multi-association and fail.

In fact, in the associative map, a unit is able to bind all possible couples of sound and object units that are in the two cortical stripes. Reciprocally, two given stripes of sound and object units can only be bound through a single associative unit in the associative map, the one that stands at the intersection of these stripes. Therefore, since our model uses a logical "AND" between the different inputs of a unit, that single associative unit must be active each time one of these sound and one of these objects are active together.

Thus, if a unit can only handle one association, each couple of sound and object stripes must contain, at most, one associated object and sound couple of units. The units that represent different object making the same sound, for instance the dynamite and the gun, must then be on different cortical stripes (see fig. 4a). The same is true for the two "BANG" units, which represent the same sound. Actually, the ability of units to handle multi-associations depends solely on the cortical learning rule used in the model, as we will now explain. This ability is crucial in the model, as explained further.

3.2   Cortical Learning Rules Requirements for Multi-association

It is the cortical learning rule that enables, or not, a unit to be strongly connected to many remote units in the same cortical stripe. Therefore, the cortical learning rule is the key to solving the multi-association problem.

Let us consider first the Hebb/anti-Hebb learning rule. Using this rule, if unit i is connected to unit j, the weight w_{ij} of the cortical connection from i to j is updated through :

$$\delta w_{ij} = \delta A_i^\star \times (A_j^\star - w_{ij})$$

where δ is the update rate of the cortical connection weights. Thus, the connection strength w_{ij} between the local unit i and a remote unit j grows when both i and j are active together (i.e. they are both in an $A^\star$ activity bubble). w_{ij} decreases each time the remote unit j is inactive while the local unit i is active.

Therefore, if unit i and j are *always* active together, w_{ij} will grow to reach the maximum weight. Consider now the case where unit j is active, for instance, half of the time when unit i is active, and remote unit j' is active the other half of that time. Both w_{ij} and $w_{ij'}$ will only reach half the maximum weight. Since cortical activation is computed on the base of the $w_{ik} \times A_k^\star$, the cortical input from j (or j') can be quite weak when $A_j^\star$ (or $A_{j'}^\star$) is strong, just because many $A_{j'}^\star$ were correlated to the $A_i^\star$ activity.

Since only the most active units are inside the $A^\star$ activity bubble, a local unit i can be active only if its cortical activities are high enough. Therefore, because of competition, local unit i can only be active if its connection weight to some remote unit j is high. This is the reason why, using the Hebb/anti-Hebb learning rule, a unit i can be activated by a unit j if $A_j^\star$ is the only one that is correlated to $A_i^\star$.

This result is actually due to the following: The global connection strength of the local unit i for a given cortical stripe is at every time distributed among all connected remote units j. Since that connection strength must be high for some connection, it is concentrated on a single remote unit, which means that all other remote units are very weakly connected to the local unit. The end result is that the local unit binds together a single remote unit per stripe connection.

As a consequence, the model cannot represent a situation where a unit in a map should be bound with multiple units in the other remote maps: It cannot handle *multi-associations*. The only way to effectively associate a given thalamic input in a map to two different thalamic inputs in another map is to have it represented by two different patches units (see fig. 4b).

This can be avoided if the cortical learning rule allows a unit to be strongly associated with multiple units for each cortical stripe. A cortical learning rule that allows this is the Widrow-Hoff learning rule.

3.3   Widrow-Hoff Learning Rule and Consequences for Multi-association

Using a learning rule adapted from the Widrow-Hoff learning rule, if unit i is connected to unit j, the weight w_{ij} of the cortical connection from i to j is updated through :

$$\delta w_{ij} = \delta(A_i^\star - \omega) \times (A_i^\star - A_i^c) \times A_j^\star$$

where δ is the update rate of the cortical connection weights, A_i^c is the cortical activity of a unit i, and ω is the decay rate of cortical connections. Here, cortical activity $A_i^c = \sum_j w_{ij} A_j^\star$ is seen as a predictor of $A_i^\star$. When both the local unit i and the remote unit j are active together, if A_i^c is lower than $A_i^\star$, w_{ij} grows, and if A_i^c is higher than $A_i^\star$, w_{ij} decreases. w_{ij} also decreases slowly over time.

Here, the global connection strength of the local unit i for a given cortical stripe is distributed among all *active* remote units j, and *not* among *all* remote units. As with the Hebb/anti-Hebb rule, because of the local competition, the connection strength w_{ij} between i and a unit j must be high. However, here, raising w_{ij} doesn't imply lowering all w_{ik} for all k in the remote connection stripe. Raising w_{ij} will only lower w_{ik} if j and k are active at the same time.

However, since only a small $A^\star$ activity bubble is present on each map, most remote units in the connection stripe cannot be active at the same time. Thus, the local unit i can bind together multiple units in a given stripe connection to a unit in another stripe connection. This is the reason why the use of the Widrow-Hoff learning rule in our model leads to multi-map organization as in fig. 4a.

Solving the multi-association problem has one main benefit: The maps need fewer units to represent a certain situation than when multi-association between unit is impossible. Moreover, since instances of a given thalamic input are not duplicated in different parts of a cortical map, it is easier for the model to perform a compromise between the local organization and the cortical connectivity requirements, i.e. *joint organization* is less constrained.

4   Model Behavior on a Simplified Example

4.1   The Phonetic-Action Association Problem

Several brain imaging studies [15] have shown that word encoding within the brain is not only organized around purely phonetic codes but is also organized around action. How this is done within the brain has not yet been fully explained but we would like to present how these action based representations naturally emerge in our model by virtue of solving constraints coming from motor maps.

We therefore applied our model to a simple word-action association. A part of the word set from the European MirrorBot project, which is a 3 year EU-IST-FET project, was used in a "phonetic" map, and we tried to associate these words to the body part that performs the corresponding action. One goal of this project is to define multimodal robotic experiments and the corresponding protocols are consequently well suited for this task.

4.2   Phonetic and Motor Coding

The phonetic coding used in our model is taken from the MirrorBot project. A word is separated into its constituting phonemes. Each phoneme is then coded by a binary vector of length 20. Since the longest word that is used has 4 phonemes, each word is coded by 4 phonemes, and if they have less, they are completed by empty phonemes.

The distance between two different phonemes is the Cartesian distance between the coding vectors. The distance between two words is the sum of the distances between their constituting phonemes. While we are well aware that this is a very functional way to represent the phonetic distance between two words, it is sufficient in order to exhibit the joint organization properties discussed in this paper.

The actions are coded in the same way as the words: There are 3 different input actions (head action, body action and hand action), and each action is coded as a binary vector of length 3. The distance between two actions is, once again, the Cartesian distance between their representing vectors.

Each word is semantically associated to a specific action. The word-action relationship is shown on figure 5.

The thalamic prototypes (i.e. external inputs) of the motor and the phonetic units are, respectively, coded actions and coded words. However, these do not necessarily correspond to real input words or actions: These prototypes are vector of float values, not binary ones. The prototype of a unit, in the figures of

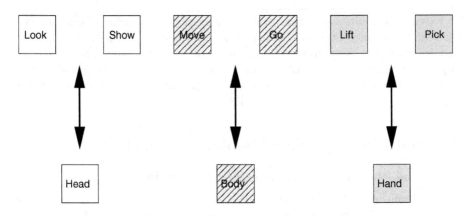

Fig. 5. Word-Action Relationship

this section, is represented as the nearest "real" input, in term of the distance previously discussed.

4.3   Interest of Associative Learning

Our model fundamentally differs from a classical Kohonen map since this latter one is somehow topologically organizing information against the sole notion of distance between inputs and prototypes. Thus if we were to use a Kohonen map to represent words from the MirrorBot grammar (encoded as a phonetic sequence), a consequence of the Kohonen algorithm and existing lateral interaction between units would be an organization toward similarity relation of word codes only (i.e. two words having similar code would be represented by the same prototype or neighbor prototypes) as illustrated in figure 6. This kind of representation is not satisfactory in the sense that it is totally disconnected from other maps and does not take any semantics of words into account.

4.4   Emergence of Action Oriented Representation

Let us consider three maps, one for word representation, one for action representation and finally an associative one that links word to action (cf. figure 7).

The central point of our model is that coherent learning within a map depends on some other maps, so that the inter-map connectivity biases the convergence to a particular self-organized state, when self-organization alone would have allowed for many more possible ones. The final state of organization in each map must allow the bubbles to be set up at intersecting cortical connection stripes, solving inter-map constraints as the one illustrated on fig 2. The cortical maps perform an effective compromise between the local and remote constraints. Remote constraints, coming from the architecture, makes activity bubbles have strong cortical connections to each other. Local constraints, coming from the thalamic layers, requires bubbles of activity to raise where the phonetic or action

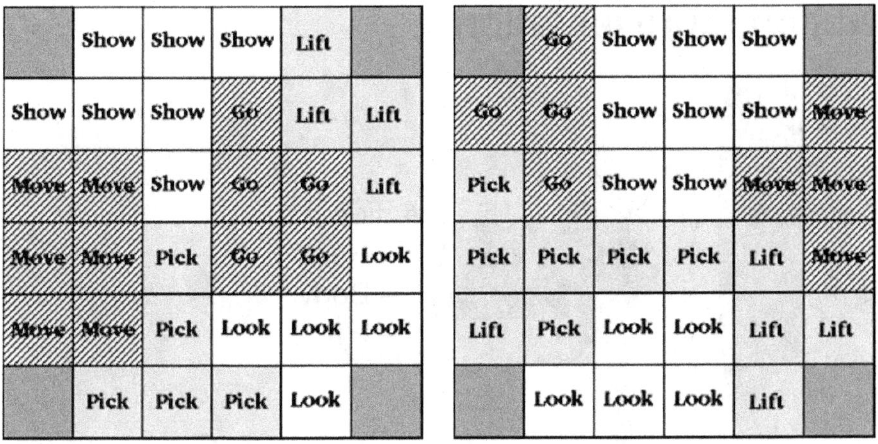

	Show	Show	Show	Lift	
Show	Show	Show	Go	Lift	Lift
Move	Move	Show	Go	Go	Lift
Move	Move	Pick	Go	Go	Look
Move	Move	Pick	Look	Look	Look
	Pick	Pick	Pick	Look	

	Go	Show	Show	Show	
Go	Go	Show	Show	Show	Move
Pick	Go	Show	Show	Move	Move
Pick	Pick	Pick	Pick	Lift	Move
Lift	Pick	Look	Look	Lift	Lift
	Look	Look	Look	Lift	

Fig. 6. Two different results of word classification by a Kohonen map based on purely phonetic representations. Words representing eye action (white), hand action (gray) or body action (stripes) are spread all over the map without paying any attention to the underlying semantic of words

prototypes best match the phonetic or action input. This compromise is poor at the beginning, but it gets better as learning proceeds.

In the current word-action association, we have chosen to impose a frozen organization to the action map, in order to illustrate how the phonetic map self-organizes when keeping coherence with the action map. As an illustration, let us consider the words "look" and "show". The phonetic representations of these words are completely different, so that a Kohonen map classifies them in different parts of the map (cf. fig. 6). In our model, however, the higher level associative map linking auditory representation with motor action will use close units to represent these words, since they both relate to the same action (head action), see "look" and "show" positions on fig. 8. As our model deals with an implicit global coherence, it is able to reflect this higher level of association and to overcome the simpler phonetic organization.

The interesting point to consider here is that word representations (e.g. phonetic map) are constrained by some topology that mimics to some extent physical properties of effectors, i.e. a word unit is dedicated to one action (e.g. hand) and cannot trigger another one (e.g. head). In order to solve this constraint and to ensure a global coherence, the model must then organize word representation in such a way that, for example, any "body" word should be linked to a body action.

As illustrated in figure 8, we can clearly see that the topological organization found by the model meets these criteria. Within the word map, words are grouped relatively to the body part they represent: Body action words are grouped together (stripes) as well as hand action words (gray) and head action words (white).

Action representation

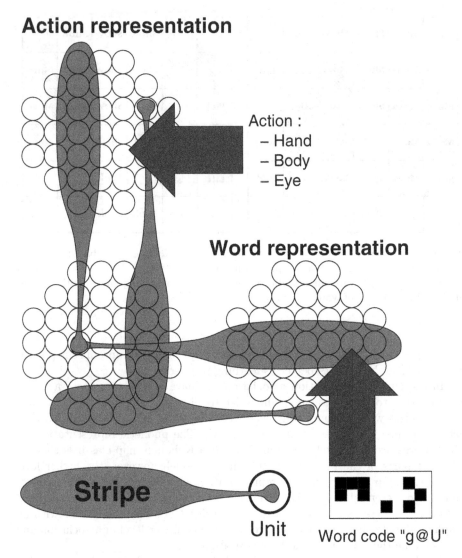

Action :
- Hand
- Body
- Eye

Word representation

Fig. 7. Schematic view of the model architecture: The word representations and the action representations are presented in separate maps, that are both connected to an associative map by reciprocal cortical connection stripes

However, the phonetic distribution of words remains the most important factor in the phonetic map organization. Each word is represented by a "cluster" of close units, and the words whose phonetic representation is close tend to be represented in close clusters of units. For instance, while "Go" and "Show" correspond to different motor actions, their phonetic representations are close, so that

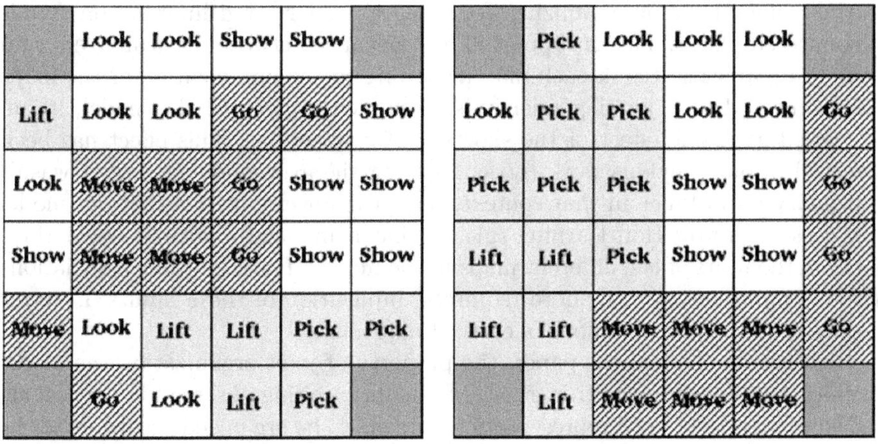

	Look	Look	Show	Show	
Lift	Look	Look	Go	Go	Show
Look	Move	Move	Go	Show	Show
Show	Move	Move	Go	Show	Show
Move	Look	Lift	Lift	Pick	Pick
	Go	Look	Lift	Pick	

	Pick	Look	Look	Look	
Look	Pick	Pick	Look	Look	Go
Pick	Pick	Pick	Show	Show	Go
Lift	Lift	Pick	Show	Show	Go
Lift	Lift	Move	Move	Move	Go
	Lift	Move	Move	Move	

Fig. 8. Two simulation results of word representation map after coherent learning has occurred with our model. Word representations are now constrained by the motor map via the associative map and, as a result, words that correspond to the same action are grouped together. Nevertheless, phonetic proximity is still kept

their representing clusters are adjacent (cf. fig. 8). This illustrates the fact that the model is actually doing a successful compromise between the local demands, which tend to organize the words phonetically, and the motor demands, which tend to put together the words that correspond to the same action. The joint organization does not destroy the local self-organization, but rather modulates it so that it becomes *coherent* with the other map organization.

Finally, having this model based on the self-organization of information prototypes leads implicitly to an organization that can be interpreted since it is easy to see what a unit is tuned on. This might be useful for further qualitative comparisons with real fMRI activations.

5   Discussion

The model presented in this paper is designed for a general cortically-inspired associative learning. It is based on the cooperation of several self-organizing maps, that are connected one with the other. The design of this model stresses some computational points, that keeps the model functionally close to the biology. The first one is locality, since each unit computes its status from the units it is connected to, without any superior managing process. This leads to the set up of a distributed competition mechanism, whose emerging effect is the rise of a bubble of activity at locally relevant places in the map. The second computational point the model stresses is stripe connectivity between maps. From a strictly computational point of view, this keeps the number of connections under combinatorial explosion. Moreover, the consequent computation has more interesting properties. Using stripes actually constraints the model to overcome partial connectivity by organizing the maps so that related information stands at connected places. This is supported by resonance between cortical layers, and leads to organize states in each map according to the organization of the maps it is connected to. This dependency isn't explicitly given to the model, it can be viewed as a side effect of the shortage of connections. This effect has been observed in our previous work [17] concerning the arm guidance, but it wasn't of primary importance in that context. Last, the novelty here is that our model now uses a Widrow-Hoff learning rule, so that it manages multiple associations between the units of the different maps in the model. Thus, multiple associations between inputs do not anymore require a duplication of these inputs, in terms of different units, on the model's cortical maps.

However, in the present paper, the property of joint organization the model exhibits is reported in the framework of semantic coding observed in cortical areas, since high level word representation appears to be organized according to the body part the word refers to. The ability of our model to generate such kind of organization without any supplementary specification supports its relevance as a functional model of cortical computation, in spite of sometimes less plausible computational mechanisms that keeps the model tractable for a large amount of units.

Considering self-organization of many interconnected self-organizing modules leads to discuss the organization of representations at a global level, that may appear rather more abstract than the organization resulting from the mapping of a mono-modal distribution, as performed by usual unsupervised learning techniques. In the context of autonomous robotics, that this model addresses, anything that is learned is obviously a situated representation. Moreover, the model makes the organization of a particular module in the architecture, dealing with one specific modality, be understandable according to the other modules, and more generally according to the global purpose of such an architecture to address situated behavior. This raises the hypothesis that the influence of the global behavioral purpose at the level of each modality representation is the very property that endows this representation with a semantic value. Therefore, this view of semantics, inspired from biological facts about cortical areas involved in language, appears to be tractable by a joint organizing model, and to be more generally suitable for any situated multimodal processing in robotics.

Acknowledgments

The authors wish to thank the Lorraine region, the Robea program of the CNRS and the European MirrorBot project for their contribution.

References

1. Rolls, E.: The Brain and Emotion. Oxford University Press: Oxford (1999)
2. Burnod, Y.: An adaptive neural network : the cerebral cortex. Masson (1989)
3. Ballard, D.H.: Cortical connections and parallel processing : Structure and function. The Behavioral and Brain Sciences **9** (1986) 67–129

4. Ito, M.: The cerebellum and neural control. New-York, Raven (1984)
5. Mountcastle, V.B.: An organizing principle for cerebral function. The unit module and the distributed system. In: The mindful brain, Cambridge, MIT Press. (1978)
6. Doya, K.: What are the computations in the cerebellum, the basal ganglia, and the cerebral cortex. Neural Networks **12** (1999) 961–974
7. Kohonen, T.: Self-Organization and Associative Memory. Springer-Verlag (1988)
8. D.J.Willshaw, von der Malsburg, C.: How parrerned neural connections can be set up by self-organization. In: Proceedings of the royal society of London. Volume B 194. (1976) 431–445
9. Kohonen, T., Oja, E.: Visual feature analysis by the self-organising maps. Neural Computing and Applications **7** (1998) 273–286
10. Kohonen, T.: The neural phonetic typewriter. Computer **21** (1988) 11–22
11. R. Miikkulainen, J.A. Bednar, T.C., Sirosh, J.: Self-organization, plasticity, and low-level visual phenomena in a laterally connected map model of the primary visual cortex. In R.L. Goldstone, P.S., Medin, D., eds.: Psychology of Learning and Motivation (36: perceptual learning). San Diego, CA: Academic Press (1997) 257–308
12. Ritter, H., Martinetz, T., Schulten, K.: Neural Computation and Self-Organizing Maps: An Introduction. Addison-Wesley Longman Publishing Co. (1992)
13. T. Allard, S.A. Clark, W.J., Merzenich, M.: Reorganization of somatosensory area 3b representations in adult owl monkeys after digital syndactyly. J Neurophysiol. **66** (1991) 1048–58
14. Bach-y-Rita, P.: Tactile sensory substitution studies. Ann NY Acad Sci. **1013** (2004) 83–91
15. Pulvermüller, F.: The Neuroscience of Language. Cambridge University Press (2003)
16. Ménard, O., Frezza-Buet, H.: Rewarded multi-modal neuronal self-organization: Example of the arm reaching movement. In: Proc. AISTA. (2004)
17. Ménard, O., Frezza-Buet, H.: Multi-map self-organization for sensorimotor learning: a cortical approach. In: Proc. IJCNN. (2003)
18. Amari, S.I.: Dynamical study of formation of cortical maps. Biological Cybernetics **27** (1977) 77–87
19. Taylor, J.G.: Neural networks for consciousness. Neural Netowrks **10** (1997) 1207–1225
20. Guigon, E., Dorizzi, B., Burnod, Y., Schultz, W.: Neural correlates of learning in the prefrontal cortex of the monkey: A predictive model. Cerebral Cortex **5** (1995) 135–147
21. Grossberg, S.: Adaptative pattern classification and universal recoding, i:parallel development and coding of neural feature detectors. Biological Cybernetics **23** (1976) 121–134

Grounding Neural Robot Language in Action

Stefan Wermter[1], Cornelius Weber[1], Mark Elshaw[1], Vittorio Gallese[2],
and Friedemann Pulvermüller[3]

[1] Hybrid Intelligent Systems, University of Sunderland,
School of Computing and Technology,
St Peter's Way, Sunderland, SR6 0DD, UK
{Stefan.Wermter, Cornelius.Weber, Mark.Elshaw}@sunderland.ac.uk
www.his.sunderland.ac.uk
[2] Institute of Neurophysiology, University of Parma,
Via Volturno, 39/E I-43100 Parma, Italy
vittorio.gallese@unipr.it
[3] Medical Research Council, Cognition and Brain Sciences Unit,
15 Chaucer Road, Cambridge, UK
friedemann.pulvermuller@mrc-cbu.cam.ac.uk

Abstract. In this paper we describe two models for neural grounding
of robotic language processing in actions. These models are inspired by
concepts of the mirror neuron system in order to produce learning by
imitation by combining high-level vision, language and motor command
inputs. The models learn to perform and recognise three behaviours, 'go',
'pick' and 'lift'. The first single-layer model uses an adapted Helmholtz
machine wake-sleep algorithm to act like a Kohonen self-organising network that receives all inputs into a single layer. In contrast, the second,
hierarchical model has two layers. In the lower level hidden layer the
Helmholtz machine wake-sleep algorithm is used to learn the relationship between action and vision, while the upper layer uses the Kohonen
self-organising approach to combine the output of the lower hidden layer
and the language input.

On the hidden layer of the single-layer model, the action words are
represented on non-overlapping regions and any neuron in each region
accounts for a corresponding sensory-motor binding. In the hierarchical
model rather separate sensory- and motor representations on the lower
level are bound to corresponding sensory-motor pairings via the top level
that organises according to the language input.

1 Introduction

In order to ground language with vision and actions in a robot we consider two
models, a single-layer and a hierarchical approach based on an imitation learning. Harnad 1990 and Harnad 2003 [10, 11] devised the concept of the symbol
grounding problem in that abstract symbols must be grounded or associated to
objects and events in the real world to know what they actually mean. Hence,
in order to actually attribute meaning to language there must be interaction

S. Wermter et al. (Eds.): Biomimetic Neural Learning, LNAI 3575, pp. 162–181, 2005.
© Springer-Verlag Berlin Heidelberg 2005

with the world to provide relevance to the symbolic representation. In terms of robotics there is a need to ground actions and visual information with symbolic information provided by language to meaningfully portray what is meant [11]. For instance, the action verb 'lift' could be grounded in the real-world robot behaviour of closing the gripper on an object, moving backward and turning around. The importance of grounding abstract representations can be seen from Glenberg and Kaschak 2002 [9] who found that the understanding of language is grounded in the action, how the action can be achieved and the likelihood of the action occurring.

Although the grounding problem is fundamental to achieve the development of social robots, Roy [29] states that there has not been the grounding of language in actions but abstract representations whose meaning must be interpreted by humans. As a result limited progress has been made in the development of truly social robots that can process multimodal inputs in a manner that grounds language in vision and actions. For instance, robots like the tour-guide robots Rhino [5] and Minerva [32] do not consider grounding of language with vision and actions.

We pursue an imitation learning approach as it allows the observer robot to ground language by creating a representation of the teacher's behaviour, and an understanding of the teacher's aims [14]. As a result of the role played by imitation learning in animal and human development there has been a great deal of interest from diverse fields such as neuroscience, robotics, computation and psychology. Imitation learning offers the ability to ground language with robot actions by taking an external action and relating it with the student robot's internal representation of the action [30]. It is a promising approach for grounding robots in language as it should allow them to learn to cope with complex environments and reduces the search space and the number of training examples compared with reinforcement learning [7].

In our language grounding approach we used the concepts of the mirror neuron system by using multimodal inputs applied to predictive behaviour perception and imitation. Mirror neurons are a class of neurons in the F5 motor area of the monkey cortex which not only fire when the monkey performs an action but also when it sees or hears the action being performed by someone else [23]. Mirror neurons in humans [8] have been associated with Broca's area which indicates their role in language development [23]. Their sensory property justifies our use of models designed for sensory systems that use self-organising learning approaches such as the Helmholtz machine wake-sleep algorithm and the Kohonen algorithm.

2   Robot Approaches to Grounding Language in Actions

Other approaches based on imitation learning have been developed to ground robot language in neural action. For instance, Billard 1999 [1] used the Dynamic Recurrent Associative Memory Architecture approach when grounding a proto-language in actions through imitation. This approach uses a hierarchy of neural networks and provides an abstract and high-level depiction of the neurological structure that are the basis of the visuo-motor pathways. By using this recurrent approach the student is able to learn actions and labels associated with them. Experiments were performed using a doll-like robot. The robot can imitate the arms and head movements of the human teacher after being trained to perform a series of actions performed by the teacher and to label this series with a name. The name is entered by using a keyboard attached to the robot. This was also expanded to use proto-sentences such as 'I touch left arm' to describe the actions. The experiments showed that the hierarchical imitation architecture was able to ground a 'proto-language' in actions performed by the human teacher and recreated by the robot.

Vogt 2000 [34] considered the grounding of language in action using imitation in robots through playing games. In the experiment two robots play various language games while one follows the other. The robots are required to develop various categories and a lexicon so they are able to ground language in actions such as 'turn left' or 'go forward'. The robots share the roles of teacher and student, and language understanding is not preprogrammed. The experiments consist of two stages. In the development stage the task is to acquire categories and a lexicon related to the categories. In the test stage the aim is to determine how well the robot performs the task when only receiving the lexicon. In this phase the teacher and student swap roles after each language game. In this imitation learning language approach only the motor signals are categorised as the teacher and student robots have different sensory-motor signals to control their actions. The categorisation achieved is found to be much more successful than the naming.

In addition, a non-imitation approach to grounding language with robot actions developed by Bailey et al. 1998 [3] investigates the neurally plausible grounding of action verbs in motor actions, such that an agent could execute the action it has learnt. They develop a system called VerbLearn that could learn motor-action prototypes for verbs such as 'slide' or 'push' that allows both recognition and execution of a learnt verb. Verb Learn learns from examples of verb word/action pairs and employs Bayesian Model Merging to accommodate different verb senses where representations of prototypical motor-actions for a verb are created or merged according to a minimum description length criterion. However, Bailey's approach makes use of discrete values that rely on opinion rather than on real world values.

3   Neurocognitive Evidence as Basis for Robot Language Neural Grounding

The robot language grounding model developed in this chapter makes use of neurocognitive evidence on word representation. The neurocognitive evidence of Pulvermüller states that cortical assemblies have been identified in the cortex that activate in response to the performance of motor tasks at a semantic level [21, 23, 24]. Accordingly, a cognitive representation is distributed among cortical neuronal populations. Using MRI and CT scans it was found that these semantic word categories elicit different activity patterns in the fronto-central areas of the cortex, in the areas where body actions are known to be processed [24, 12].

Pulvermüller and his colleagues have performed various brain imaging experiments [22, 12] on the processing of action verbs to test their hypothesis on a distributed semantic word representation. From these experiments it has been possible to identify a distributed representation where the activation was different between action verbs based on the body parts they relate to. It was found that there were clustered activation patterns for the three types of action verbs (arm, leg and face) in the left hemispheric inferior-temporal and inferior-frontal gyrus foci. There were also however differences between these three types of action verbs in terms of the average response times for lexical decisions. For instance, the response time is faster for head-associated words than for arm-associated words, and the arm-associated words are faster processed than leg words. Consistent with the somatotopy of the motor and premotor cortex [20], leg-words elicited greater activation in the central brain region around the vertex, with face-words activating inferior-frontal areas, thereby suggesting that the relevant body-part representations are differentially activated when words that denote actions are being comprehended.

These findings suggest the word semantics is represented in different parts of the brain in a systematic way. Particularly, the representation of the word is related to the actual motor and premotor regions of the brain that perform the action. This is evidence for distributed cortical assemblies that bind acoustic, visual and motor information and stresses the role of fronto-central premotor cortex as a prominent binding site for creating neural representations at an abstract semantic level. Fig. 1 shows a schematic view of this distributed representation of regions in the brain activated by leg, arm and face based on the brain imaging experiments.

Previously, we have developed a computational model of the somatotopy of action words model that recreates the findings on action word processing [37, 38]. This neural model shown in Fig. 2 grounds language in the actual sensor readings

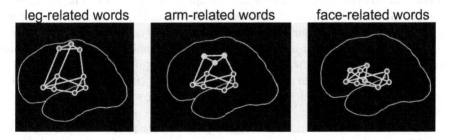

Fig. 1. Based on the brain imaging studies a schematic of the distributed semantic representation in the brain of action verb processing based on the body-parts performing them

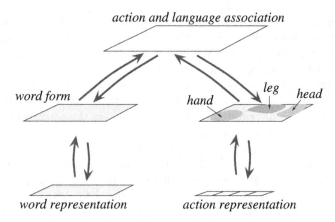

Fig. 2. A neural model of the somatotopy of action words model

from an autonomous robot. In particular, the actual sensor readings represent semantic features of the action verbs. The approach provides a computational implementation of distributed cell assemblies representing and processing action words along with the actions they can refer to [22]. In the novel architecture presented in this paper, the link between perception and production and between action and language is set up by one single map.

4   Mirror Neuron Grounding of Robot Language with Actions

The mirror neuron approach offers a biological explanation for the grounding of language with vision and actions. Rizzolatti and Arbib 1998 [23], Gallese and Goldman 1998 [8] and Umilta et al. 2001 [33] found that neurons located in the rostral region of a primate's F5 motor area were activated by the movement of the hand, mouth or both. It was found that these neurons fire as a result of the goal-oriented action but not the movements that make up this action. The recognition of motor actions depends on the presence of a goal and so the motor system does not solely control movement [8, 25]. Hence, the mirror neuron system produces a neural representation that is identical for the performance and recognition of the action [2]. Fig. 3 shows neuronal responses during recognition and performance of object-related actions. The neurons are active during performance of the action (shown for neurons 3 and 4) and during recognition where recognition can be either visual or auditory.

These mirror neurons do not fire in response to the presence of the object or mimicking of the action. Mirror neuron responses require the action to interact with the actual object. They differentiate not only between the aim of the action but also how the action is carried out [33]. What turns a set of movements into an action is the goal, with the belief that performing the movements will achieve a specific goal [2]. Such a system requires the recognition of the grasping hand

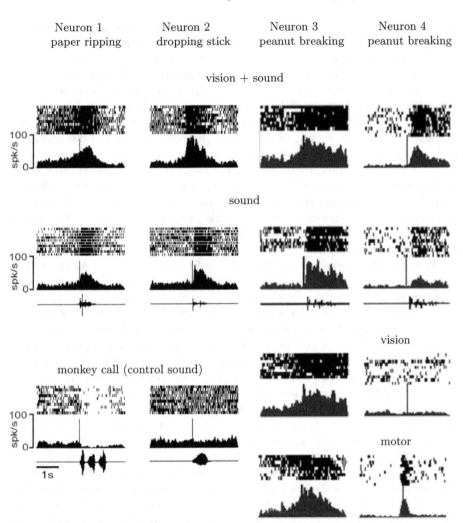

Fig. 3. Responses of macaque F5 mirror neurons to actions. From left to right, the four selected neurons and the chosen stimulus which is in each case a strong driving stimulus. From top to bottom, their responses to vision plus sound, to sound only, and for neurons 3 and 4 to vision of the action only and to the monkey's own performance of the action. For neurons 1 and 2, their reaction to a control sound is shown instead. In each little figure, the above rastergram shows spikes during 10 trials, below which a histogram is depicted (a vertical line indicates the time of the onset of the sound or at which the monkey touches the object). In the case of sound stimuli, an oscillogram of the sound is depicted below the histogram. It can be seen that neurons respond to their driving action via any modality through which they perceive the action, but not to the control stimuli. As an exception, neuron 4 does not respond if the monkey only visually perceives the action (from Kohler et al. 2002 [17] and Keysers et al. 2003 [15])

and examination of its movement and an examination of the association of the hand parameters to the position and affordance to reach the object [2].

The role of mirror neurons was to depict actions so they are understood or can be imitated [23]. Furthermore, the mirror neuron system is held to have a major role in the immediate imitation if an action exists in the observer's repertoire [4]. According to Schaal et al. 2003 [31] and Demiris and Hayes, 2003 [7] imitation learning is common to everyday life and is able to speed up the learning process. Imitation can take the form of mimicking the behaviour of the demonstrator or learning how the demonstrator behaves, responds or deals with unexpected events. Complex imitation not only has the capacity to recognise the actions of another person as familiar movements and to produce them, but also to identify that the action contains novel movements that can be approximated by using movements already known. Imitation learning requires learning and the ability to take the seen action and produce the appropriate motor action to recreate the observed behaviour [4].

An explanation proposed by Rizzolatti and Luppino 2001 [26] for the ability to imitate through the mirror neuron system is an internal vocabulary of actions that are recognised by the mirror neurons. Normally the action is recognised even when the final section is hidden [25]. Understanding comes through the recognition of the action and the intention of the indiviudal. This allows the observer to predict the future actions of the action performers and so determine if they are helpful, unhelpful, threatening and to act accordingly [8]. Such understanding of others' actions also allows primates to cooperate, perform teamwork and deal with threats. The mirror neuron system was a critical discovery as it shows the role played by the motor cortex in action depiction [27]. Hence, the observing primate is put in the same internal state as the one performing the action.

The mirror neuron system also exists in humans [8]. Increased excitation was found in the regions of the motor cortex that was responsible for performing a movement even when the subject was simply observing it. Motor neurons in humans are thus excited when both performing and observing an action [8]. The F5 area in monkeys corresponds to various cortical areas in humans including the left superior temporal sulcus of the left inferior parietal lobule and of the anterior region of Broca's area. The association of mirror neurons with Broca's area in human and F5 in primates points to their role in grounding of language in vision and actions [17]. The ability to recognise an action is required for the development of communication between members of a group and finally speech. It is possible that the mirror neuron system was firstly part of an intentional communication system based on hand and face gestures [17] and then in a language based system [23]. Once language became associated with actions it was no longer appropriate for it to be located in the emotional vocalisation centre. It would emerge in the human Broca's area from an F5-like region that had mirror neuron features and a gesture system. The importance of gestures reduced until they were seen as an accessory to language [23].

Arbib 2004 [2] examined the emergence of language from the mirror neuron system by considering the neural and functional basis of language and the development of the recognition ability of primates to the full language in humans. In doing so Arbib produced a notion of language development over 7 stages: (i) grasping; (ii) a mirror system for grasping; (iii) a simple imitation system for object grasping; (iv) a complex imitation system that allows the recognition of a grasping action and then repeat; (v) a gesture based language system; (vi) proto-speech and (vii) language that moves from action-object frames to a semantic syntax based approach. Hence, evolution has enabled the language system to develop from the basic mirror neuron system that recognises actions to a complex system that allowed cultural development. This concept of the mirror neurons forms the basis of our models for the grounding of robot language in neural actions. In the remainder of this paper we will consider two models that neurally learn to perform the grounding of language with actions.

5   Methods and Architectures

A robot simulator was produced with a teacher robot performing 'go', 'pick' and 'lift' actions. The actions were performed one after another in a loop in an environment (Fig. 4). The student robot observed the teacher robot performing the behaviours and was trained by receiving multimodal inputs. These multimodal inputs were *(i)* high-level visual inputs which were the x and y coordinates and the rotation angle φ of the teacher robot relative to the front wall, *(ii)* the motor directions of the robot ('forward', 'backward', 'turn left' and 'turn right') and *(iii)* a symbolic language description stating the behaviour the teacher is performing ('go', 'pick' or 'lift').

The first behaviour, 'go', involves the robot moving forward in the environment until it reaches a wall and then turns away from it. The coordinates x and φ ensure that the robot avoids the wall, irrespective of y. The second behaviour, 'pick', involves the robot moving toward the target object depicted in Fig. 4 at the top of the arena. This "docking" procedure is produced by a reinforcement approach as described in [36] and uses all, x, y and φ coordinates. The final behaviour, 'lift', involves moving backward to leave the table and then turning around to face toward the middle of the arena. Coordinates x and φ determine how far to move backward and in which direction to turn around. These coordinates which are shared by teacher and learner are chosen such that they could be retrieved once the imitation system is implemented on a real robot.

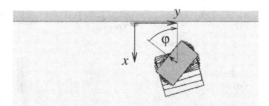

Fig. 4. The simulated environment containing the robot at coordinates x, y and rotation angle φ. The robot has performed ten movement steps and currently turns away from the wall in the learnt 'go' behaviour

When receiving the multimodal inputs corresponding to the teacher's actions the student robot was required to learn these behaviours so that it could recognise them in the future or perform them from a language instruction. Two neural architectures were considered.

5.1   Single-Layer and Hierarchical Architectures

Both imitation models used an associator network based on the Helmholtz machine approach [6]. The Helmholtz machine generates representations of data using unsupervised learning. Bottom-up weights W^{bu} generate a hidden representation r of some input data z. Conversely, top-down weights W^{td} reconstruct an approximation of the data $\tilde{z}$ from the hidden representation. Both sets of weights are trained by the unsupervised wake-sleep algorithm which uses the local delta rule. Parameterised by a sparse coding approach the Helmholtz machine creates biologically realistic edge detectors from natural images [35] and unlike a pure bottom-up recognition model [18] produces also the generative model of the data via neural connections. This is used during testing when we regard either the language area or the motor area as the model's output.

These two models' multimodal inputs included the higher-level vision which represents the x and y coordinates and rotation angle φ of the teacher robot, a language input consisting of a 80-dimensional binary phoneme representation and the motor directives of the four motor units as input.

For the single-layer model all inputs are fed into the hidden layer at the same time during training. The hidden layer of the associator network in Fig. 5 that acted as the student robot's "computatioal cortex" had 16 by 48 units. The sparse coding paradigm of the wake-sleep algorithm leads to the extraction of independent components in the data which is not desired since many of these components would not span over multiple modalities. Therefore we augmented the sparsity toward a winner-take-all mechanism as used in Kohonen networks [18]. The drawback, however, of this winner coding is that the activation of just one unit must account for all input modalities' activations. So if there is a variation in just one modality, for example if an action can be described by two

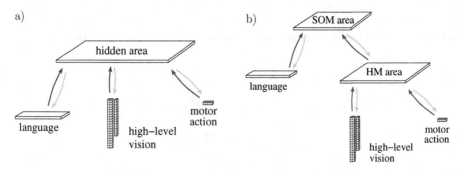

Fig. 5. a) A single-step (3-to-1) architecture. b) A two-layer hierarchical architecture. Bottom-up weights are depicted dark, top-down weights light

different words, then twice as many units are needed to represent this action. This inefficiency motivates the hierarchical model.

In the hierarchical model there is the association of the motor and high-level vision inputs using the first hidden layer, denoted HM area, which uses sparse but distributed population coding. The activations of the first hidden layer are then associated with the language region input at the second hidden layer, denoted SOM area. The first hidden layer uses a Helmholtz machine learning algorithm and the second hidden layer uses Kohonen's self-organising map learning algorithm. Such an architecture allows the features created on the Helmholtz machine hidden layer to relate a specific action to one of the three behaviours given the particular high-level visual information to "flexible" associations of pairs/patterns of activations on the hidden area.

5.2   Processing of Data

On the language region representations of phonemes were presented. This approach used a feature description of 46 English phonemes based on the phonemes in the CELEX lexical databases (http://www.kun.nl/celex/). Each of the phonemes was represented by 20 phonetic features, which produced a different binary pattern of activation in the language input region for each phoneme. These features represent the phoneme sound properties for instance voiced or unvoiced, so similar phonemes have similar structures. The input patterns representing the three used words are depicted in Fig. 6.

The higher-level vision represents the x and y coordinates and rotation angle φ of the teacher robot. The x, y and φ coordinates in the environment were represented by two arrays of 36 units and one array of 24 units, respectively. For a close distance of the robot to the nearest wall, the x position was a Gaussian of activation centred near the first unit while for a robot position near the middle of the arena the Gaussian was centred near the last unit of the first column of 36 units. The next column of 36 units represented the y coordinates so that a Gaussian centred near the middle unit represented the robot to be in the centre of the environment along the y axis. Rotation angles φ from -180° to 180° were represented along 24 units with the Gaussian centred on the centre unit if $\varphi = 0^\circ$.

As final part of the multimodal inputs the teacher robot's motor directives were presented on the 4 motor units (forward, backward, turn right and turn left) one for each of the possible actions with only one active at a time. The activation values in all three input areas were between 0 and 1.

g @ U   p I k   l I f t

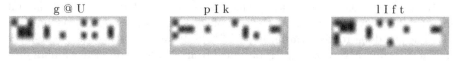

Fig. 6. The phonemes and the corresponding 4×20-dimensional vectors representing 'go', 'pick' and 'lift'

During training the models received all the inputs, however when testing, either the language area or the motor inputs were omitted. The language input was omitted when the student network was required to take the other inputs that would be gained from observing the teacher robot and recognise the behaviour that was performed. Recognition was verified by comparing the units which are activated on the language area via the top-down weights W^{td} (Fig. 5) with the activation pattern belonging to the verbal description of the corresponding behaviour. The motor input was omitted when the student robot was required to perform the learnt behaviours based on a language instruction. It then continuously received its own current x, y and φ coordinates and the language instruction of the behaviour to be performed. Without motor input it had to produce the appropriate motor activations via W^{td} which it had learnt from observing the teacher to produce the required behaviour.

The size of the HM hidden layer is 32 by 32 units and the SOM layer has 24 by 24 units. The number of training steps was around 500000. The duration of a single behaviour depended on the initial conditions and may average at around 25 consecutive steps before the end condition (robot far from wall or target object reached) was met.

5.3   Training Algorithms

The algorithms used the Helmholtz machine [6] and the self-organising map (SOM) algorithm [18] generate internal representations of their training data using unsupervised learning. Bottom-up weights W^{bu} (Fig. 5) generate a hidden representation r of some input data z. Conversely, top-down weights W^{td} an used to reconstruct an approximation $\tilde{z}$ of the data from the hidden representation.

The characteristic of the SOM is that each single data point is represented by a single active ("winning") unit on the hidden area, thus only one element of r is non-zero. The network approximates a data point by this unit's weights. In contrary, the canonical Helmholtz machine's internal representation r contains a varying number of inactive and active, binary stochastic units. A data point is thus reconstructed by a linear superposition of individual units' contributions. Their mean activation can be approximated using a continuous transfer function instead of binary activations. Furthermore, by changing transfer function parameters the characteristics can be manipulated such that units are predominantly inactive which leads to a sparse coding paradigm. At the extreme (that would involve lateral inhibition) one unit might only be allowed to become active at a time.

The learning algorithm for the single-layer model and the HM layer of the hierarchical model is described in the following and consists of alternating wake- and sleep phases to train the top-down and the bottom-up weights, respectively.

In the wake phase, a full data point z is presented which consists of the full motor, higher-level vision and in the case of the single-layer model also language. The linear hidden representation $r^l = W^{bu}z$ is obtained first. In the single-layer model, a competitive version r^c is obtained from this by taking the winning unit of r^l (given by the strongest active unit) and assigning activation values under a Gaussian envelope to the units around the winner. Thus, r^c is effectively a smoothed localist code. On the HM area of the hierarchical model, the linear activation is converted into a sparse representation r^s using the transfer function $r_j^s = e^{\beta x_j}/(e^{\beta r_j^l} + n)$, where $\beta = 2$ controls the slope and $n = 64$ the sparseness of firing. The reconstruction of the data is obtained by $\tilde{z} = W^{td} r^{c/s}$ and the top-down weights from units j to units i are modified according to

$$\Delta w_{ij}^{td} = \eta \, r_j^{c/s} \cdot (z_i - \tilde{z}_i) \tag{1}$$

with an empirically determined learning rate $\eta = 0.001$. The learning rate was increased 5-fold whenever the active motor unit of the teacher changed. This was critical during the 'go' behaviour when the robot turned for a while in front of a wall until it would do its first step forward. Without emphasising the 'forward' step, the student would learn only the 'turn' command which dominates this situation. Behaviour changes are significant events [13] and neuroscience evidence supports that the brain has a network of neurons that detect novel or significant behaviour to aid learning [16, 19].

In the sleep phase, a random hidden code r^r is produced to initialise the activation flow. Binary activation values were assigned under a Gaussian envelope centred on a random position on the hidden layer. Its linear input representation $z^r = W^{td} r^r$ is obtained, and then the reconstructed linear hidden representation $\tilde{r}^r = W^{bu} z^r$. From this, in the single-layer model we obtain a competitive version $\tilde{r}^c$ by assigning activation values under a Gaussian envelope centred around the winner. In the HM area of the hierarchical model, we obtain a sparse version $\tilde{r}^s$ using the above transfer function and parameters on the linear representation. The bottom-up weights from units i to units j are modified according to

$$\Delta w_{ji}^{bu} = \epsilon \, (w_{ji}^{bu} - z_i^r) \cdot \tilde{r}_j^{c/s} \tag{2}$$

with an empirically determined learning rate $\epsilon = 0.01$.

The learning rates η and ϵ were decreased linearly to zero during the last quarter of training in order to reduce noise. All weights W^{td} and W^{bu} were rectified to be non-negative at every learning step. In the single-layer model, the bottom-up weights W^{bu} of each hidden unit were normalised to unit length. In the HM area of the hierarchical model, to ensure that weights did not grow too large, a weight decay term of $-0.015 \cdot w_{ij}^{td}$ is added to Eq. 1 and $-0.015 \cdot w_{ji}^{bu}$ to Eq. 2.

Only the wake phases of training involved multimodal inputs from the motor, higher visual and language regions z based on observing the actions of the teacher robot performing the three behaviours. The sleep phases on the other hand use only random initial activations.

The SOM area of the hierarchical model was trained by the classical self-organising map algorithm [18]. The hidden representation o is in our model the activation vector on the SOM area while its input data i is the concatenated vector from the language input together with the HM area activation r. Only the bottom-up weights, depicted dark in Fig. 5 b), are trained. Top-down weights are not modelled but can formally be obtained from the bottom-up weights by taking the transpose of the weight matrix. Training of the SOM area weights was done after the HM area weight learning was completed.

The representation o_k of unit k is established by determining the Euclidean distance of the weight vector to its inputs, given by: $o_k = \|w_k - i\|$. The weights are originally randomised and hence a unit of the network will react more strongly than others to a specific input representation. The winning unit is the unit k' where the distance $o_{k'}$ is smallest. The weight vector of this winning unit k' as well as the neighbouring units are altered based on the following equation which leads to the weight vectors resembling more the data:

$$\Delta w_{kj} = \alpha T_{kk'} \cdot (i_j - w_{kj}).$$

The learning rate α was set to 0.01. The neighbour function was a Gaussian: $T_{k,k'} = \exp(-d^2_{k,k'}/2\sigma^2)$, where $d_{k,k'}$ is the distance between unit k and the winning unit k' on the SOM area grid.

At the beginning of training, a larger neighbourhood ($\sigma = 12$) achieved broad topologic learning following a reduction during training to ($\sigma = 0.1$). Additional finer training was done with smaller neighbourhood interaction widths by reducing σ from 0.1 to 0.01.

6   Single-Layer Model Results

The single-layer associator network imitation learning robot performed well when recognising the behaviour being performed by the teacher robot and performing the behaviour based on a language instruction. Recognition was tested by the network producing a phonological representation on the language area which was compared to the appropriate language instruction.

Furthermore, when considering if the trained student robot was able to produce a certain behaviour requested by a language input, the movement traces in Fig. 7 on the next page show that when positioned in the same location the robot performs these different behaviours successfully.

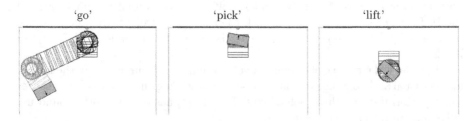

Fig. 7. The simulated trained student robot performance when positioned at the same point in the environment but instructed with different language input. The robot was initially placed in the top middle of the arena facing upward. In the 'go' behaviour it moves around the arena; during 'pick' it approaches the middle of the top wall (target position) and then alternates between left- and right turns; during 'lift' it moves back and then keeps turning

'go'   'pick'   'lift'   all

Fig. 8. Trained weights W^{td} to four selected language units of the student robot. Each rectangle denotes the hidden area, dark are strong connections from the corresponding regions. Each of the three left units is active only at one input which is denoted above. The rightmost unit is active at all language words

forward   backward   right   left

Fig. 9. The trained weights W^{td} to the four motor units of the student robot. As in Fig. 8 the regions from which strong connections originate in the hidden area are depicted dark

Fig. 8 indicates that the three behaviours are represented at three separate regions on the hidden area. In contrast, the four motor outputs are represented each at more scattered patches on the hidden area (Fig. 9). This indicates that language has been more dominant in the clustering process.

7   Hierarchical Model Results

First, we have trained a HM area to perform a single behaviour, 'pick', without the use of a higher-level SOM area. The robot thereby self-imitates a behaviour

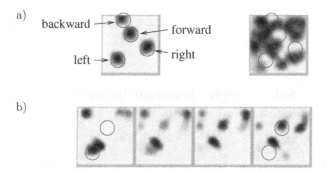

Fig. 10. a) Left, the projections of the four motor units onto the HM area. Right, the projections of all high-level vision inputs on to the HM area. b) Four neighbouring SOM units' RFs in the HM area. These selected units are active during the 'go' behaviour. Circles indicate that the leftmost units' RFs overlap with those of the 'left' motor unit while the rightmost unit's RF overlaps with the RF of the 'forward' motor unit

it has previously learnt by reinforcement [36]. Example videos of its movements can be seen on-line at: www.his.sunderland.ac.uk/supplements/AI04/.

Fig. 10 a) shows the total incoming innervation originating from the motor units (left) and the high-level vision units (right) on the HM area. The figure has been obtained by activating all four motor units or all high-level vision units, respectively, with activation 1 and by displaying the resulting activation pattern on the HM area.

It can be seen that the patches of motor innervation avoid areas of high-density sensory innervation, and vice versa. This effect is due to competitive effects between incoming innervation. This does not mean that motor activation is independent of sensory activation: Fig. 10 b) shows the innervation of SOM area units on the HM area which bind regions specialised on motor- and sensory input.

The leftmost of the four units binds the "left" motor action with some sensory input while the rightmost binds the "forward" motor action with partially different sensory input. In the cortex we would expect such binding not only to occur via another cortical area (such as the SOM area in our model) but also via horizontal lateral inner-area connections which we do not model.

The action patterns during recognition of the 'go' behaviour action sequence depicted in Fig. 4 and during its performance are shown in Figs. 11 and 12, respectively. At first glance, the activation patterns on the HM- and SOM areas are very similar between recognition and performance which suggests that most neurons display mirror neuron properties.

The largest difference can be seen within performance between the two activation steps of the HM area: in the first step it is activated from vision alone

HM area activation based on high-level vision and motor input

SOM area winner-based classification based only on HM area input

language classification by the SOM winning unit

Fig. 11. Activation sequences during observation of a 'go' behaviour, without language input. Strong activations are depicted dark, and shown at ten time steps from left to right. Circles mark the bottom-up input of the active motor unit of the teacher which changes from 'forward' in the first 6 steps to 'turn left' during the last 4 steps (cf. Fig. 10 b)). Language classification is correct except for the last time step which is classified as 'pick' (cf. Fig. 6))

HM area activation based only on high-level vision input

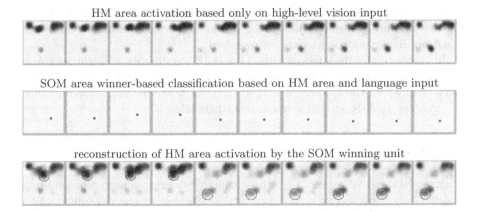

SOM area winner-based classification based on HM area and language input

reconstruction of HM area activation by the SOM winning unit

Fig. 12. Activation sequences during performance of a 'go' behaviour, i.e. without motor input. The performed sequence is visualised in Fig. 4. Circles mark the region on the HM area at each time step which has the decisive influence on the action being performed (cf. Fig. 10)

(top row of Fig. 12) in order to perceive the robot state and in the second step it is activated from the SOM area (bottom row of Fig. 12) in order to relay activation to the associated motor unit. The difference between these two steps comes from the lack of motor input in the first step and the completion of the pattern to include the motor induced activation as would come during full observation in the second step. Naturally, the second step's activation pattern resembles the pattern during recognition in the top row of Fig. 11, since patterns reconstructed from SOM units resemble the training data.

The differences in HM area unit activation patterns during recognition and performance are thus localised at the RF site of the active motor unit. If during training, the input differs only by the motor input (which happens if in the same situation a different action is performed according to a different behaviour) then the difference must be large enough to activate a different SOM unit, so that it can differentiate between behaviours. During performance, however, the absence of the motor input is not desired to have a too strong effect on the HM area representation, because the winner in the SOM area would become unpredictable and the performed action a random one.

The last row in Fig. 11 shows the activations of the language area as a result of the top-down influence from the winning SOM area unit during recognition. An error is made at the last time step which as far as the input is concerned (HM area activation in top row) is barely distinguishable from the second last time step. Note that the recognition error is in general difficult to quantify since large parts of some behaviours are ambiguous: for example, during 'go' and 'pick', a forward movement toward the front wall is made in large areas of the arena at certain orientation angles φ, or a 'turn' movement near the wall toward the centre might also be a result of either behaviour. Inclusion of additional information like the presence of a goal object or the action history could disambiguate many situations, if a more complex model was used.

In the examples depicted in Figs. 11 and 12, the teacher and learner robots are initialised at the same position. Both then act similar during the first 4 time steps after which the learner decides to turn, while the teacher turns only after 6 time steps (see the circled areas in these figures).

8   Discussion

Each model recreates some of the neurocognitive evidence on word representation and the mirror neuron system. While for the single layer model a single unit is active at a specific time-step, for the hierarchical model multiple units are active. In terms of the neurocognitive evidence it can be argued that the hierarchical model is closer to the brain as it involves a distributed representation.

The ability of the single-layer and hierarchical model controlled robot to both recognise an observed behaviour and perform the behaviour that it has learnt by imitating a teacher shows the models were able to recreate one core concept of the mirror neuron system. For instance, in the single-layer model the student robot displays mirror neuron properties by producing similar regional unit activation patterns when observing the behaviour and performing it, as seen on some examples in Fig. 13. Furthermore, the achieved common "action understanding" between the teacher and student on the behaviour's meaning through language corresponds to the findings in the human mirror neuron system expressed by Arbib [2] whereby language would be allowed to emerge.

With regard to the hierarchical model it is suggested to identify the HM area of the model with area F5 of the primate cortex and the SOM area with F6. F5 represents motor primitives where the stimulation of neurons leads to involuntary limb movements. F6 rather acts as a switch, facilitating or suppressing the effects of F5 unit activations but it is itself unable to evoke reliable and fast motor responses. In our model, the HM area is directly linked to the motor output and identifiable groups of neurons activate specific motor units while the SOM area

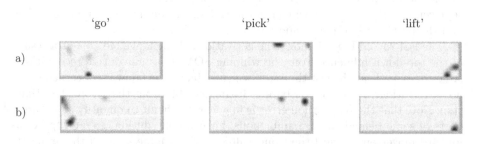

Fig. 13. Activations for the associator network summed up during short phases while the student robot **a)** correctly predicts the behaviours and **b)** performs them based on a language instruction

represents the channel through which a verbal command must pass in order to reach the motor related HM units.

Mirror neurons have so far been reported in F5. By design, the hierarchical model uses the HM area for both, recognition and production, so an overlap in the activation patterns as observed in mirror neurons is expected. This overlap is mainly due to those neurons which receive high-level vision input. This perceptual input is tightly related to the motor action as it is necessarily present during the performance of an action and contributes to the "motor affordances" [8]. The decisive influence on the motor action, however, is localised in our model on smaller regions on the HM area, as defined by the motor units' receptive fields (Fig. 10 a)). The units in these regions would correspond to the canonical motor neurons which make up one third of F5 neurons. These non-mirror neurons have only motor control function and are not activated by action observation alone.

A prediction of our model would then be that if the visually related mirror neurons alone are activated, e.g. by electrode stimulation, then neurons downstream would not be directly excited and no motor action would take place. It is, however, difficult to activate such a distinguished group of neurons since horizontal, lateral connections in the cortex are likely to link them to the canonical motor neurons.

9   Conclusion

We have developed both a single-layer and an hierarchical approach to robot learning by imitation. We considered an approach to ground language with vision and actions to learn three behaviours in a robot system. The single-layer model relies on a competitive winner-take-all coding scheme. However, the hierarchical approach combines a sparse, distributed coding scheme on the lower layer with winner-take-all coding on the top layer. Although both models offer neural based robot language grounding by recreating concepts of the mirror neuron system in region F5, the hierarchical model suggests analogies to the organisation of motor cortical areas F5 and F6 and to the properties of mirror neurons found in these areas. In doing so it provides insight to the organisation and activation of sensory-motor schemata from a computational modelling perspective. Considering functional processing logics it explains the position of mirror neurons connecting multiple modalities in the brain. This hierarchical architecture based on multi-modal inputs can be extended in the future to the inclusion of reward values that are also represented in cortical structures [28] to achieve goal driven teleological behaviour.

Acknowledgments. This work is part of the MirrorBot project supported by the EU in the FET-IST programme under grant IST- 2001-35282. We thank Fermín Moscoso del Prado Martín at the Cognition and Brain Science Unit in Cambridge for his assistance with developing the language phoneme representation.

References

1. Billard A. and Hayes G. Drama, a connectionist architecture for control and learning in autonomous robots. *Behavior Journal*, 7(1):35–64, 1999.
2. M. Arbib. From monkey-like action recognition to human language: An evolutionary framework for neurolinguistics. *Behavioral and Brain Science*, pages 1–9, 2004.
3. D. Bailey, N. Chang, J. Feldman, and S. Narayanan. Extending embodied lexical development. In *Proceedings of the Nineteenth Annual Meeting of the Cognitive Science Conference*, 1998.
4. G. Buccino, S. Vogt, A. Ritzi, G. Fink, K. Zilles, H.-J. Freund, and G. Rizzolatti. Neural circuits underlying imitation learning of hand actions: An event-related fMRI study. *Neuron*, 42:323–334, 2004.
5. W. Burgard, A.B. Cremers, D. Fox, D. Hahnel, G. Lakemeyer, D. Schulz, W. Steiner, and S. Thrun. Experiences with an interactive museum tour-guide robot. *Artificial Intelligence*, 114(1-2), 2000.
6. P. Dayan. Helmholtz machines and wake-sleep learning. In M. Arbib, editor, *Handbook of Brain Theory and Neural Network*. MIT Press, Cambridge, MA, 2000.
7. Y. Demiris and G. Hayes. Imitation as a dual-route process featuring prediction and learning components: a biologically-plausible computational model. In K. Dautenhaln and C. Nehaniv, editors, *Imitation in animals and artifacts*. MIT Press, 2002.
8. V. Gallese and A. Goldman. Mirror neurons and the simulation theory of mind-reading. *Trends in Cognitive Science*, 2(12):493–501, 1998.
9. A. Glenberg and M. Kaschak. Grounding language in action. *Psychonomic Bulletin and Review*, 9:558–565, 2002.
10. S. Harnad. The symbol grounding problem. *Physica D*, 42:335–346, 1990.
11. S. Harnad. The symbol grounding problem. In *Encyclopedia of Cognitive Science*, Dublin, 2003.
12. O. Hauk and F. Pulvermüller. Neurophysiological distinction of action words in the frontal lobe: An ERP study using minimum current estimates. *Human Brain Mapping*, pages 1–9, 2004.
13. G. Hayes and J. Demiris. A robot controller using learning by imitation. In *Proceedings of the 2nd International Symposium on Intelligent Robotic Systems, Greijcnnnoble, France*, 1994.
14. I. Infantino, A. Chella, H. Dzindo, and I. Macaluso. A posture sequence learning system for an anthropomorphic robotic hand. In *Proceedings of the IROS-2003 Workshop on Robot Programming by Demonstration*, 2003.
15. C. Keysers, E. Kohler, M.A. Umilt, L. Nanetti, L. Fogassi, and V. Gallese. Audio-visual mirror neurons and action recognition. *Exp. Brain Res.*, 153:628–636, 2003.
16. R. Knight. Contribution of human hippocampal region to novelty detection. *Computer Speech and Language*, 383(6597):256–259, 1996.
17. E. Kohler, C. Keysers, M. Umilta, L. Fogassi, V. Gallese, and G. Rizzolatti. Hearing sounds, understanding actions: Action representation in mirror neurons. *Science*, 297:846–848, 2002.
18. T. Kohonen. *Self-Organizing Maps*. Springer Verlag, Heidelberg, 1997.
19. B. Opitz, A. Mecklinger, A.D. Friederici, and D.Y. von Cramon. The functional neuroanatomy of novelty processing: Integrating erp and fMRI results. *Cerebral Cortex*, 9(4):379–391, 1999.
20. W. Penfield and T. Rasmussen. *The cerebral cortex of man*. Macmillan, Cambridge, MA, 1950.

21. F. Pulvermüller. Words in the brain's language. *Behavioral and Brain Sciences*, 22(2):253–336, 1999.
22. F. Pulvermüller, R. Assadollahi, and T. Elbert. Neuromagnetic evidence for early semantic access in word recognition. *European Journal of Neuroscience*, 13:201–205, 2001.
23. G. Rizzolatti and M. Arbib. Language within our grasp. *Trends in Neuroscience*, 21(5):188–194, 1998.
24. G. Rizzolatti, L. Fogassi, and V. Gallese. Neurophysiological mechanisms underlying the understanding and imitation of action. *Nature Review*, 2:661–670, 2001.
25. G. Rizzolatti, L. Fogassi, and V. Gallese. Motor and cognitive functions of the ventral premotor cortex. *Current Opinion in Neurobiology*, 12:149–154, 2002.
26. G. Rizzolatti and G. Luppino. The cortical motor system. *Neuron*, 18(2):889–901, 1995.
27. G. Rizzolatti, G. Luppino, and M. Matelli. The organization of the cortical motor system: New concepts. *Electroencephalography and Clinical Neurophysiology*, 106:283–296, 1998.
28. E.T. Rolls. The orbitofrontal cortex and reward. *Cereb. Cortex*, 10(3):284–94, 2000.
29. D. Roy. Learning visually grounded words and syntax of natural language sk. *Computer Speech and Language*, 16(3), 2002.
30. S. Schaal. Is imitation learning the route to humanoid robots. *Trends in Cognitive Science*, 3(6):233–242, 1999.
31. S. Schaal, A. Ijspeert, and A. Billard. Computational approaches to motor learning by imitation. *Transaction of the Royal Society of London: Serial B, Biological Sciences*, 358:537–547, 2003.
32. S. Thrun, M. Bennewitz, W. Burgard, F. Dellaert, D. Fox, D. Haehnel, C. Rosenberg, N. Roy, J. Schulte, and D. Schulz. Minerva: A second generation mobile tour-guide robot. In *Proceedings of the IEEE international conference on robotics and automation (ICRA'99)*, 1999.
33. M. Umilta, E. Kohler, V. Gallese, L. Fogassi, L. Fadiga, and G. Keysers, C.and Rizzolatti. I know what you are doing: A neurophysical study. *Neuron*, 31:155–165, 2001.
34. P. Vogt. Grounding language about actions: Mobile robots playing follow me games. In J. Meyer, A. Berthoz, D. Floreano, H. Roitblat, and S. Wilson, editors, *SAB00, Honolulu, Hawaii*. MIT Press, 2000.
35. C. Weber. Self-organization of orientation maps, lateral connections, and dynamic receptive fields in the primary visual cortex. In *Proceedings of the ICANN Conference*, 2001.
36. C. Weber, S. Wermter, and A. Zochios. Robot docking with neural vision and reinforcement. *Knowledge-Based Systems*, 17(2-4):165–72, 2004.
37. S. Wermter and M. Elshaw. Learning robot actions based on self-organising language memory. *Lecture Notes in Artificial Intelligent*, 16(5-6):661–669, 2003.
38. S. Wermter, C. Weber, M. Elshaw, C. Panchev, H. Erwin, and F. Pulvermüller. Towards multimodal neural robot learning. *Robotics and Autonomous Systems Journal*, 47:171–175, 2004.

A Spiking Neural Network Model of Multi-modal Language Processing of Robot Instructions

Christo Panchev

School of Computing and Technology,
St. Peter's Campus, University of Sunderland,
Sunderland SR6 0DD, United Kingdom
`christo.panchev@sunderland.ac.uk`

Abstract. Presented is a spiking neural network architecture of human language instruction recognition and robot control. The network is based on a model of a leaky Integrate-And-Fire (lIAF) spiking neurone with Active Dendrites and Dynamic Synapses (ADDS) [1, 2, 3]. The architecture contains several main modules associating information across different modalities: an auditory system recognising single spoken words, a visual system recognising objects of different colour and shape, motor control system for navigation and motor control and a working memory. The main focus of this presentation is the working memory module whose function is sequential processing of word from a language instruction, task and goal representation and cross-modal association of objects and actions. We test the model with a robot whose goal is to recognise and execute language instructions. The work demonstrates the potential of spiking neurons for processing spatio-temporal patterns and the experiments present spiking neural networks as a paradigm which can be applied for modelling sequence detectors at word level for robot instructions.

1   Introduction

Being the foremost attribute of human intelligence, language is a particularly challenging and interesting task. It emerged within a rich problem space and the task of language processing in the brain is performed by mapping inherently non-linear patterns of thought onto linear sequence of signals, under a severe set of processing constraints from human perception, motor co-ordination and production, and memory [4].

There have been two major directions for studying the representational and functional processing of language in the brain: the emergent language skills of humans, e.g. innate vs. learnt [5]; and the effects of language impairment such as those due to brain injury, e.g. aphasia [6, 7]. Recent advances in neuroscience and brain imaging have allowed us to open a third route into understanding the language mechanisms in the brain. Constraints over the computational and representational mechanisms in the brain involved in language can be derived

S. Wermter et al. (Eds.): Biomimetic Neural Learning, LNAI 3575, pp. 182–210, 2005.
© Springer-Verlag Berlin Heidelberg 2005

from many studies on single neuron function and plasticity paradigms, neuronal ensembles organisation and processing, working memory, etc. [8]

In parallel with the neuro-psychological and linguistic aspects, computational simulations constitute a very important part of language studies - helping to understand the representational and functional language processes in the brain. There is a broad range of open questions that need an adequate answer before we reach any significant success in our computational models. Such questions include very precise biophysics or biochemistry problems of single neurons, as well as many interdisciplinary ones within the Nature v Nurture debate, localist and distributed representational/functional paradigms, language localisation and plasticity paradigms and finally many questions arise in the application and theoretical levels of linguistics, psychology and philosophy.

Within recently published work on language, cognition and communicative development in children with focal brain injury [9, 10], the favourite viewpoint of brain organisation for language has changed. Many scientists have taken a new consensus position between the historical extremes of equipotentiality [11] and innate pre-determination of the adult pattern of brain organisation for language [12]. However, there is still not a definite understanding of the levels of innateness and plasticity in the brain. In our approach we consider possible partial innate architectural or processing constraints in the language system, but emphasise on the development of the language skills.

Modularity

Early studies on language included mainly work in linguistics and psychology. In the late nineteenth century contribution into the field came from the neurologists' findings of "affections of speech from disease of the brain" [13], a condition known as aphasia. After a stroke or tumour, adults who have been fluent in speaking and understanding their native language, were sometimes found to acquire severe and specific deficiency in language abilities. Major brain areas have been identified as a result of aphasia studies, including the speech production and language comprehension areas of Broca [14] with lesions resulting in motor aphasia, and Wernicke [15] where possible trauma can lead to sensory aphasia. Some proposals have even argued for quite specific aphasia types and therefore some possibly very specialised language-related areas in the brain [16]. It has been accepted that the main language areas normally reside in the dominant (usually the left) hemisphere of the human brain. While some early language models even assign the two main language functions syntax and semantic processing to the Broca's and Wernicke's regions, and some others distribute the functionality of these two areas for language production and comprehension respectively, later neural models of human language processing appeal for a much wider functional and encoding network [8].

The converging evidence shows that it is unlikely that there are areas in the brain specialised only for specific language functions independent of other language areas, e.g. speech production centre (inferior frontal area of Broca) and entirely separate and independent language comprehension centre (superior temporal area of Wernicke). These two areas are the most critical for language processing in the brain and appear to be functionally interdependent. A fluent and unimpaired language processing requires a wider network of regions in the brain. The functionality and knowledge necessary for language comprehension or production, are not entirely available in these two regions, and thereby the language skills cannot be exclusively assigned to that region. For example, patients with left-dominant language hemisphere and brain lesions in the right hemisphere, would also exhibit some selective language deficiencies with words from specific categories, e.g. words related to actions or words related to visual perceptions [17]. In summary, the language skills are dominantly implemented in one hemisphere (usually the left), with the pre-frontal areas playing critical part, and a network of additional regions in the brain including some from the non-dominant hemisphere are necessary for complete language functionality [18].

Neuroimaging studies confirm that both classical language areas Broca's and Wernicke's become active during language task and both areas are necessary for both language production and language comprehension tasks. At the same time, as mentioned earlier, neither of these areas alone is sufficient for language processing. It is clear though that both areas play a central role in all language tasks. As Ojemann [19] points out, the language system involves both essential areas, as well as neurons widely distributed across the cortex (including event the non-dominant hemisphere), while at the same time, the evidence for clear distinction of separate systems for speech production and speech perception is less clear.

Further support of the distributed nature of language areas in the brain come from brain imaging studies of bilingual subjects [20, 21, 22, 23, 24] showing that the language mechanism in the left temporal lobe is more activated by the native language than be any lesser known secondary language. Such studies have shown that only bilinguals who learnt their second language early enough and had enough practice in both languages have overlaying areas of both languages in contrast to subjects who have learnt their second language later in their life and have been found to activate two non-overlapping subregions in Broca's area [25]. In some recent studies for the task sentence comprehension, the critical factor has been shown to be the level of fluency in both languages rather that the time (age) they have been learnt. Studies with higher spatial resolution revealed that even with one brain region for the languages there might be smaller-scale circuits specialised for each particular language [23, 25].

Distributed word representations

The question of the areas in the brain where words are represented and processed has been address in many studies. Although there is still some dispute over the exact areas involved [26], significant advances have been made recently. Many brain imaging studies reveal that a number of areas outside the classical language area become active during a word processing task. Thus many brain areas in addition to the classical Broca's and Wernicke's areas are now considered related to and important for language processing. One of the first models covering a number of brain areas was proposed by Freud [27] suggesting that words are represented in the brain by multi-modal associations between ensembles of neurons in different cortical areas. Freud's idea was not very well received at the time, but now it is in the core of many recent computational and experimental neuroscience based theories of language. Recent metabolic and neurophysiological imaging studies have shown that words are organised as *distributed neuron ensembles that differ in their cortical topographies* [28]. Evidence of merging miltimodal sensory perceptions [29, 30], sensory information of one modality being able to modulate the responses of cortical circuits in the areas of another [31, 32, 33, 34] and of cortical cells with multi-modal response properties [35, 36, 37] supports the idea that the cortex is an information merging computational device which uses neurons representing and processing information from various sensory and motor modalities. Furthermore, the neuroanatomical cortical connectivity patterns show that links between the primary areas are not direct and often include more than one route involving several non-primary areas. The merging of multi-modal information is therefore not necessarily done through direct links between the primary areas of the modalities involved [38, 39, 40], but rather via intervening cortical areas [41] allowing complex mapping of sensory/motor patterns.

Simultaneous speech input naming an object and visual input from observing that object can activate the auditory and visual representations of the word and the object itself [42]. This will lead to associations between the auditory and visual representations of the word, forming a distributed, cross-modal cell assembly representing the word/object. Similarly, associations between the auditory representation of an action verb and motor representations of that action can be associatively combined to form cell assemblies for actions/verbs. Continuing further, objects which are strongly related to some action could be represented by cross-modal cell assemblies including the auditory representation of the word, the visual representation of the object and the motor representation of the actions associated with the object. Similar argument can be put forward for words which do not map directly to visual appearances or motor actions. Such words are usually learnt based on the context within which they appear. In such cases, the building of the word cell assembly representing such word can be formed based on the statistical grounds. There are two possible cases here. The first type of words are learnt from the context within which they appear. The meaning of such words can be defined by a number of other words with which they frequently appear [43]. The auditory representation of such a word will often be simultaneously active with the multi modal (vision and/or motor) representations of the words which built the context (if such exist) and consequently the frequently active neuronal assemblies in these areas will become associated. Thereby, we can assume the such words will also be represented by cell assemblies occupying different cortical areas, and possibly different modalities. The second type of word, so called functional words is represented by cell assemblies covering cortical areas in the left Broca's region [44]. The learning of such words will acquire neuronal ensembles which are derived from the syntax and in some cases the semantics of the context within which they occur.

Considering the gradual language learning in children and incremental acquisition of the lexicon, it can also be argued that such a mechanism of associative learning and word cell assemblies will lead to shared representations, i.e. common neurons in the cell assemblies representing word with similar semantics, functional role or pronunciation. It particular that will be the case with semantically similar words which have been learnt in the same context, or have served to build the new meaning of one another.

How infants/children learn the phonetic construction of words can be explained by the correlation learning principles. During early language acquisition, only a relatively small number of word forms are learnt from single word repetitions [45]. The phoneme-to-word associations for nearly all learnt word has to be derived from a continuous stream of phonemes, usually without acoustic clues of inter-word intervals. The reoccurring phoneme sequences constituting a word can be isolated/distinguished from the rest of the acoustic/phoneme stream based on statistical discrimination [46, 47, 48]. Young infants can distinguish phoneme sequences which constitute familiar words from the surrounding sounds [49] and it is quite plausible that they use the co-occurrence correlation statistics and mutual information of phonemes and syllable sequences in this process [50].

One of the earliest elements of learning word representation in the brain appear with the babbling - the earliest sign of language-like articulation performed by infants old six months or more [51]. Soon after acquiring this ability, the infant is able to repeat words spoken by others [52]. Early babbling and word-like production are believed to activate cortical areas in the inferior frontal lobe, including inferior motor cortex and adjacent pre-frontal areas. At the same time, these sounds activate the neurons in the auditory system, including areas in the superior temporal lobe. The parallel activation of these cortical areas allows the association of the cortical representation of motor programmes for the production of particular sounds and the auditory representation of the perception of such sounds. Such association constitutes a building block in a word cell assembly including the two modality specific representations of the same word.

Another interesting question is when are the words being recognised by the brain. Many studies have worked on this question by contrasting brain responses to word and pseudo-word stimulus. The word recognition point is the earliest point in time at which the subject is presented with enough information to make decision/distinction [53]. There are strong indications for particular set of high frequency responses with start/end and peak timings relative to the word recognition point.

Some neurophysiological studies have found word-related high frequency (gamma band) cortical responses around 400 ms after the recognition point of spoken words [42]. Other studies have found word/pseudo-word differences of event-related potentials of cortical activity around 100-200 ms after the onset of visually presented stimuli [54]. These results indicate relatively early and fast word discrimination in the brain.

A number of neuroimaging studies have shown that auditory stimulus can activate at least part of the cell assemblies representing the word in the absence of auditory attention [55, 56, 57, 58], indicating that word/pseudo-word distinction can use sensory information alone, and perhaps be performed in the early perceptual stages. While considering these results, it should be noted that the word/pseudo-word discrimination is a relatively simple task in the initial stages of language processing, and that further word recognition and language comprehension task would require attention. Furthermore, since attention modulates the activity patterns in the primary areas, learning cross-modal associations will require the attentive process.

Interesting constraints on the representational and processing language mechanisms in the brain can be seen in the theories presenting the potential evolutionary development of the brain structure related to language developed by Aboitiz and Garicia[59] and Rizzolatti and Arbib [60]. A number of these constraints have been incorporated in the model presented in this work.

Aboitiz and Garicia see the Wernicke's area in the superior temporal lobe as the zone for cross-modal associations which link the phonological loop with object/action representations in other sensory/motor areas. Their model is based on phonological-rehearsal device (phonological loop) across Broca's and Wernicke's regions which could express simple syntactic regularities. Another main hypothesis in their work is that language processing is closely linked to working memory in two main aspects: (1) in terms of anatomical arrangement of the neural circuits involved, e.g. the mapping of multi-modal concepts in Wernicke's and phonological sequences; and (2) in terms of operating within an efficient working memory system, e.g. gradual context building and maintenance.

Rizzolatti and Arbib suggest that the development of the human lateral speech circuit in the Broca's area is a consequence of the fact that this area was previously associated with a mechanism of recognising observed actions. They argue that this mechanism is the neural prerequisite for the development of inter-individual communication and finally speech [61, 62, 63, 64, 65]. The rostal part of ventral pre-motor cortex (area F5) in monkeys is seen as the analog of human Broca's area [66, 60, 67]. This area contains a class of neurons (called *mirror neurons*) that do not become active just when an action is performed, but also respond when the same specific action is observed [68, 69, 70, 71]. Most of the neurons are highly selective and respond only to a particular action, with some coding not only the action but also how it is performed. Furthermore, the same mirror neurons respond to the same action also when it is internally generated. Such an observation/execution matching system is suggested to be the bridge from "doing" to "communicating", i.e. the link between sender and receiver. Significantly, mirror neurons in monkeys have also been found to associate between an action and the acoustic perceptions related to the actions [72, 73, 74], e.g. the sound of breaking a nut. The authors stress that the mirror neurons are not innate, but in fact they are learnt representations of recognised actions or methods of performing acquired actions. Another important point which the authors note is that the full system of communicating and understanding between human is based on a far richer set of brain mechanisms than the core "mirror neuron system for grasping" which is shared by monkeys and human [75]. One further conclusion from their work is that the Broca's areas would encode "verb phrases" and constraints about the noun phrases that can fill the slot, but not information about the noun phrases themselves. This knowledge (objects or noun phrases) could be completely outside Broca's area, for example in the temporal lobe.

Some related computational models for language processing

Most of the related computational models of language processing have been based on associative learning, following the view that language should not be processed in isolation [76, 77, 78]. Similar to the brain where a language processing is related to actual interaction with the environment in conjunction with internal constraints, computational language systems should be able to ground objects and actions to their perceptions and affordances.

Recent experimental and modelling work has taken a direction of embodiment of a cognitive system as a device capable of perceiving the environment as well as reacting based on interpretation of sensory information and internal drives. Such paradigm incorporates substantial part of the field of Intelligent Robotics, and represents a new opportunity for development of novel types of cognitive systems understanding natural language.

In their work on the Talking Heads experiment and the AIBO's communication skills, Steels and Kaplan successfully present the hypothesis that basic language communication skills can be "bootstrapped in a social learning process under the strong influence of culture" [79, 80, 81]. In the earlier work, interacting agents evolved a language sufficient to perform a descriptive communication task. In subsequent experiments a robot played a game with a human and was learning associations between the visual perception of objects and their names. The authors argued for three main prerequisites for early language acquisition: (1) the ability to engage in social integrations, which in turn requires abilities like recognition of others, attentions, talking, moving, etc.; (2) the presence of a mediator; and (3) incremental learning/acquisition of concepts.

The development of the neural architectures presented here is based on the above evidence for language representations and processing in the brain. Mainly, it follows the views that concepts are represented by distributed cell assemblies across multiple areas of different modalities, with objects activating neurons in both auditory and visual areas, and actions activating neurons in auditory and motor areas. Furthermore, following [60], the presented architecture implements two separate subareas of distributed representations of actions and objects: one which encodes the action and object related constraints and another which represents the objects.

2   Overall Model Architecture

Despite the recent advances in brain imaging and recent empirical results, there is a clear need for further developments in the theoretical framework, as well as on computational simulations being able to approximate the existing empirical results, suggest further experimental directions and make predictions for phenomena and paradigms which are currently difficult or impossible to obtain

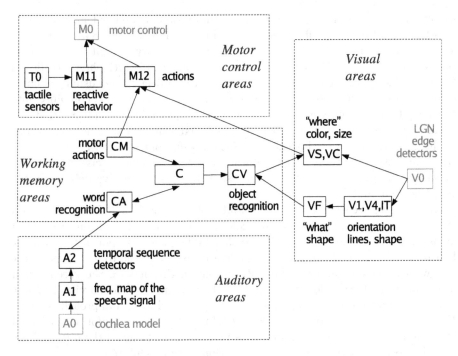

Fig. 1. The overall multi-modal spiking neural network architecture. It consists of four main interconnected modules: The auditory (A0, A1 and A2), the visual (V0, V1, V12 and V22), the motor control (M0, M11, M12 and T0) and the central associative and working memory modules (CA, CV, CM and W). All areas are implemented using ADDS spiking neurones. All areas except the primary input/output (A0, V0, M0 and T0) have been trained using the Hebbian learning algorithm presented in [2, 3]

experimentally [82]. The modelling work presented here aims to contribute toward such computational models. It focused on building a model base for making theoretical predictions about particular functionalities, as well as to provide an architecture and integrated system for developing and testing future hypotheses.

The goal of this model and experiments is to build and test the control mechanisms using ADDS spiking neurons [1, 2, 3]. The work involves modelling of primary sensory areas (auditory, visual and tactile), higher cognitive functions areas (language processing/understanding, working memory) and motor control areas. The overall architecture is presented in figure 1.

3   Multi-modal Sensory-Motor Integration and Working Memory

Sound and vision are processed in different parts of the brain. Nevertheless, a spoken word describing an object and an object present in the visual field would raise activity in the brain which leads to a coherent representation of the object associating the interpretation of the sound and vision, as well as features related to other modalities, e.g. affordances related to possible actions [83, 65, 63]. Such associations are not necessary within the same modality as well. For example, the description of an object might contain several words which should be linked together. The perceived visual properties of the object would usually include colour, shape, texture, size, position, etc. These also have to be combined into a coherent visual representation. What are the possible mechanisms that the brain could use, and how this can be modelled using spiking neurons is one of the central themes in this chapter. Such associations seem to be deceptively trivial and effortless to us, but the question of what are the precise mechanisms which our brain uses to perform the task still causes extensive debates in the scientific community.

Lets discuss the situation where a child is playing football and is about to kick the ball. The brain processes shape, colour and spatial information in different areas of the visual cortex. How would the child's visual system distinguish the ball from other objects on the field, having to associate the round shape of the ball with its colour, and recognise it without mixing it with the green colour of the field or other children on the field. How would the child's visual system inform it about the location of the ball so that the child can approach the ball. While approaching the ball, the brain would have to receive information about other objects or children standing in its path, and try to avoid them. It will not mix the position of the ball with the position of those children.

Lets extend the scenario to one in which the training coach gives an instruction to the child to kick the ball. The brain would have had to hear the words in the coach's instruction and interpret their meaning. Part of the instruction will refer to the kick action, so the brain would have to associate it with the program of the action itself. Another part of the instruction will be a description of the ball, possibly involving one or more words (e.g. specifying the colour or the position of the ball). The brain would have to combine and associate the meaning of these words with the visual perception of the ball which matches the description. In such a situation, the child's brain will have to performs several different levels of association, i.e. its brain will simultaneously solve many instances of the binding problem. The model presented in this chapter attempts to simulate a computational and suitably simplified interpretation of brain mechanisms underlying the above functionality. As a cognitive system it should have the capability to bind features together at several levels and across spatial and temporal domains.

Furthermore, a computational system supporting the above scenario will have to address an additional cognitive task. In the above scenario, there would usually be a short time period during which the coach's instruction is being executed. During that period, the child will need the maintain the task and the goal. The brain would employ a working memory mechanism capable of maintaining arbitrary pattern for a short time period. The exact neural mechanisms which the brain uses are still open questions for a debate. This chapter will review some of the main theoretical ideas for such models and suggest one particular solution. The model of a working memory is another major theme in the current chapter.

3.1   The Binding Problem

The problem of feature binding is sometimes a challenge even for the real brain. Discriminating objects is more difficult is they share the same features. Illusionary conjunctions may appear under several conditions, such as in very short visual presentation features from one object are perceived as belonging to another [84, 85, 86].

The term (*feature*) *binding problem* refers to a class of problems related to how a distributed representation of information across multiple spatial and temporal dimensions results in coherent representations [87]. It can be seen at different levels of complexity, spatial and temporal scales. One of the most often explored and modelled form of binding, that of perceptual binding and binding in the visual system in particular, is concerned with the problem of correct association of one visual feature, such as objects shape, to another, such as location, in order to build a unified and coherent description of an actual appearance of the object [88]. The visual system also needs to perform some form of binding across time in order to interpret object motion. Binding exists in other modalities as well. For example, the auditory system may need binding in order to discriminate the sound from a single voice in a crowd. Binding can also span across different modalities, e.g. the sound of breaking a nut with the visual percept of it, so that they constitute a single event. In addition to the above perceptual types of binding, one could speak of a higher level cognitive binding [89]. For example binding might be necessary when relating perceptions to concepts, e.g. the visual perception of an apple to all the knowledge about it (e.g. could eat it, its taste, etc.). Another major area where binding could play a central role is the understanding of natural language. Binding could be necessary to associate the phonetic and semantic representation of a word with its syntactic role in a sentence - the slot/filler problem. Furthermore, binding might be the mechanism of linking the different words of a sentence into a coherent meaning/interpretation.

The two main types of mechanisms which have been suggested as the possible models of the brain are combinatorial and temporal binding [89].

Although, computationally expensive, combinatorial binding has been suggested as a possible mechanism being implemented in the brain as well as a solution to the binding problem in cognitive systems [90, 91]. The idea is suggests and is based on neurons with highly specialised response properties - the "cardinal cells" [92]. The problem of gradual binding of complex features into coherent representations can be solved by a hierarchical structure of interconnected cells with increased complexity. Neurons in the early stages of the visual system can be seen as representing low level features such as local colour and orientation, whereas those in the higher levels can represent complex patterns and shapes [93, 94, 95]. The concept could be further extended from perceptual, to cross-modal and cognitive binding by considering the latest results on *mirror neurons*. These neurons have been found to be highly selective to complex auditory, visual and motor patterns [60, 72, 74, 73, 63].

The temporal correlation hypothesis has been proposed as a possible solution of the combinatorial problem in neural coding [87] and is based on neurophysiological studies showing that neurons driven by a single stimulus respond synchronously with oscillations in the 30-70 Hz range. Such synchronous oscillation have been observed in the visual areas [96] and between sensory and motor regions [97, 98]

Although in most theoretical models and implementations the combinatorial and temporal binding are seen as alternatives, there is no reason why they should not coexist in a large scale cognitive system. In fact, considering the variety of different types of binding which the brain has to perform, it will be unlikely if a single mechanism is used throughout. In reality, a complex cognitive system would require a variety of mechanisms and implementations feature binding solutions. For example, as suggested by psychological and neurological data [84, 99] attention should also be closely linked to solving the binding problem [100].

3.2   Working Memory

The Working Memory (WM) implements mechanisms for holding information for a short period of time, usually until an action for which that information is necessary can be executed and is often refereed to as Short-Term Memory (STM) or Active Memory (AM) [101]. The information which is held could be about objects or events which are either recently perceived, recalled from the Long-Term Memory (LTM) or inferred. The neural architectures implemented here follow the view that the WM/STM is the activated part of the LTM. Furthermore, similar to the positions taken by [102], [59] and [63], the pre-frontal cortex is seen as a collection of interconnected special-purpose working memory areas, e.g. separate WMs for spatial location of objects and for object characteristics [103] and perhaps linguistic working memory [59]. The computational analog of the pre-frontal cortex presented here, i.e. the WM of the model, is implemented as multiple special-purpose STM systems, interconnected and working in parallel, mainly WM for actions and WM for object characteristics, interconnected according to the affordances associating particular actions and objects.

3.3   Implementation

The Working Memory (Central) modules implement three main functions: (1) The networks in the Central module (the AW-CA-C circuit in particular) implement the simple grammar of the robot's instruction set and the integration of a sequence of words into a phrase (object and/or instruction); (2) The neural circuits from the Working Memory maintain the instructions given to the robots for the period necessary for their execution; and (3) Neurons from the central modules are the points of integration and transfer of information across modalities.

One view on the major part of the functionality of the neural circuits in the Central module is to achieve a number of different types of binding across the temporal and spatial domains.

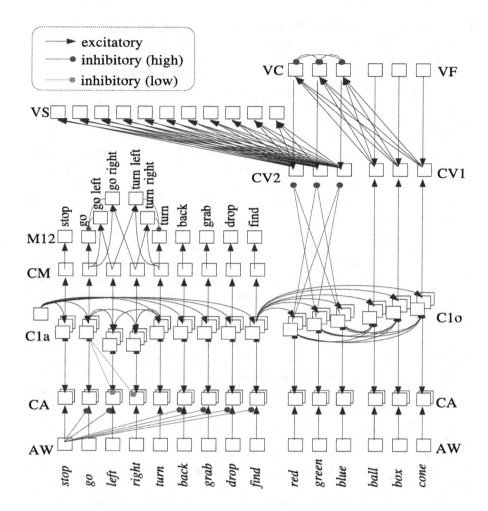

Fig. 2. Architecture of the Central (Working Memory) module. The Central layers interact with the other modules via CA for auditory, CV for the visual and CM for the motor areas. One of the main functionalities is implemented in the C1 layer which includes small clusters of neurons representing each of the words in the robot's lexicon (C1a sub-area for action words and C1o sub-area for objects). Lateral and feed-back connections implement the relationship between the words in an instruction. Each concept recognised by the robot is represented by distributed cell assemblies across different modalities. For example the representation of *go* includes neurons from the AW, CA, C1, CM and M12 areas, and entities such as *box* activate neurons from AW, CA, C1, CV and VF (possibly also including VC and VS) areas. For clarity only some inhibitory connections from AW to CA and from C1 to CA are shown. The inhibitory connections are setup so that high inhibition prevents the neurons from firing whereas low inhibition delays their firing. The strong inhibition from AW to CA connects neurons representing words which have the same order position in the instructions. The weak inhibition from C to CA connects neurons representing words in successive position in the instruction phrases

Language understanding: Recognising sequences of words. One of the main goals of this work is to build a system which is able to recognise and respond to instructions given using natural language. The first stage of this system was presented earlier as the auditory modules recognising single words of spoken commands. Further challenges posed by the task include the necessity to integrate the sequence different words into a coherent instruction to be recognised and executed by the robot.

Different neural mechanisms have been proposed as the architectures which the real brain might use in the areas handling language processing and understanding. Some of the main requirement considered during the design and implementation of the computational architecture presented here are: (1) The words forming an instruction would come in a sequence, and the mechanism should support gradual building of the semantic information contained in the sequence under a set of syntactic and semantic constraints; (2) The temporal structure of a spoken sequence of words contains relatively high levels of noise, more specifically, the interval between two consecutive words ranges from a few hundred milliseconds up to a few seconds. The proposed neural mechanism should cope with such fluctuations; (3) The mechanism should allow for insertions such as adjectives. For example, the instructions *Find box* and *Find red box* should lead to the same behaviour of the robot if there is a red box in front of it.

The architecture presented here derives from a model presented by Lisman [104, 105, 106, 107, 108]. It is implemented in the CA-C1 circuit running two nested oscillations. The oscillation representing a phrase runs in the theta range (10 Hz for the current implementation). Within each cycle of the theta oscillation, the cell assemblies representing each of the currently active concepts spike in a sequence forming a gamma oscillation (about 30 Hz in the current implementation).

Each cycle of the theta oscillation can include one phrase (instruction). The start of the cycle of the theta oscillation is marked by a Central Pattern Generator (CPG) neuron in area C1 which spikes at 10 Hz and sends signals to all C1 neurons representing words which can appear at the beginning of a phrase. The signals generate sub-threshold membrane potentials at the C1 neurons and alone are not sufficient to activate the assemblies. Additional input from the neurons in CA is required for these neurons to spike. The CA and C1 neurons in a cell assembly representing a particular concept have bidirectional connections and formulate a reverberating oscillatory activity between the two areas. Thus, activation of a single word constituting an instruction (or being the first word of an instruction), e.g. *go*, would be as follows (figure 3: upon recognition of the auditory input stream as the word *go* the AW neuron for that word will respond with a spike burst causing several spikes in the CA neurons for *go*. In parallel the inhibition from AW to CA will shut down any instruction currently held in CA-C. One or more of the spikes generated in CA will appear sufficiently close to the spikes in CPG. The nearly synchronous spikes from the CPG and CA neurons will activate the neurons representing *go* in C1a. In return, the neurons from C1a will activate (with some delay) the CA again as well as motor control

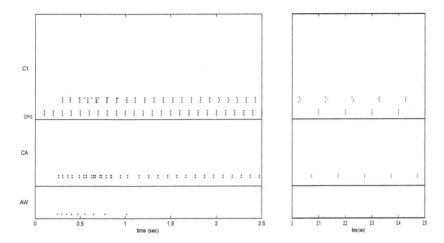

Fig. 3. Working memory activity in processing the instruction *Go* given at time 0. Left: At approximately 250 milliseconds the word is recognised and activates the AW neuron representing *go*. The spike burst from AW activates the neurons in CA and thereby the working memory oscillation C1a-CA. Right: Zoom in the oscillation after the activation pattern has stabilised

neurons in CM and M12. Following the propagation of activity, the CA-C1a neurons for the word *go* will oscillate with a precise frequency led by the spikes from CPG and maintain this activity while subsequent words come in and the instruction is being executed.

The processing of words taking second and third position in the instruction phrases follows a similar activation patter (figure 4). The main difference is that instead of receiving inputs from CPG, the neurons representing such words in area C1 receive lateral inputs from the C1 neurons representing words which can precede them in a valid instruction. For example the C1a neurons representing *left* receive lateral excitatory connections from the C neurons representing *go* and *turn*. In addition, the CA neurons for the word *left* will also receive low strength fast inhibition from the C1a neurons of *go* and *turn*. This inhibition is not sufficient to prevent the CA neurons from firing but rather to delay their firing and ensure proper order of activation in the CA-C1 oscillations. Most critically, the weak inhibition from C1 to CA ensures that when new words come as input, they enter the CA-C1 oscillation at the appropriate place, i.e. after the word which should precede them.

The architecture supports a gradual build up of the current context which allows a wide range of fluctuations in the intervals between the consecutive words in the input stream, including insertions. The CA-C1 oscillation maintains the current context until the next word arrives. Upon arriving of a word from the auditory stream (that is activation in AW), the new entity is included at the appropriate place in the current context in accordance with the semantic and syntactic constraints in the robot's dictionary. This is implemented by the low

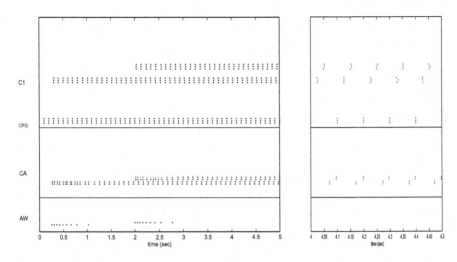

Fig. 4. Working memory activity in processing the instruction *Turn left* given as a sequence of the words *turn* at time 0 sec and *left* at 1.66 sec. Left: At approximately 250 milliseconds the first word is recognised and activates the AW neuron representing *turn*. The spike burst from AW activates the neurons in CA and thereby the working memory oscillation C1a-CA. At approximately 2 sec the word *left* is recognised an enters the working memory oscillation after *turn* Right: Zoom in the oscillation after the activation pattern has stabilised

inhibition from C to CA. For example, if the current context is *find red* and the new input word is *box*, the new oscillation will be a sequential firing of the assemblies for *find red box*, whereas if the current context is *find box*, a subsequent input word *red* will be included in just before *box* and again lead to the representation of *find red box* (figure 5). Furthermore, the architecture allows corrections in the instruction, e.g. change in the colour or shape of the object. Figure 6 presents an example where the colour of the object has been change from blue to green.

Associating Language Descriptions and Visual Perceptions of Objects. This section describes the neural mechanisms employed by the robot to understand and execute instructions on finding and approaching an object. Such instructions are based on the *find* command followed by a description of the target object. The description might fully describe the target object as having a specific colour and shape, or might include only the shape, in which case the robot will seek an object of any colour. The execution of the commands of the type *find blue ball* or *find ball* requires the robot to maintain in its working memory the description of the object which it has to find and respond once the object is recognised in the visual field. The recognition of the target object involves two stages of matching first the shape and then the colour of the target (as maintained in the working memory) and perceived (as recognised by the visual module) objects. The activation of the instruction *find* drives the robot forward,

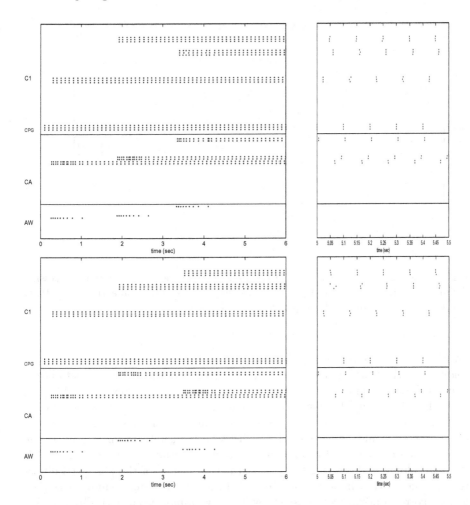

Fig. 5. Working memory activity in processing the instruction *Find red box* (Top) and *Find box red* (Bottom). In both cases the pattern of the final stable oscillation represents the sequence *find red box*

while it is seeking the target object. Initially the robot move freely and avoids any obstacles in its path. All objects which do not match the description given in the instruction are treated as obstacles and the robot should steer around them. Once the object whose description is being held in the working memory has been recognised in the visual field, the robot should approach and stop in front of it.

Distributed feature-based representations are used to encode the objects in both working memory and visual areas. The CA-Clo neural circuits of the central module can maintain a lexical description of the objects as part of the *find* instruction. In both CA and Clo areas, each of the 3 colours and 3 shapes is represented by a small cluster of neurons. Once activated by a language input,

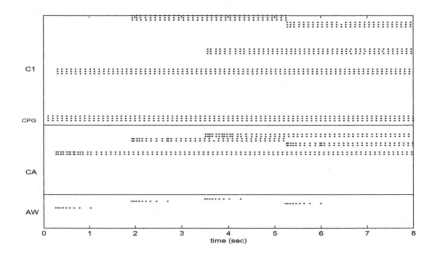

Fig. 6. Correcting the target in the instruction: Working memory activity in processing the instruction *Find blue ball* followed by the word *green*. The final activation pattern of the working memory is the sequence *find green ball*

reverberating cell assemblies containing neural clusters of a shape and possibly colour from the working memory maintain a representation of the target object. Each of these clusters will potentiate/inhibit target neurons in CV. At the same time, the robot would be navigating around the environment, and while moving around different objects from the scene will fall into the visual field. Each of these objects would be recognised by the robot's visual system activating the relevant neurons in VF (for the shape), and potentiating the neurons in VC (for the colour) and VS (for the position). If the shape of the perceived object is the same as the shape of the object held in the working memory, the CV1 neuron corresponding to that shape will receive inputs from both C1o and VF which is strong enough to cause a spike. That in turn will activate the VC neuron representing the colour of the perceived object which has already been potentiated via the dorsal stream. If the instruction does not specify the colour of the object which the robot should find, but only the shape, then the working memory will maintain a representation of an object which does not activate the colour neurons in area C1o. Consequently, there will be no inhibition from C1o to CV2 colour neurons. In such cases the activation of a VC neuron is sufficient to fire the corresponding neuron in CV2. As a result, regardless of the colour of the perceived object, a CV2 neuron will fire every time the shape of the perceived and target objects match. If however, the language instruction specifies the colour, a cluster of neurons in area C1o representing that colour will be part of the active cell assembly representing the target object. Consequently, the neurons in that cell assembly will inhibit the input streams from VC which represents the other two colours and allow a CV2 neuron to fire only if the colours represented in C1o and VC are the same. As a result, CV2 neurons will respond only if the colour and the shape of the perceived object match the colour and the shape of the object described in the language instruction. Finally, spikes in CV2 will activate the already potentiated neuron in VS which, as described in the previous section, represents the horizontal position of the perceived object in the visual scene. The VS neurons have direct connection to the M12 motor control neurons and can steer the robot so that the perceived object is in the centre of the visual field. As a result the robot will navigate toward an object which has been recognised to match the one given by the current language instruction. An example of the Working Memory activity in approaching a target is given in the later experimental section.

Once the robot arrives in close proximity to the target, the object does not fit in the camera image and object recognition is no longer possible. The robot can no longer navigate toward the object based on visual information. However, before this situation is reached, the robot can "sense" the target object using the front distance sensors and can be "attracted" to the object instead of treating it as an obstacle. In a condition where the robot recognises the target object on the visual field, and the front sensors are activated, the attractive neurons in area M11 are activated. These neurons inhibit the obstacle avoiding repulsive cells in M11 and control the relative speed of the wheels so that the robot moves forward toward the object in front. Based on the input from the attractive neurons, the robot can move within a grasping distance from the target object.

The presented working memory and navigation architecture is partially derived from known mechanisms which the brain could be using for temporal storage of concepts and navigation based on multi-modal sensory information. The architecture developed here is nevertheless, a suitably simplified and computationally optimised model of the brain mechanisms. Further extension could possibly include models of hippocampal place and head direction cells, which will provide the robot with much finer information about its relative position to the target object or a landmark.

4   A Robot Executing Language Instructions

The model and experiments presented in this chapter aim at exploring the computational performance of the ADDS neurons in a real environment, processing real data and controlling a mobile robot. More specifically, this chapter presents a models of a "remote brain" architecture of a mobile robot receiving and executing language instructions. The overall setup is presented in figure 7. The environment includes a mobile robot (Khepera or PeopleBot) equipped with camera and gripper. The robots are moving using differential wheels. The robots send real-time images to the cluster via image pre-processing workstation. Spoken language instructions are also send to the cluster, after recording and pre-processing at the workstation. Each robot is connected to the "remote brain" (a multi-CPU Beowulf cluster) via wireless or tether, and receives direct control signals for the wheels and the gripper. The robot is moving in an environment with objects of

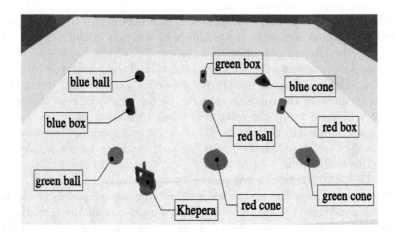

Fig. 7. The robot's environment contains nine objects of different colour and shape

different shape and colour. Given a particular instruction, e.g. "find blue ball" or "lift red box", the task of the robot is to understand and execute it.

The system was implemented in both real Khepera robot equipped with a drippier and a camera and the Webots simulation environment with a realistic Khepera model. The experiments reported is this chapter were performed using Webots where more precise control on the environment and observation of the robot's position and status was available. The experiments were also performed with a real robot and qualitatively equivalent results were obtained.

5   Navigation and Language Command Execution

The first experiment tests the behavioural responses of the robot to direct commands. Figure 8 shows the trajectory of the robot executing a sequence of direct commands. The robot begins at position 0 where the command *go* is issued. After a while, at position 1, the command is extended with the word *left*. Since the sequence *go left* constitutes a valid instruction recognisable by the robot, to word *left* does not replace the existing active instruction, but is rather added to it. Further, at position 2, the word *right* is send to the robot. Since this instruction is alternative to the *left* one, it replaces it and the active instruction for the robot now is *go left*. At position 3, the word *go* is sent to the robot. Being a primary word in the instruction set, the connectivity of the WM for this word is such that it will stop the oscillatory activity of any other word in the WM. At this point the active instruction for the robot becomes *go*, and the robot moves in a straight line. Further, at position 4, the robot is *stop*ped and instructed to *turn right*. After some delay, the instruction *go* is sent. Similar to point 3, this command cancels all current instructions. As a result, the robot stops turning and starts moving in a straight line until it is stopped at position 5.

0	*go*
1	*left*
2	*right*
3	*go*
4	*stop*
	turn right
	go
5	*stop*

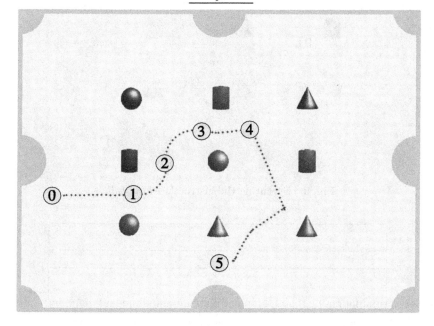

Fig. 8. Navigating the robot

While moving without a fixed target in the visual field, the obstacle avoiding behaviour is active, hence the robot bounces from the green cone.

6   Finding and Manipulating Objects

Figure 9 shows the trajectory of the robot executing the instruction *find ball* given at position 0. Initially, the robot follows a simple exploring behaviour, i.e. it moves freely in the environment, avoiding walls and objects which are not of interest, until the target object falls in the visual field and is recognised by the robot. For the present experiment, the target object is described as a ball. Therefore, the robot will identify as a target object the first ball which falls into the visual field, that is the green ball visible from position 1. At that point the robot's movements are guided by the visual information on the relative position of the target object and approaching behaviour. The robot moves toward the object guided by visual information until the object falls into the range of the

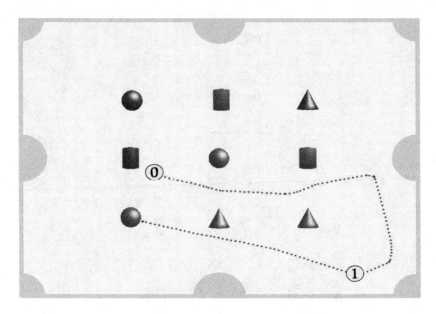

Fig. 9. Executing the instruction *Find ball*

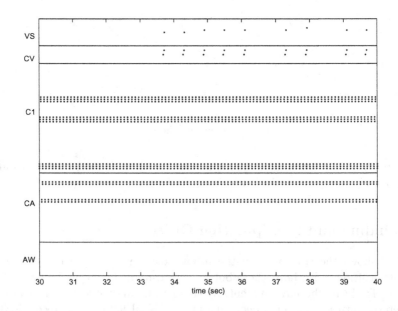

Fig. 10. Working memory activity in executing the instruction *Find ball* at the time when the robot first recognises the target object (position 1 on figure 9). Every time when the robot receives an image containing the green ball, the CV neurons for shape and colour respond. Consequently, the VS neuron corresponding to the relative position of the current object fires steering the robot toward the target

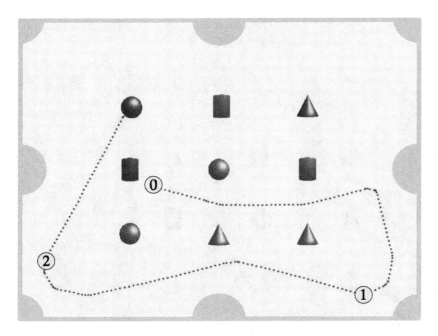

Fig. 11. Executing the instruction *Find blue ball*

front distance sensors. Shortly after this point the object's projection of the retina becomes too big, and it can no longer be recognised by the visual system. Guided by sensory information received from the front distance sensors, the robot can further move closer to the object in front, reaching a distance at which it can use the gripper. At this point the ball can be grabbed and moved to a different place. Figure 10 shows the spike plot of the Working Memory activity when the robot first recognises the green ball. Following the activation of the CV1 for the ball shape and CV2 for the green colour, the VS neurons respond steering the robot toward the target.

Figure 11 shows the execution of the instruction *find blue ball*. The robot starts from the same position as in the previous experiment. However, at position 1, the robot does not identify the green ball as the target given in the instruction, and continues with the exploring behaviour. The blue ball is recognised at position 2 and the robot approaches is directly.

Figure 12 shows the behaviour of the robot while executing instructions with changing targets. At position 0, the robot is given instruction *find red box* and starts moving around until the target object falls into the visual field. At position 1, the robot recognises the red box and starts approaching it. At position 2, the instruction is changed. The robot receives the word sequence *green box*. As a result, the instruction active in the working memory is *find green box*. The robot no longer recognises the red box as a target and avoids it as an obstacle. Later on, at position 3, the robot identifies the green box and approaches it.

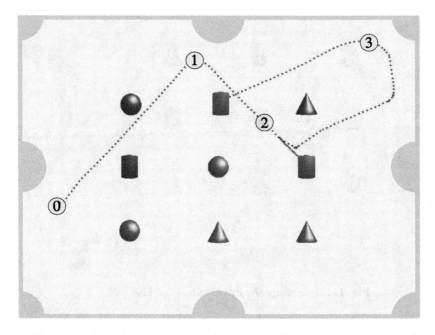

| 0 *find red box* |
| 2 *green box* |

Fig. 12. Executing an instruction with changing targets

7   Discussion

This chapter presented a multi-modal architecture based on ADDS spiking neurons for robot control using language instructions. The central modules of the network integrate sequences of words constituting an instruction, and associated their interpretation with relevant neural assemblies in the visual and motor modalities. The architecture incorporates a working memory model based on oscillatory activity of neural assemblies from different modalities. The model implements a gradual build up of context and language instructions as subsequent word are sent to the robot. Furthermore, the working memory allows the current representation of the instruction to be maintained for as long as required, that is until a target is found, or the instruction is replaced by another one.

Being the first robot control architecture built entirely on spiking neurons, the current model would allow a number of new behavioural and computational experiments which provide results that could be related much closer to the processes in the real brain. Further work with the architecture would focus on behavioural properties of the model. For example, the current set up allows the task and objects in the environment to be used in a reinforcement learning development. The colours of the objects could be used for encoding the "taste" of the robot, e.g. green and blue objects could be attractive, while red ones are repulsive. In addition, the shapes of the objects could be used for encoding affordances, e.g. balls and boxes could be lifted, while cones cannot. In the longer term, the experimental setup can be extended with multiple robots with imitating and swarming behaviours, which in addition will provide a rich environment requiring more complex language instructions.

References

1. Panchev, C., Wermter, S.: Spike-timing-dependent synaptic plasticity from single spikes to spike trains. Neurocomputing (2004) to appear.
2. Panchev, C., Wermter, S., Chen, H.: Spike-timing dependent competitive learning of integrate-and-fire neurons with active dendrites. In: Lecture Notes in Computer Science. Proceedings of the International Conference on Artificial Neural Networks, Madrid, Spain, Springer (2002) 896–901
3. Panchev, C., Wermter, S.: Temporal sequence detection with spiking neurons: towards recognizing robot language instructions. Connection Science **17**(1) (2005)
4. Elman, J.L., Bates, E.A., Johnson, M.H., Karmiloff-Smith, A., Parisi, D., Plunkett, K.: Rethinking Innateness. MIT Press, Cambridge, MA (1996)
5. Bates, E.: Plasticity, localization and language development. In Broman, S., Fletcher, J., eds.: The changing nervous system: Neurobehavioral consequences of early brain disorders. New York: Oxford University Press (1999)
6. Goodglass, H.: Understanding aphasia. Technical report, University of California, San Diego, San Diego: Academic Press (1993)
7. Bates, E., Vicari, S., Trauner, D.: Neural mediation of language development: Perspectives from lesion studies of infants and children. In Tager-Flusberg, H., ed.: Neurodevelopmental disorders. Cambridge, MA: MIT Press (1999) 533–581
8. Pulvermuller, F.: The Neuroscience of Language: On Brain Circuits of Words and Serial Order. Cambridge University Press, Cambridge (2002)
9. Bates, E., Thal, D., Trauner, D., Fenson, J., Aram, D., Eisele, J., Nass, R.: From first words to grammar in children with focal brain injury. Special issue on Origins of Communication Disorders, Developmental Neuropsychology (1997) 275–343
10. Stiles, J., Bates, E., Thal, D., Trauner, D., Reilly, J.: Linguistic, cognitive and affective development in children with pre- and perinatal focal brain injury: A ten-year overview from the san diego longitudinal project. In C. Rovee-Collier, L.L., Hayne, H., eds.: Advances in infancy research. Norwood, NJ: Ablex (1998) 131–163
11. Lenneberg, E.H.: Biological foundations of language. New York: Wiley (1962)
12. Stromswold, K.: The cognitive and neural bases of language acquisition. In Gazzaniga, M.S., ed.: The Cognitive Neuroscience. Cambridge, MA: MIT Press (1995)
13. Jackson, J.H.: On affections of speech from disease of the brain. Brain **1** (1878) 304–330
14. Broca, P.: Remarques sur le siége de la faculté du langage articulé, suivies d'une observation d'aphemie (perte de la parole). Bulletin de la Société d'Anatomie **36** (1861) 330–357
15. Wernike, C.: Der Aphasische Symtomenkomplex. Eine Psychologische Studie auf Anatomischer Basis. Breslau: M. Cohn und Weigart (1874)
16. Caplan, D.: Neurolinguisitcs and linguistic aphasiology: An introduction. Cambridge University Press (1987)

17. Neininger, B., Pulvermüller, F.: The right hemisphere's role in action word processing: a double case study. Neurocase **7** (2001) 303–320
18. Pulvermüller, F., Mohr, B.: The concept of transcortical cell assemblies: a key to the understanding of cortical lateralization and interhemispheric interaction. Neuroscience and Biobehavioral Reviews **20** (1996) 557–566
19. Ojemann, G.: Cortical organisation of language. Journal of Neuroscience **11**(8) (1991) 20–39
20. Price, C., Green, D., Studnitz, R.: A functional imaging study of translation and language switching. Brain **122** (1999) 2221–2235
21. Mazoyer, B.M., Tzourio, N., Frak, V., Syrota, A., Murayama, N., Levrier, O.: The cortical representation of speech. Journal of Cognitive Neuroscience **5** (1993) 467–479
22. Perani, D., Dehaene, S., Grassi, F., Cohen, L., Cappa, S., Dupoux, E.: Brain processing of native and foreign languages. Neuroreport **7** (1996) 2439–2444
23. Dehaene, S., Dupoux, E., Mehler, J., Cohen, L., Paulesu, E., Perani, D.: Anatomical variability in the cortical representation of first and second languages. Neuroreport **8** (1997) 3809–3815
24. Bavelier, D., Corina, D., Jezzard, P., Clark, V., Karni, A., Lalwani, A.: Hemispheric specialization for english and asl: left invariance-right variability. Neuroreport **9** (1998) 1537–1542
25. Kim, K.H., Relkin, N.R., Lee, K.M., Hirsch, J.: Distinct cortical areas associated with native and second languages. Nature **388** (1997) 171–174
26. Posner, M.I., DiGirolamo, G.J.: Flexible neural circuitry in word processing. Behavioral and Brain Sciences **22** (1999) 299–300
27. Freud, S.: Zur Auffassung der Aphasien. Wien: Franz Deuticke (1891) English tanslation: On Aphasia: A Critical Study. Translated by E. Stengel, International Universities Press, 1953.
28. Pulvermüller, F.: A brain perspective on language mechanisms: from discrete neuronal ensembles to serial order. Progress in Neurobiology **67**(2) (2002) 85–111
29. Stein, B.E., Meredith, M.A.: Merging of the senses. MIT Press, Cambridge, MA (1993)
30. Meredith, M.A.: On the neuronal basis for multisensory convergence: A brief overview. Cognitive Brain Researc **14** (2002) 31–40
31. Shimojo, S., Shams, L.: Sensory modalities are not separate modalities: plasticity and interactions. Current Opinion in Neurobiology **11** (2001) 505–509
32. von Melchner, L., Pallas, S.L., Sur, M.: Visual behaviour mediated by retinal projections directed to the auditory pathway. Nature **404** (2000) 871–876
33. Blake, R., Grossman, E.: Neural synergy between seeing and hearing. (draft) (2000)
34. Lappe, M.: Information transfer between sensory and motor networks. In Moss, F., Gielen, S., eds.: Handbook of bilogical physics. Volume 4. Elsevier Science (2000)
35. Fuster, J.M., Bodner, M., Kroger, J.K.: Cross-modal and cross-temporal association in neurons of frontal cortex. Nature **405** (2000) 347–351
36. Zhou, Y.D., Fuster, J.M.: Visio-tactile cross-modal associations in cortical somatosensory cells. In: Proceedings of National Academy of Sciences U.S.A. Volume 97. (2000) 9777–9782
37. Rizzolatti, G., Luppino, G., Matelli, M.: The organization of the cortical motor system: new concepts. Electroencephalography and Clinical Neurophysiology **106**(4) (1998) 283–296

38. Barbas, H.: Connections underlying the synthesis of cognition, memory, and emotion in primate prefrontal cortices. Brain Research Bulletin **52**(5) (2000) 319–330

39. Lewis, J.W., van Essen, D.C.: Corticocortical connections of visual, sensimotor, and multimodal processing areas in the parietal lobe of the macaque monkey. Journal of Computational Neurology **428** (2000) 112–137

40. Pandya, D.N., Yeterian, E.H.: Architecture and connections of cortical association areas. In Peters, A., Jones, E., eds.: Cerebral Cortex, Association and Auditory Cortices. Volume 4. Plenum Press, London (1985) 3–61

41. Deacon, T.W.: Cortical connections of the inferior arcuate sulcus cortex in the macaque brain. Brain Reseach **573** (1992) 8–26

42. Pulvermüller, F.: Brain reflections of words and their meaning. Trends in Cognitive Sciences **5**(12) (2001) 517–524

43. Landauer, T.K., Dumais, S.T.: A solution to plato's problem: the latent semantic analysis theory of acquisition, induction, and representation of knowledge. Psychological Review **104** (1997) 211–254

44. Pulvermüller, F.: Words in the brain's language. Behavioral and Brain Sciences **22** (1999) 253–336

45. Pulvermüller, F.: Words in the brain's language. Behavioral and Brain Sciences **22** (1999) 253–336

46. Brent, M.R., Cartwright, T.A.: Distributional regularity and phonotactic constraints are useful for segmentation. Cognition **61**(1-2) (1996) 93–125

47. Harris, Z.S.: From phonemes to morphemes. Language **28** (1955) 1–30

48. Redlich, A.N.: Redundancy reduction as a strategy for unsupervised learning. Neural Computation **3** (1993) 289–304

49. Shaffran, J.R., Aslin, R.N., Newport, E.L.: Statistical learning by 8-month-old infants. Science **274**(5294) (1996) 1926–1934

50. Shannon, C.E., Weaver, W.: The Mathematical Theory Of Communication. University of Illinois Press (1963)

51. Locke, J.L.: Babbling and early speech: continuity and individual differences. First Language **9** (1989) 191–206

52. Locke, J.L.: The child's path to spoken language. Harvard University Press, Cambridge, MA (1993)

53. Marslen-Wilson, W., Tyler, L.K.: The temporal structure of spoken language understanding. Cognition **8** (1980) 1–71

54. Rugg, M.D.: Further study of electrophysiological correlates of lexical decision. Brain and Language **19** (1983) 142–194

55. Price, C.J., Wise, R.J.S., Frackowiak, R.S.J.: Demonstrating the implicit processing of visually presented words and pseudowords. Cerebral Cortex **6** (1996) 62–70

56. Korpilahti, P., Krause, C.M., Holopained, I., Lang, A.H.: Early and late mismatch negativity elicited by words and speech-like stimuli in children. Brain and Language **76** (2001) 332–371

57. Pulvermüller, F., T., K., Shtyrov, Y., Simola, J., Tiitinen, H., Alku, P., Alho, K., Martinkauppi, S., Ilmoniemi, R.J., Näätänen, R.: Memory traces for words as revealed by the mismatch negativity. Neuroimage **14** (2001) 607–623

58. Shtyrov, Y., Pulvermüller, F.: Neurophysiological evidence for memory traces for words in the human brain. Neuroreport **13** (2002) 521–525

59. Aboitiz, F., García, V.R.: The evolutionary origin of the language areas in the human brain. a neuroanatomical percpective. Brain Research Reviews **25** (1997) 381–396

60. Rizzolatti, G., Arbib, M.: Language within our grasp. Trends in Neurosciences **21** (1998) 188–194
61. Arbib, M.A.: The mirror neuron hypothesis for language ready brain. In Angelo, C., Domenico, P., eds.: Computational approaches to the evolution of language and communication. Springer, Berlin (2001) 229–254
62. Arbib, M.A.: The mirror system, imitation, and the evolution of language. In Christopher, N., Kerstin, D., eds.: Imitation in animals and artifacts. MIT Press, Cambridge, MA (2002) 229–280
63. Arbib, M., Bota, M.: Language evolution: neural homologies and neuroinformatics. Neural Networks **16**(9) (2003) 1237–1260
64. Arbib, M.: Rana computatrix to human language: towards a computational neuroethology of language evolution. Philosophical Transactions: Mathematical, Physical and Engineering Sciences **361**(1811) (2003) 2345–2379
65. Arbib, M.: From monkey-like action recognition to human language: An evolutionary framework for neurolinguistics. Behavioral and Brain Sciences (2005)
66. Gallese, V., Fadiga, L., Fogassi, L., Rizzolatti, G.: Action recognition in the premotor cortex. Brain **119** (1996) 593–609
67. Gallese, V., Keysers, C.: Mirror neurons: a sensorimotor representation system. Behavioral Brain Sciences **24**(5) (2001) 983–984
68. Rizzolatti, G., Fadiga, L., Matelli, M., Bettinardi, V., Paulesu, E., Perani, D., Fazio, F.: Localization of grasp representations in human by pet: 1. observation versus execution. Exp Brain Res. **111** (1996) 246–252
69. Rizzolatti, G., Fadiga, L., Matelli, M., Bettinardi, V., Paulesu, E., Perani, D., Fazio, F.: Parietal cortex: from sight to action. Experimental Brain Research **11**(2) (1996) 246–252
70. Gallese, V., Goldman, A.: Mirror neurons and the simulation theory of mindreading. Trends in Cognitive Science **2** (1998) 493–501
71. Rizzolatti, G., Fogassi, L., Gallese, V.: Neurophysiological mechanisms underlying the understanding and imitation of action. Nature Review **2** (2001) 661–670
72. Gallese, V.: Echo mirror neurons: Recognizing action by sound. In: Human Frontiers Science Program Workshop "Mirror System: Humans, Monkeys and Models", University of Southern California, Los Angelis, CA (2001)
73. Kohler, E., Keysers, C., Umilta, M., Fogassi, L., Gallese, V., Rizzolatti, G.: Hearing sounds, understanding actions: action representation in mirror neurons. Science **297** (2002) 846–848
74. Keysers, C., Kohler, E., Umiltà, M., Fogassi, L., Nanetti, L., Gallese, V.: Audiovisual mirror neurones and action recognition. Experimental Brain Research **153** (2003) 628–636
75. Zukow-Goldring, P., Arbib, M., Oztop, E.: Language and the mirror system: A perception/action based approach to cognitive development. in preparation (2002)
76. Burns, B., Sutton, C., Morrison, C., Cohen, P.: Information theory and representation in associative word learning. In Prince, C., Berthouze, L., Kozima, H., Bullock, D., Stojanov, G., Balkenius, C., eds.: Proceedings Third International Workshop on Epigenetic Robotics: Modeling Cognitive Development in Robotic Systems, Boston, MA, USA (2003) 65–71
77. Oates, T.: Grounding Knowledge in Sensors: Unsupervised Learning for Language and Planning. PhD thesis, University of Massacgusetts, Amherst (2001)
78. Siskind, J.M.: Learning word-to-meaning mappings. In Broeder, P., Murre, J., eds.: Models of Language Acquisition: Inductive and Deductive Approaches. Oxford University Press (2000) 121–153

79. Steels, L.: The Talking Heads Experiment. Volume 1. Words and Meanings. Antwerpen (1999)
80. Steels, L., Kaplan, F.: Bootstrapping grounded word semantics. In Briscoe, T., ed.: Linguistic evolution through language acquisition: formal and computational models. Cambridge University Press, Cambridge, UK (1999)
81. Steels, L., Kaplan, F.: AIBO's first words. the social learning of language and meaning. preprint (2001)
82. Arbib, M.A., Grethe, J.S., eds.: Computing the Brain: A Guide to Neuroinformatics. San Diego Academic Press (2001)
83. Shtyrov, Y., Hauk, O., Pulvermüller, F.: Distributed neuronal networks for encoding catergory-specific semantic information: The mismatch negative to action words. European Journal of Neuroscience **19** (2004) 1–10
84. Wolfe, J., Cave, K.: The psychophysical evidence for a binding proble in human vision. Neuron **24**(1) (1999) 11–17
85. Triesman, A.: Solutions to the binding problem: Progress through controversy and convergence. Neuron **24**(1) (1999) 105–110
86. Treisman, A., Schmidt, H.: Illusory conjunctions in the perception of objects. Cognitive Psychology **14** (1982) 107–141
87. von der Malsburg, C.: The correlation theory of brain function. Technical report, Max-Planck-Institute for Biophysical Chemistry (1981) Internal Report 81-2.
88. Rosenblatt, F.: Principles of Neurodynamics: Perceptions and the Theory of Brain Mechanisms. Spartan Books, Washington (1961)
89. Roskies, A.: The binding problem. Neuron **24**(1) (1999) 7–9
90. Ghose, G., Maunsell, J.: Specialized representations in the visual cortext: A role for binding. Neuron **24**(1) (1999) 79–85
91. Riesenhuber, M., Poggio, T.: Are cortical models really bound by the "binding problem". Neuron **24**(1) (1999) 87–93
92. Barlow, H.B.: Single units and sensation: A neuron doctrine for perceptual psychology? Perception **1** (1972) 371–394
93. Van Essen, D., Gallant, J.: Neural mechanisms of form and motion processing in the primate visual system. Neuron **13** (1994) 1–10
94. Kobatake, E., Tanaka, K.: Neuronal selectivities to complex object features in the ventral visual pathway of the macaque cerebral cortext. Journal of Neurophysiology **71** (1994) 856–857
95. Gallant, J.L., Connor, C.E., Rakshit, S., Lewis, J., Van Essen, D.: Neural responses to polar, hyperbolic, and cartesian gratings in area v4 of the macaque monkey. Journal of Neurophysiology **76** (1996) 2718–2739
96. Tallon-Baudry, C., Bertrand, O.: Oscillatory gamma activity in humans and its role in object representation. Trends in Cognitive Sciences **3** (1999) 151–162
97. Roelfsema, P.R., Engel, A.K., König, P., Singer, W.: Visio-motor integration is associated with zero time-lag synchronization among cortical areas. Nature **385** (1997) 157–161
98. Bressler, S.L., Coppola, R., Nakamura, R.: Episodic multiregional cortical coherence at multiple frequencies during visual task performance. Nature **366** (1993) 153–156
99. Moran, J., Desimone, R.: Selective attention gates visual processing in the extrastriate cortex. Science **229** (1985) 782–784
100. Reynolds, J., Desimone, R.: The role of neural mechanisms of attention in solving the binding problem. Neuron **24**(1) (1999) 19–29
101. Fuster, J.M.: Memory in the Cerebral Cortex. MIT Press, Cambridge, MA (1995)

102. Goldman-Rakic, P.S.: Architecture of the prefrontal cortex and the central executive. In: Annals of the New York Academy of Science. Volume 769. New York (1995) 71–83
103. Wilson, F.A., Ó Scalaide, S.P., Goldman-Rakic, P.S.: Dissociation of object and spatial processing domains in primate prefrontal cortex. Science **260** (1993) 1955–1958
104. Lisman, J., Idiart, M.A.P.: Storage of 7+/-2 short-term memories in oscillatory subcycles. Science **267** (1995) 1512–1515
105. Idiart, M.A.P., Lisman, J.: Short-term memory as a single cell phenomenon. In Bower, J.M., ed.: The Neurobiology of Computation: Proceedings of the third annual computational and neural systems conference. Kluwar Academic Publishers (1995)
106. Jensen, O., Lisman, J.E.: Novel lists of 7 2 known items can be reliably stored in an oscillatory short-term memory network: interaction with long-term memory. Learning and Memory **3** (1996) 257–263
107. Jensen, O., Lisman, J.E.: Theta/gamma networks with slow nmda channels learn sequences and encode episodic memory: role of nmda channels in recall. Learning and Memory (1996)
108. Jensen, O., Lisman, J.E.: Hippocampal sequence-encoding driven by a cortical multi-item working memroy buffer. Trends in Neurosciences **28**(2) (2005) 67–72

A Virtual Reality Platform for Modeling Cognitive Development

Hector Jasso[1] and Jochen Triesch[2]

[1] Department of Computer Science and Engineering,
University of California San Diego, La Jolla CA 92093, USA
hjasso@cs.ucsd.edu
[2] Cognitive Science Department, University of California San Diego,
La Jolla CA 92093, USA
triesch@cogsci.ucsd.edu

Abstract. We present a virtual reality platform for developing and evaluating embodied models of cognitive development. The platform facilitates structuring of the learning agent, of its visual environment, and of other virtual characters that interact with the learning agent. It allows us to systematically study the role of the visual and social environment for the development of particular cognitive skills in a controlled fashion. We describe how it is currently being used for constructing an embodied model of the emergence of gaze following in infant-caregiver interactions and discuss the relative benefits of virtual vs. robotic modeling approaches.[1]

1 Introduction

Recently, the field of cognitive science has been paying close attention to the fact that cognitive skills are unlikely to be fully specified genetically, but develop through interactions with the environment and caregivers. The importance of interactions with the physical and social environment for cognitive development has been stressed by connectionist [7] and dynamical systems [17] approaches.

Developmental schemes are also being proposed in the field of intelligent robotics [1, 3, 18]. Instead of building a fully working robot, a body capable of interacting with the environment is given general learning mechanisms that allows it to evaluate the results of its actions. It is then "set free" in the world to learn a task through repeated interactions with both the environment and a human supervisor.

Our motivation is to develop embodied models of cognitive development, that allows us to systematically study the emergence of cognitive skills in naturalistic settings. We focus on visually mediated skills since vision is the dominant modality for humans. The kinds of cognitive skills whose development we would

---

[1] This paper was presented at the 3rd International Conference for Development and Learning, ICDL'04, La Jolla, California, USA.

S. Wermter et al. (Eds.): Biomimetic Neural Learning, LNAI 3575, pp. 211–224, 2005.
© Springer-Verlag Berlin Heidelberg 2005

ultimately like to model range from gaze and point following and other shared attention skills over imitation of complex behaviors to language acquisition. Our hope is that embodied computational models will help to clarify the mechanisms underlying the emergence of cognitive skills and elucidate the role of intrinsic and environmental factors in this development.

In this chapter, we present a platform for creating embodied computational models of the emergence of cognitive skills using computer-generated *virtual environments*. These virtual environments allow the semi-realistic rendering of arbitrary visual surroundings that make it easy to relate model simulations to experimental data gathered in various settings. Our platform facilitates structuring of the graphical environment and of any social agents in the model. Typically, a single developing infant and a single caregiver are modeled, but arbitrary physical and social settings are easily accommodated. To illustrate the features of our platform, we show how it can be used to build an embodied model of the emergence of gaze following in infant-caregiver interactions. This effort is a component of a larger research project studying the emergence of shared attention skills within the MESA (Modeling the Emergence of Shared Attention) project at the University of California San Diego[2].

The remainder of the chapter is organized as follows. Section 2 describes our modeling platform and the underlying software infrastructure. Section 3 shows how it is currently being used to build an embodied model of the emergence of gaze following in mother-infant interactions. Finally, we discuss our work and the relative benefits of virtual vs. robotic modeling approaches in Section 4.

2   The Platform

2.1   Platform Overview

The platform allows the construction of semi-realistic models of arbitrary visual environments. A virtual room with furniture and objects can be set up easily to model, say, a testing room used in a controlled developmental psychology experiment, or a typical living room. These visual environments are populated with virtual characters. The behavior and learning mechanisms of all characters can be specified. Typically, a virtual character will have a vision system that receives images from a virtual camera placed inside the character's head. The simulated vision system will process these images and the resulting representation will drive the character's behavior [15]. Figure 1 shows an example setting.

An overview of the software structure is given in Figure 2. The central core of software, the "Simulation Environment," is responsible for simulating the learning agent (infant model) and its social and physical environment (caregiver model, objects, ...). The Simulation Environment was programmed in C++ and will be described in more detail below. It interfaces with a number of 3rd party libraries for animating human characters (BDI DI-Guy), managing and

---

[2] http://mesa.ucsd.edu

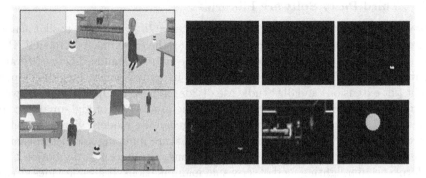

Fig. 1. Left: various views of a virtual living room used to model the emergence of gaze following. From top left, clockwise: caregiver's view, lateral view, birds eye view, and infant's view. Right: Saliency maps generated by analyzing the infant's visual input (lower left image in left half of figure). Top row, left to right: red, green, blue. Bottom row, left to right: yellow, contrast, face position

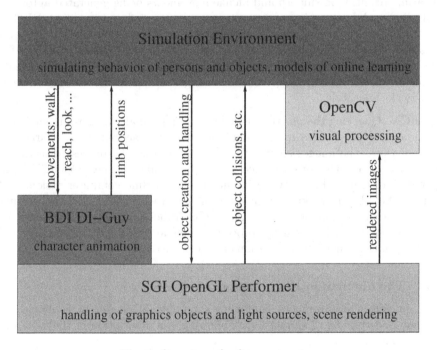

Fig. 2. Overview of software structure

rendering of the graphics (SGI OpenGL Performer), and visual processing of rendered images to simulate the agents' vision systems (OpenCV).

The platform currently runs on a Dell Dimension 4600 desktop computer with a Pentium 4 processor running at 2.8GHz. The operating system is Linux. An NVidia GeForce video graphics accelerator speeds up the graphical simulations.

2.2   Third Party Software Libraries

OpenGL Performer. The Silicon Graphics *OpenGL Performer*[3] toolkit is used to create the graphical environment for running the experiments. OpenGL Performer is a programming interface built atop the industry standard *OpenGL* graphics library . It can import textured 3D objects in many formats, including Open-Flight (.flt extension) and 3D Studio Max (.3ds extension). OpenGL is a software interface for graphics hardware that allows the production of high-quality color images of 3D objects. It can be used to build geometric models, view them interactively in 3D, and perform operations like texture mapping and depth cueing. It can be used to manipulate lighting conditions, introduce fog, do motion blur, perform specular lighting, and other visual manipulations. It also provides virtual cameras that can be positioned at any location to view the simulated world.

DI-Guy. On top of OpenGL Performer, Boston Dynamics's *DI-Guy* libraries[4] provide lifelike human characters that can be created and readily inserted into the virtual world. They can be controlled using simple high-level commands such as "look at position (X, Y, Z)," or "reach for position (X, Y, Z) using the left arm," resulting in smooth and lifelike movements being generated automatically. The facial expression of characters can be queried and modified. DI-Guy provides access to the character's coordinates and link positions such as arm and leg segments, shoulders, hips, head, etc. More than 800 different functions for manipulating and querying the characters are available in all. Male and female characters of different ages are available, configurable with different appearances such as clothing style.

OpenCV. Querying the position of a character's head allows us to dynamically position a virtual camera at the same location, thus accessing the character's point of view. The images coming from the camera can be processed using Intel's *OpenCV* library[5] of optimized visual processing routines. OpenCV is an open-source, extendable software intended for real-time computer vision, and is useful for object tracking, segmentation, and recognition, face and gesture recognition, motion understanding, and mobile robotics. It provides routines for image processing such as contour processing, line and ellipse fitting, convex hull calculation, and calculation of various image statistics.

2.3   The Simulation Environment

The Simulation Environment comprises a number of classes to facilitate the creation and running of simulations. Following is a description of the most important ones.

The Object Class. The OBJECT class is used to create all inanimate objects (walls, furniture, toys, etc.) in the simulation. Instances of the OBJECT class are

---

[3] http://www.sgi.com/products/software/performer/

[4] http://www.bdi.com

[5] http://www.intel.com/research/mrl/research/opencv/

created by giving the name of the file containing the description of a 3D geometrically modeled object, a name to be used as a handle, a boolean variable stating whether the object should be allowed to move, and its initial scale. The file must be of a format readable by OpenGL Performer, such as 3D Studio Max (.3ds files) or OpenFlight (.flt files). When an OBJECT is created, it is attached to the Performer environment. There are methods for changing the position of the OBJECT, for rotating it, and changing its scale. Thus, it can easily be modeled that characters in the simulation can grasp and manipulate objects, if this is desired.

The Object Manager Class. The OBJECT MANAGER class holds an array of instances of the OBJECT class. The OBJECT MANAGER has methods for adding objects (which must be previously created) to the scene, removing them, and querying their visibility from a specific location. The latter function allows to assess if, e.g., an object is within the field of view of a character, or if the character is looking directly at an object.

The Person Class. The PERSON class is used to add any characters to the simulation. These may be rather complicated models of, say, a developing infant simulating its visual perception and learning processes, or they may be rather simplistic agents that behave according to simple scripts. To create an instance of the PERSON class, a DI-Guy character type must be specified, which determines the visual appearance of the person, along with a handle to the OpenGL Performer camera assigned to the character. The BRAIN type and VISION SYSTEM type (see below) must be specified. If the character's actions will result from a script, then a filename with the script must be given. For example, such a script may specify what the character is looking at any given time. One BRAIN object and one VISION SYSTEM object are created, according to the parameters passed when creating the PERSON object. The PERSON object must be called periodically using the "update" method. This causes the link corresponding to the head of the character to be queried, and its coordinates to be passed to the virtual camera associated with the character. The image from the virtual camera in turn is passed to the character's VISION SYSTEM, if the character has any. The output of the VISION SYSTEM along with a handle to the DI-Guy character is passed to the BRAIN object, which will decide the next action to take and execute it in the DI-Guy character.

The Brain class. The BRAIN class specifies the actions to be taken by an instance of the PERSON class. The space of allowable actions is determined by the DI-Guy character type associated with the person. The simplest way of how a BRAIN object can control the actions of a PERSON is by following a script. In this case the PERSON will "play back" a pre-specified sequence of actions like a tape recorder. More interestingly, a BRAIN object can contain a simulation of the person's nervous system (at various levels of abstraction). The only constraint is that this simulation has to run in discrete time steps. For example, the BRAIN object may instantiate a reinforcement learning agent [14] whose state information is derived from a perceptual process (see below) and whose action space is the space of allowable actions for this character.

An "update" method is called every time step to do any perceptual processing, generate new actions, and possibly simulate experience dependent learning.

The actions used to control a character are fairly high-level commands such as "look to location (X,Y,Z)," "walk in direction Θ with speed v," or "reach for location (X,Y,Z) with the left arm," compared to direct specification of joint angles or torques. Thus, this simulation platform is not well suited for studying the development of such motor behaviors. Our focus is on the development of higher-level skills that use gaze shifts, reaches, etc. as building blocks. Thus, it is assumed that elementary behaviors such as looking and reaching have already developed and can be executed reliably in the age group of infants being modeled — an assumption that of course needs to be verified for the particular skills and ages under consideration. The positive aspect of this is that it allows to focus efforts on modeling the development of higher level cognitive processes without having to worry about such lower-level skills. This is in sharp contrast to robotic models of infant development, where invariably a significant portion of time is spent on implementing such lower level skills. In fact, skills like two-legged walking and running, or reaching and grasping are still full-blown research topics in their own right in the area of humanoid robotics.

The Vision System class. The VISION SYSTEM class specifies the processing to be done on the raw image corresponding to the person's point of view (as extracted from a virtual camera dynamically positioned inside the person's head). It is used to construct a representation of the visual scene that a BRAIN object can use to generate behavior. Thus, it will typically contain various computer vision algorithms and/or some more specific models of visual processing in human infants, depending on the primary goal of the model.

If desirable, the VISION SYSTEM class may also use so-called "oracle vision" to speed up the simulation. Since the simulation environment provides perfect knowledge about the state of all objects and characters in the simulation, it is sometimes neither necessary nor desirable to infer such knowledge from the rendered images through computer vision techniques, which can be difficult and time consuming. Instead, some property, say the identity of an object in the field of view, can simply be looked up in the internal representations maintained by the simulation environment — it functions as an oracle. This simplification is desirable if the visual processing (in this case object recognition) is not central to the developmental process under consideration, and if it can be assumed that it is sufficiently well developed prior to the developmental process being studied primarily. In contrast, in a robotic model of infant development, there is no "oracle" available, which means that all perceptual processes required for the cognitive skill under consideration have to be modeled explicitly. This is time-consuming and difficult.

Main Program and Control Flow. The main program is written in C++ using object-oriented programming. OpenGL Performer is first initialized, and a scene with a light source is created and positioned. A window to display the 3D world is initialized, and positioned on the screen. Virtual cameras are created and positioned in the world, for example as a birds eye view or a lateral view. Cameras corresponding to the characters are created but positioned dynamically as the characters move their heads. Each camera's field of view can be set (characters would usually have around a 90° field of view), and can be configured to eliminate objects that are too close or too far. All cameras created are linked to the window that displays the 3D world. Environment settings such as fog, clouds, etc. can be specified. The DI-Guy platform is then initialized, and a scenario is created. The scenario holds information about all the characters, and must be used to create new characters. New instances of the PERSON class are created, and their activities are specified by periodically giving them new actions to perform. The level of graphical detail of the characters can be specified to either get fairly realistically looking characters or to speed up processing.

Statistics gathering. Throughout the session, statistics are gathered by querying the different libraries: DI-Guy calls can be used to extract the position of the different characters or the configuration of their joints. The OBJECT MANAGER can be used to query the position of objects and their visibility from the point of view of the different characters. In addition, the internal states of all characters' simulated nervous systems are perfectly known. This data or arbitrary subsets of it can easily be recorded on a frame by frame basis for later analysis. These statistics are useful for analyzing long-term runs, and allow to evaluate whether the desired behavior is being achieved and at what rate. We point out that every simulation is perfectly reproducible and can be re-run if additional statistics need to be collected.

3   A First Example: Gaze Following

The motivation for constructing the platform was to facilitate the development of embodied models of cognitive and social development. To illustrate how the platform can be used through a concrete example, we will outline how we are currently developing an embodied model of the emergence of gaze following [5]. Gaze following is the capacity to redirect visual attention to a target when it is the object of someone else's attention. Gaze following does not occur at birth, but instead develops during a child's first 18 months of life.

The model we are developing is aimed at testing and refining the *basic set hypothesis* [8], which states that the following conditions are sufficient for gaze following to develop in infants: a) a reward-driven general purpose learning mechanism, b) a structured environment where the caregiver often looks at objects or events that the infant will find rewarding to look at, c) innate or early defined preferences that result in the infant finding the caregiver's face pleasant to look at, and d) a habituation mechanism that causes visual reward to decay over time while looking at an object and to be restored when attention is directed to a different object. Recently, Carlson and Triesch [4] demonstrated with a very abstract and simplified computational model, how the basic set may lead to the emergence of gaze following and how plausible alterations of model parameters lead to deficits in gaze following reminiscent of developmental disorders such as autism or Williams syndrome.

In our current work, we want to investigate if the basic set hypothesis still holds for a more realistic situation, where learning takes place in a complex naturalistic environment. The platform is configured for an experimental setup consisting of a living room with furniture and a toy, all of them instantiations of the OBJECT class and built from 3D Studio Max objects. Two instantiations of the PERSON class are created, one for the caregiver and one for the baby. The caregiver and learning infant are placed facing each other. The caregiver instantiates a BRAIN object controlling its behavior. A single toy periodically changes location within a meter of the infant, and its position is fed to the caregiver's BRAIN. In a first version of the model, the caregiver's BRAIN will simply cause the character to look at the position of the interesting toy with fairly high probability (75%). No visual system is given to the caregiver.

The baby instantiates a VISUAL SYSTEM object that models a simple infant vision system. In particular, it evaluates the *saliency* of different portions of the visual field [9], it recognizes the caregiver's head, and it discriminates different head poses of the caregiver. Saliency computation is based on six different features, each habituating individually according to Stanley's model of habituation [13]. The feature maps (see Figure 1) are: red, green, blue and yellow color features based on a color opponency scheme [12], a contrast feature that acts as an edge detector by giving a high saliency to locations in the image where the intensity gradient is high, and finally a face detector feature that assigns a high saliency to the region of the caregiver's face, which is localized through orace vision. The saliency of the face can be varied depending on the pose of the caregiver's face with respect to the infant (infant sees frontal view vs. profile view of the caregiver). A similar scheme for visual saliency computation has been used by Breazeal [2] for a non-developing model of gaze following, using skin tone, color, and motion features.

The infant's BRAIN object consists of a two-agent reinforcement learning system similar to that used in [4]. The first agent learns to decide when to simply look at the point of highest saliency (reflexive gaze shift) or whether to execute a planned gaze shift. The second agent learns to generate planned gaze shifts based on the caregiver's head pose. The infant should learn to direct gaze to the caregiver to maximize visual reward, and habituation will cause him/her to look elsewhere before looking back to the caregiver. With time, the infant learns to follow the caregiver's line of regard, which increases the infant's chance of seeing the interesting toy. However, the caregiver's gaze does not directly index the position of the object, but instead only specifies a direction with respect to the caregiver but not the distance from the caregiver. One goal of the current model is to better understand such spatial ambiguities and how infants learn to overcome them [11].

3.1 Platform Performance

To illustrate the performance of the platform given our current hardware, we made a number of measurements to establish the computational bottlenecks for this specific model. The time spent for each frame was divided into three separate measures for analysis: the time to calculate the feature maps (Vision), the time to display them (Map Display), and the time for the DI-Guy environment to calculate the next character positions and display them (Animation). Table 1 shows how the times vary with the resolution of the infant's vision system. As can be seen, most time is spent on simulating the infant's visual processing. Real time performance is achievable if the image resolution is not set too high.

Table 1. Simulation times (sec.)

Image Scale	Vision	Map Display	Animation
80×60	0.0226	0.0073	0.0476
160×120	0.0539	0.0092	0.0431
240×180	0.0980	0.0121	0.0522
320×240	0.1507	0.0113	0.0422
400×300	0.2257	0.0208	0.0507
480×360	0.3025	0.0276	0.0539

4 Discussion

The platform presented here is particularly useful for modeling the development of *embodied* cognitive skills. In the case of the emergence of gaze following discussed above, it is suitable because the skill is about the inference of mental states from bodily configurations, such as head and eye position, which are realistically simulated in our platform.

4.1 Virtual vs. Robotic Models

Recently, there has been a surge of interest in building robotic models of cognitive development. Compared to the virtual modeling platform presented here, there are a number of important advantages and serious disadvantages of robotic models that we will discuss in the following. A summary of this discussion is given in Table 2.

Physics. The virtual simulation is only an approximation of real-world physics. The movements of the characters do not necessarily obey physical laws but are merely animated to "look realistic." For the inanimate objects, we currently do not simulate any physics at all. In a robotic model, the physics are real, of course. The justification of neglecting physics in the virtual model is that the cognitive skills we are most interested in are fairly high-level skills, i.e., we simply do not want to study behavior at the level of muscle activations, joint torques, and

Table 2. Robotic vs. virtual models of infant cognitive development

Property	Robotic Model	Virtual Model
physics	real	simplified or ignored
agent body	difficult to create	much easier to simulate
motor control	full motor control problem	substantially simplified
visual environment	realistic	simplified computer graphics
visual processing	full vision problem	can be simplified through oracle vision
social environment	real humans	real humans or simulated agents
real time requirements	yes	no, simulation can be slowed down or sped up
data collection	difficult	perfect knowledge of system state
reproducibility of experiments	difficult	perfect
ease-of-use	very difficult	easy
development costs	extremely high	very modest

frictional forces, but at the level of primitive actions such as gaze shifts, reaches, etc., and their coordination into useful behaviors.

Agent body. In the virtual modeling platform, we can choose from a set of existing bodies for the agents. These bodies have a high number of degrees of freedom, comparable to that of the most advanced humanoid robots. Further, since physics is not an issue, we are not restricted by current limitations in robotic actuator technology. Our characters will readily run, crawl, and do many other things.

Motor control. Our interface to the agents in the model allows us to specify high-level commands (walk here, reach for that point, look at this object). The underlying motor control problems do not have to be addressed. In contrast, for a robotic model the full motor control problem needs to be solved, which represents a major challenge. Clearly, the platform should not be used to study the specifics of human motor control but it makes it much easier to focus on higher level skills. At the same time, perfect control over individual joint angles is possible, if desired.

Visual environment. The simulated computer graphics environment is of course vastly simpler than images taken by a robot in a real environment. For example, shadows and reflections are not rendered accurately, and the virtual characters are only coarse approximations of human appearance. Clearly, again, such a modeling platform should not be used to, say, study the specifics of human object recognition under lighting changes. The skills we are most interested in, however, use object recognition as a basic building block (e.g., the ability to distinguish different head poses of the caregiver with a certain accuracy). We believe that the details of the underlying mechanism are not crucial as long as the level of competence is accurately captured by the model.

Visual processing. In the virtual modeling platform we can vastly simplify perceptual processes through the use of oracle vision. In a robotic model, this is not possible and the perceptual capabilities required for some higher level cognitive skills may simply not have been achieved by contemporary computer vision methods.

Social environment. A robotic model can interact with a real social environment, i.e., one composed of real human beings. In our virtual modeling platform we could achieve this to some extent by using standard Virtual Reality interfaces such as head mounted displays in conjunction with motion tracking devices. In such a setup a real person would control a virtual person in the simulation, seeing what the virtual person is seeing through the head mounted display. However, the ability to experiment with vastly simplified agents as the social environment allows us to systematically study what aspects of the social environment, i.e., which behaviors of caregivers, are really crucial for the development of specific social skills [16]. This degree of control over the social environment cannot be achieved with human subjects. Also, the social agents may be programmed to exhibit behavior that replicates important statistics of caregiver behavior observed in real infant caregiver interactions. For example, Deák et al. are collecting such statistics from videos of infant-caregiver dyad interactions [6]. We are planning on developing caregiver models that closely replicate the observed behaviors.

Real time requirements. A robotic model must be able to operate in real time. This severely limits the complexity of the model. Perceptual processes in particular are notoriously time consuming to simulate. In the virtual model, we are not restricted to simulating in real time. Simulations may be slowed down or sped up arbitrarily. In addition, the availability of oracle vision allows to save precious computational resources.

Data collection. In the virtual model it is trivial to record data about every smallest detail of the model at any time. This is much harder to achieve in a robotic model interacting with real human caregivers. In particular, the exact behavior of the caregiver is inherently difficult to capture. Useful information about the caregiver behavior can be recovered by manually coding video records of the experiment, but this information is not available at the time of the experiment.

Reproducibility of experiments. Along similar lines, the virtual modeling platform allows perfect reproducibility of experiments. Every last pixel of the visual input to the learning agent can be recreated with fidelity. This is simply impossible in a robotic model.

Ease-of-use. Not having to deal with robotic hardware shortens development times, reduces maintenance efforts to a minimum, and makes it much easier to exchange model components with other researchers. Also, recreating the specific setup of a real-world behavioral experiment, only requires changing a configuration file specifying where walls and objects are, rather than prompting a renovation.

Development costs. Finally, robotic models are much more expensive. Most of the software components used in our platform (Linux OS, SGI OpenGL Performer, Intel OpenCV) are freely available to researchers. The lion share of the costs is the price of the BDI DI-Guy software.

All these benefits may make a virtual model the methodology of choice. Even if a robotic model is ultimately desirable, a virtual model may be used for rapid proto-typing. We see the use of virtual and robotic models as complementary. In fact, we are pursuing both methodologies at the same time in our lab [10].

4.2   Possible Extensions

There are several extensions to our platform that may be worth pursuing. First, we have only considered monocular vision. It is easy to incorporate binocular vision by simply placing two virtual cameras side by side inside a character's head. Foveation could also be added to the characters' vision systems. Second, in order to model language acquisition, a simulation of vocal systems and auditory systems of the characters could be added. Even in the context of non-verbal communication, a caregiver turning his head to identify the source of a noise may be a powerful training stimulus for the developing infant. Third, the platform is not restricted to modeling human development, but could be extended to model, say, the development of cognitive skills in a variety of non-human primates. To this end the appropriate graphical characters and their atomic behaviors would have to be designed. Fourth, on the technical side, it may be worth investigating in how far the simulation could be parallelized to run on a cluster of computers.

5   Conclusion

In conclusion, we have proposed a research platform for creating embodied virtual models of cognitive development. We have outlined how the platform may be used to model the emergence of gaze following in naturalistic infant-caregiver interactions. The virtual modeling platform has a number of important advantages compared to robotic modeling approaches. The relative benefits of virtual models over robotic models on the one hand or more abstract computational models on the other hand need to be evaluated on a case-by-case basis.

Acknowledgments

This work is part of the MESA project at UCSD. We acknowledge funding from the UC Davis MIND Institute[6] and the National Alliance for Autism Research[7].

---

[6] http://www.ucdmc.ucdavis.edu/mindinstitute/
[7] http://www.naar.org

Tone Milazzo developed the first version of the platform. Tone Milazzo and Jochen Triesch developed the initial object-oriented programming scheme. The following people contributed with valuable input: Gedeon Deák, Javier Movellan, Hyundo Kim, Boris Lau, and Christof Teuscher.

References

1. M. Asada, K. F. MacDorman, H. Ishiguro, and Y. Kuniyoshi. Cognitive developmental robotics as a new paradigm for the design of humanoid robots. *Robotics and Autonomous Systems*, 37:185–193, 2001.
2. C. Breazeal. *Designing Sociable Robots*. MIT Press, Cambridge, MA, USA, 2002.
3. R. A. Brooks, C. Breazeal, R. Irie, C. C. Kemp, M. Marjanovic, B. Scassellatti, and M. M. Williamson. Alternative essences of intelligence. In *Proc. of the American Association of Artificial Intelligence*, pages 961–968, 1998.
4. E. Carlson and J. Triesch. A computational model of the emergence of gaze following. In H. Bowman and C. Labiouse, editors, *Connectionist Models of Cognition and Perception II: Proceedings of the Eighth Neural Computation and Psychology Workshop, University of Kent, UK, 28–30 August 2003*, volume 15, pages 105–114. World Scientific, 2003.
5. G. O. Deák, R. Flom, and A. D. Pick. Perceptual and motivational factors affecting joint visual attention in 12- and 18-month-olds. *Developmental Psychology*, 36:511–523, 2000.
6. G. O. Deák, Y. Wakabayashi, and H. Jasso. Attention sharing in human infants from 5 to 10 months of age in naturalistic conditions. In *Proc. of the 3rd International Conference on Development and Learning (ICDL'04)*, La Jolla, California, USA, 2004.
7. J. L. Elman, E. A. Bates, M. H. Johnson, A. Karmiloff-Smith, D. Parisi, and K. Plunkett. *Rethinking Innateness*. MIT Press, Cambridge, MA, USA, 1996.
8. I. Fasel, G. O. Deák, J. Triesch, and J. Movellan. Combining embodied models and empirical research for understanding the development of shared attention. In *Proc. of the 2nd International Conference on Development and Learning (ICDL'02)*, pages 21–27, Los Alamitos, California, USA, 2002.
9. L. Itti and C. Koch. A saliency-based search mechanism for overt and covert shifts of visual attention. *Vision Research*, 40(10-12):1489–1506, 2000.
10. H. Kim, G. York, G. Burton, E. Murphy-Chutorian, and J. Triesch. Design of an antropomorphic robot head for studying autonomous mental development and learning. In *Proc. of the IEEE 2004 International Conference on Robotics and Automation (ICRA 2004)*, New Orleans, LA, USA, 2004.
11. B. Lau and J. Triesch. Learning gaze following in space: a computational model. Proceedings of the Third International Conference on Development and Learning (ICDL'04), La Jolla, California, October 20–22, 2004.
12. T. W. Lee, T. Wachtler, and T. J. Sejnowski. Color opponency is an efficient representation of spectral properties in natural scenes. *Vision Research*, 42(17):2095–2103, 2002.
13. J. C. Stanley. A computer simulation of a model of habituation. *Nature*, 261:146–148, 1976.
14. R. S. Sutton and A. G. Barto. *Reinforcement Learning: an Introduction*. MIT Press, Cambridge, MA, USA, 1998.

15. D. Terzopoulos, X. Tu, and R. Grezeszczuk. Artificial fishes: Autonomous locomotion, perception, behavior, and learning in a simulated physical world. *Artificial Life*, 1(4):327–351, 1994.
16. C. Teuscher and J. Triesch. To care or not to care: Analyzing the caregiver in a computational gaze following framework. Proceedings of the Third International Conference on Development and Learning (ICDL'04), La Jolla, California, October 20–22, 2004.
17. E. Thelen and L. Smith. *A Dynamic Systems Approach To the Development of Cognition and Action*. MIT Press, Cambridge, MA, USA, 1994.
18. J. Weng, J. McClelland, A. Pentland, O. Sporns, I. Stockman, M. Sur, and E. Thelen. Autonomous mental development by robots and animals. *Science*, 291:599–600, 2001.

Learning to Interpret Pointing Gestures: Experiments with Four-Legged Autonomous Robots

Verena V. Hafner and Frédéric Kaplan

Sony CSL Paris, 6 rue Amyot, 75005 Paris, France
{hafner, kaplan}@csl.sony.fr

Abstract. In order to bootstrap shared communication systems, robots must have a non-verbal way to influence the attention of one another. This chapter presents an experiment in which a robot learns to interpret pointing gestures of another robot. We show that simple feature-based neural learning techniques permit reliably to discriminate between left and right pointing gestures. This is a first step towards more complex attention coordination behaviour. We discuss the results of this experiment in relation to possible developmental scenarios about how children learn to interpret pointing gestures.

1   Introduction

Experiments with robots have successfully demonstrated that shared communication systems could be negotiated between autonomous embodied agents [1, 2, 3, 4, 5, 6]. In these experiments, robots draw attention through verbal means to an object of their environment. In order to bootstrap these conventional communication systems, it is crucial that the robots have a non-verbal way to influence the attention of other robots. They can for instance point to the topic of the interaction. This non-verbal form of communication is necessary as the robots have no direct access to the "meanings" used by the other robots. They must guess it using non-linguistic cues. The interpretation of pointing gestures must therefore be sufficiently reliable, at least initially when the system is bootstrapping. Once the language is in place, such kind of external feedback is less crucial and can even be absent [7].

Research in gaze or pointing interpretation is active in the context of human robot interaction (e.g. [8, 9, 10, 11, 12, 13]). By contrast, only few works explore the same issues for interaction between autonomous robots. A small number of solutions have been proposed to enable pointing and pointing interpretation in a variety of contexts (e.g. [14]). The focus of the present chapter concerns how robots can *learn* to interpret pointing gestures.

This chapter presents a model in which pointing gesture recognition is learned using a reward-based system. This model assumes, for instance, that a robot will often see something interesting from its point of view when looking in the direction where another robot is pointing to. It can be a particular salient feature of the environment, or an object which serves a current need (e.g. the charging station), or an opportunity for learning [15]. This approach is in line with Carlson and Triesch's computational model of the emergence of gaze following based on reinforcement learning [16]. Their model

S. Wermter et al. (Eds.): Biomimetic Neural Learning, LNAI 3575, pp. 225–234, 2005.
© Springer-Verlag Berlin Heidelberg 2005

has been tested in a virtual environment by Jasso et al. [17]. To the best of our knowledge, this chapter represents the first attempt to show that a robot can learn to interpret the pointing gestures of another robot.

The rest of the paper describes the robotic experiment we conducted. We then discuss the limitation and possible extensions of this preliminary investigation.

2   Robot Experiments

2.1   The Interaction Scenario

Here we describe and show robot experiments where a pointing gesture is learned to be classified as either left or right. For these experiments, two Sony AIBOs were sitting on the floor, facing each other (see figure 1). One of the robots (the adult) is randomly pointing towards an object on the left or right side of its body using its left or right front leg, respectively. The other robot (the child) is watching it. From looking at the pointing gesture of the other robot, the learning robot guesses the direction and starts looking for an object on this side. Finding the object on this side represents a reward.

Fig. 1. An example of pointing shown with two robots. The robot on the left represents the adult who is pointing, the robot on the right represents the child who is learning to interpret the pointing gesture

Since the focus of this experiment is learning of pointing recognition and not pointing, this skill is hardwired in the adult robot. The robot is visually tracking a coloured object on its left or right side, thereby facing the object. Pointing is achieved by simply copying the joint angle of the head to the joint angle of the arm. Note that the pointing robot takes on an exact pointing position and does not only distinguish between the left and the right side.

2.2   Image Processing and Feature Space

A sample camera image from the robot's point of view can be seen in figure 2 left. For the experiments, the robot took 2300 pictures focusing on its pointing partner, 1150 for each pointing direction. The situations in which the pictures have been taken varied in the distance between the two robots, the viewing angle, the lighting conditions and the backgrounds (three different backgrounds).

From the original camera image, a small number of features has to be selected to facilitate the learning of interpreting the pointing gesture. We decided to apply two main filters to the image. One filter extracts the brightness of the image, the other filter extracts horizontal and vertical edges. These choices are biologically motivated. Eyes are very sensitive to brightness levels, and edges are the independent components of natural scenes [18]. The original image I is thus transformed to I' using a filter f:

$$I \xrightarrow{f} I'$$

For both filters, the colour image is transformed into greyscale first with pixel values between 0 and 255. In the subsequent steps, the image is divided into its left part and its right part (see figure 3). This is justified by the robot always centering on the other robot's face using an independent robot tracking mechanism, thus dividing the image into the right half of the other robot and its left half.

$$I' \longrightarrow I'_L, I'_R$$

The brightness filter B_θ applies a threshold θ to the image, which sets all pixels with a value greater than θ to 255, and all others to 0. For the experiments, values of $\theta = 120$ and $\theta = 200$ have been used. For the edge filter, we chose two Sobel filters S_H and S_V (see [19]) which extracts the horizontal and the vertical edges, respectively. An example of an image transformed by the filters can be seen in figure 2.

To the filtered images I', different operators op can be applied to extract low-dimensional features. These operators are the centre of mass $\mu = (\mu_x, \mu_y)$ and the sum Σ.

$$I' \xrightarrow{op} q$$

where q is the resulting scalar feature.

The four filters B_{120}, B_{200}, S_H and S_V together with the three operators μ_x, μ_y and Σ applied to both the left and the right side of the image I result in $4 \cdot 3 \cdot 2 = 24$

Fig. 2. Left: A robot pointing to its left side as seen from another robot's camera. The child robot tracks the adult robot in order to keep it in the centre of its visual field. Centre: Feature extraction for brightness using a threshold θ. Right: Feature extraction for horizontal edges using a Sobel edge detector

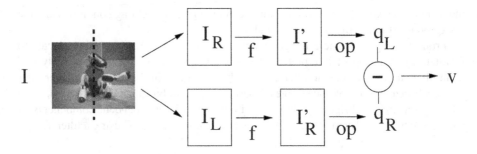

Fig. 3. Feature extraction from the original camera image

different features q_L and q_R (see figure 3). We take the differences between the left and right features resulting in 12 new features $v = q_L - q_R$.

2.3   Feature Selection

We selected a subset of the features by applying pruning methods. This is done by evaluating a subset of attributes by considering the individual predictive ability of each feature along with the degree of redundancy between them. Subsets of features that are highly correlated with the class while having low intercorrelation are preferred. The method used was greedy hillclimbing augmented with a backtracking facility provided by WEKA [20]. From the 12 features available to the robot, 3 have been selected to be the most meaningful: $B_{200} \circ \mu_y$, $S_H \circ \Sigma$ and $S_V \circ \Sigma$. Their values for all images are depicted in figure 4. Intuitively, the robot lifting its arm results in a vertical shift of brightness on this side of the image, an increase of horizontal edges and a decrease of vertical edges on this side.

For comparison, we also calculated the three least successful features. They turned out to be $B_{200} \circ \mu_x$, $B_{120} \circ \mu_x$ and $S_V \circ \mu_y$.

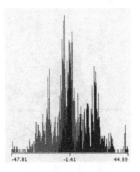 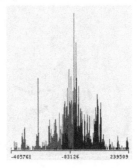 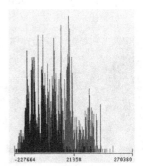

Fig. 4. Most successful scalar features for pointing gesture recognition from an image and the frequency of their values in the image data set. The red values are taken from pointing towards the left, the blue ones from pointing towards the right. Left: $B_{200} \circ \mu_y$. Centre: $S_H \circ \Sigma$. Right: $S_V \circ \Sigma$

3   Results

For learning the pointing gesture recognition, we used a multi-layer-perceptron (MLP) with the selected features as input, 3 neurons in the hidden layer, and the pointing direction (left or right) coded with two neurons as output. The learning algorithm is backpropagation with a learning rate $\lambda = 0.3$ and momentum $m = 0.2$. The evaluation is based on a 10-fold cross validation.

We chose backpropagation as a supervised learning algorithm which is comparable to a reward-based system in case of a binary decision. The choice of using MLPs and backpropagation is arbitrary and can be replaced by any other suitable machine learning technique involving reward. It is however sufficient to show that pointing gesture recognition can be easily learned between two robots.

Table 1. Learning results of different input features using 10-fold cross validation on the dataset of 2300 images

features	MLP	success rate
best 3	3-3-2	95.96%
worst 3	3-3-2	50.74%
all 12	12-7-2	98.83%

The success rate for the three chosen features (figure 4) is 95.96% (see table 1) using a 3-3-2 MLP and one epoch of training. When using all the 12 difference values v as inputs to a 12-7-2 MLP, the success rate increases to 98.83%. The success rate for the worst three features and one epoch of training is 50.74%, just slightly above chance.

In figure 5, the progress of learning can be monitored. The upper graph shows the error curve when the images of the pointing robot are presented in their natural order, alternating between left and right. The lower graph shows the error curve for images presented in a random order from a pre-recorded sequence. The error decreases more rapidly in the ordered sequence, but varies when conditions are changed.

4   Discussion

4.1   Pointing Interpretation and Intentional Understanding

We showed that with the current setup, a robot can learn to interpret another robot's pointing gesture. Although the pointing gesture of the adult robot can vary continuously depending on the position of the object, the interpretation of the pointing direction is either left or right. This corresponds to primary forms of attention detection as they can be observed in child development. Mutual gaze between an adult and a child, a special case of attentional behaviour, occurs first around the age of three months. At the age of about six months, infants are able to discriminate between a left or right position of the head and gaze of their parents, but the angle error can be as large as 60 degrees [21]. At the age of about nine months, the gaze angle can be detected correctly. Pointing gestures only start to be interpreted at the age of around one year [21] (see table 2). Children start

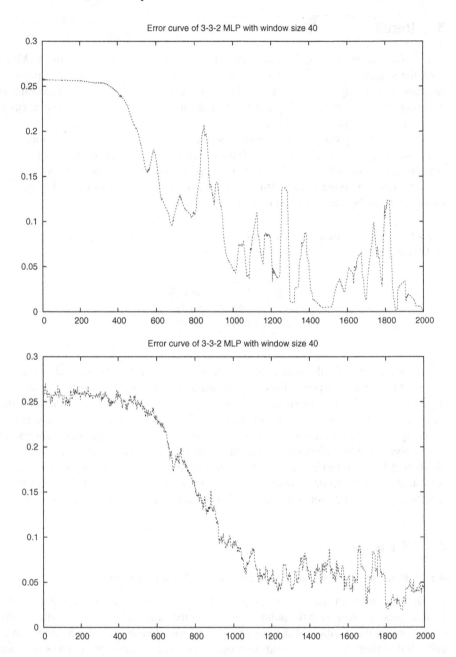

Fig. 5. Error of MLP during learning. Top: sequence of images in natural order. Bottom: random order of training images

to point first at the age of 9 months [22]. It is usually seen as a request for an object which is outside the reach of the child, and even occurs when no other person is in the room. This is called imperative pointing. At the age of 12 months, pointing behaviour

Table 2. Developmental timelines of attention detection and pointing in humans

Age from:	Attention detection	Attention manipulation
0-3 m	**Mutual gaze** - Eye contact detection	
6 m	Discrimination between **left and right position** of head and gaze	
9 m	**Gaze angle detection** - fixation on the first salient object encountered	**Imperative Pointing**: Drawing attention as a request for reaching an object (attention not monitored)
12 m	**Gaze angle detection** - fixation on any salient object encountered - Accuracy increased in the presence of a pointing gesture	**Declarative Pointing**: Drawing attention using gestures
18 m	**Gaze following** toward object outside the field of view - Full object permanence	

becomes declarative and is also used to draw attention to something interesting in the environment [23].

It is feasible to include the detection of a continuous angle of the pointing in a robotic setup. This would involve changing the current binary decision to a continuous value (possibly coded with population coding). But a higher accuracy is probably not sufficient to achieve efficient pointing interpretation. To truly learn the exact meaning of a pointing gesture, deeper issues are involved. Pointing interpretation in child development starts at an age where the infant begins to construct an intentional understanding of the behaviour of adults. This means that their actions are parsed as means towards particular goals. It could therefore be argued that pointing interpretation is much more than a geometrical analysis [23]. It involves a shared intentional relation to the world [24]. Developing some form of intentional understanding in a robot is one of the most challenging unsolved problems for developmental robotics [25].

4.2   Co-development of Pointing Gestures and Pointing Interpretation

In our robotic setup, the meaning of one gesture meaning 'left' and another gesture meaning 'right' could easily be reversed, or even completely different gestures could be used. The pointing movement of the adult robot was arbitrarily chosen to resemble a human pointing gesture. It is not clear that this gesture is the most adapted for unambiguous interpretations given the perceptual apparatus of the robots. In this perspective, it would be interesting to investigate a co-development between a pointing robot and a robot trying to understand pointing gestures. Situations of co-development between pointing and pointing gesture recognition could lead to interesting collective dynamics. Given the embodiment of the robots and the environmental conditions, some particular gestures may get selected for efficiency and learnability. Features that make them unambiguous and easy to transmit will be kept, whereas inefficient traits should be discarded. It has been argued that similar dynamics play a pivotal role for shaping linguistic systems [26].

4.3   Pointing and the Mirror Neuron System

Taking inspiration from current research in artificial mirror neuron systems [27], it would be possible to design a robot that interprets pointing gestures of others in relation with its own pointing ability. However, it is not clear whether the ability of pointing and pointing detection are correlated in human child development. Desrochers, Morisette and Ricard [28] observed that pointing seems to occur independently of pointing gesture recognition during infant development. These findings also seem to suggest that pointing does not simply arise from imitative behaviour.

4.4   Adult Robot Behaviour and Scaffolding

In the current setup, the adult robot randomly points at objects, its behaviour does not depend on the behaviour or the reaction of the child robot. Interactions between humans are very different. When pointing at something to show it to the child, a human adult carefully observes the attentional focus of the child and adjusts its behaviour to it. In some cases, the adult might even point to an object the child is already paying attention to in order to strengthen the relationship [29].

4.5   Pointing and Cross-Correlation

Nagai et al. [12] have argued in the context of human-robot interaction that simply the correlation between the presence of objects in general and gaze is sufficient for learning how to interpret gaze (without the necessity of an explicit feedback). Similar techniques based on cross-correlation could also be tried in the context of pointing interpretation between two robots. This type of learning relies on the assumption that the correlation is sufficiently strong to be discovered in practice. It is possible that a combination of both cross-correlation and reward based processes results in an efficient strategy for learning of pointing interpretation.

5   Conclusions

The interpretation of pointing is only one of the prerequisites necessary for bootstrapping human-like communication between autonomous robots. This chapter presents a first experiment showing how a robot can learn to interpret pointing gestures of another robot. In our future work, we will address the limitations of this initial prototype that have been discussed, and investigate the dynamics of social coordination and attention manipulation not yet investigated in this work.

Acknowledgements

This research has been partially supported by the ECAGENTS project founded by the Future and Emerging Technologies programme (IST-FET) of the European Community under EU R&D contract 001940. Andrew Whyte designed motor primitives for the robots used in this experiment. The authors would like to thank Anne Henning, Tricia Striano and Luc Steels for precious comments on the ideas discussed in this chapter.

References

1. Steels, L., Kaplan, F.: Situated grounded word semantics. In Dean, T., ed.: Proceedings of the Sixteenth International Joint Conference on Artificial Intelligence IJCAI'99, San Francisco, CA., Morgan Kaufmann Publishers (1999) 862–867
2. Steels, L., Kaplan, F.: Collective learning and semiotic dynamics. In Floreano, D., Nicoud, J.D., Mondada, F., eds.: Advances in Artificial Life (ECAL 99). Lecture Notes in Artificial Intelligence 1674, Berlin, Springer-Verlag (1999) 679–688
3. Steels, L., Kaplan, F.: Bootstrapping grounded word semantics. In Briscoe, T., ed.: Linguistic evolution through language acquisition: formal and computational models. Cambridge University Press, Cambridge (2002) 53–73
4. Steels, L.: The Talking Heads Experiment. Volume 1. Words and Meanings, Antwerpen (1999)
5. Vogt, P.: Lexicon grounding on mobile robots. PhD thesis, Vrije Universiteit Brussel (2000)
6. Kaplan, F.: La naissance d'une langue chez les robots. Hermes Science (2001)
7. Steels, L., Kaplan, F., McIntyre, A., Van Looveren, J.: Crucial factors in the origins of word-meaning. In Wray, A., ed.: The Transition to Language. Oxford University Press, Oxford, UK (2002) 252–271
8. Scassellati, B.: Imitation and mechanisms of joint attention: A developmental structure for building social skills on a humanoid robot. In: Computation for metaphors, analogy and agents. Vol 1562 of Springer Lecture Notes in Artificial Intelligence. Springer Verlag (1999)
9. Kozima, H., Yano, H.: A robot that learns to communicate with human caregivers. In: First International Workshop on Epigenetic Robotics (Lund, Sweden). (2001)
10. Imai, M., Ono, T., Ishiguro, H.: Physical relation and expression: Joint attention for human-robot interaction. In: Proceedings of the 10th IEEE International Workshop on Robot and Human Communication. (2001)
11. Nagai, Y., Asada, M., Hosoda, K.: A developmental approach accelerates learning of joint attention. In: Proceedings of the second international conference of development and learning. (2002)
12. Nagai, Y., Hosoda, K., Morita, A., Asada, M.: A constructive model for the development of joint attention. Connection Science 15 (2003) 211–229
13. Nickel, K., Stiefelhagen, R.: Real-time recognition of 3d-pointing gestures for human-machine-interaction. In: Proceedings of the 25th Pattern Recognition Symposium - DAGM'03. (2003)
14. Baillie, J.C.: Grounding symbols in perception with two interacting autonomous robots. In Berthouze, L., Kozima, H., Prince, C., Sandini, G., Stojanov, G., Metta, G., Balkenius, C., eds.: Proceedings of the 4th International Workshop on Epigenetic Robotics: Modeling Cognitive Development in Robotic System, Lund University Cognitive Studies 117 (2004)
15. Kaplan, F., Oudeyer, P.Y.: Maximizing learning progress: an internal reward system for development. In Iida, F., Pfeifer, R., Steels, L., Kuniyoshi, Y., eds.: Embodied Artificial Intelligence. LNAI 3139. Springer-Verlag (2004) 259–270
16. Carlson, E., Triesch, J.: A computational model of the emergence of gaze following. In: Proceedings of the 8th Neural Computation Workshop (NCPW8). (2003)
17. Jasso, H., Triesch, J., Teuscher, C.: Gaze following in the virtual living room. In Palm, G., Wermter, S., eds.: Proceedings of the KI2004 Workshop on Neurobotics. (2004)
18. Bell, A.J., Sejnowski, T.J.: Edges are the independent components of natural scenes. In: Advances in Neural Information Processing Systems (NIPS). (1996) 831–837
19. Dudek, G., Jenkin, M.: Computational principles of mobile robotics. Cambridge University Press (2000)
20. Witten, I., Eibe, F.: Data mining. Morgan Kaufmann Publishers (2000)

21. Butterworth, G.: Origins of mind in perception and action. In Moore, C., Dunham, P., eds.: Joint attention: its origins and role in development. Lawrence Erlbaum Associates (1995) 29–40
22. Baron-Cohen, S.: Mindblindness: an essay on autism and theory of mind. MIT Press, Boston, MA, USA (1997)
23. Tomasello, M.: Joint attention as social cognition. In Moore, C., Dunham, P., eds.: Joint attention: its origins and role in development. Lawrence Erlbaum Associates (1995) 103–130
24. Hobson, P.: The craddle of thought. MacMillan (2002)
25. Kaplan, F., Hafner, V.: The challenges of joint attention. In Berthouze, L., Kozima, H., Prince, C., Sandini, G., Stojanov, G., Metta, G., Balkenius, C., eds.: Proceedings of the 4th International Workshop on Epigenetic Robotics: Modeling Cognitive Development in Robotic System, Lund University Cognitive Studies 117 (2004) 67–74
26. Brighton, H., Kirby, S., Smith, K.: Cultural selection for learnability: Three hypotheses underlying the view that language adapts to be learnable. In Tallerman, M., ed.: Language Origins: Perspective on Evolution. Oxford University Press, Oxford (2005)
27. Elshaw, M., Weber, C., Zochios, A., Wermter, S.A.: A mirror neuron inspired hierarchical network for action selection. In Palm, G., Wermter, S., eds.: Proceedings of the KI2004 Workshop on NeuroBotics, Ulm, Germany (2004) 98–105
28. Desrochers, S., Morisette, P., Ricard, M.: Two perspectives on pointing in infancy. In Moore, C., Dunham, P., eds.: Joint Attention: its origins and role in development. Lawrence Erlbaum Associates (1995) 85–101
29. Liszkowski, U., Carpenter, M., Henning, A., Striano, T., Tomasello, M.: Twelve-month-olds point to share attention and interest. Developmental Science 7 (2004) 297 – 307

Reinforcement Learning Using a Grid Based Function Approximator

Alexander Sung[1], Artur Merke[2], and Martin Riedmiller[3]

[1] Institute of Computer Science, University of Tübingen,
Sand 1, 72076 Tübingen, Germany
[2] Institute of Computer Science, University of Dortmund,
Otto Hahn Str. 16, 44227 Dortmund, Germany
[3] Institute of Computer Science, University of Osnabrück,
Albrechtstraße 28, 49069 Osnabrück, Germany

Abstract. Function approximators are commonly in use for reinforcement learning systems to cover the untrained situations, when dealing with large problems. By doing so however, theoretical proves on the convergence criteria are still lacking, and practical researches have both positive and negative results. In a recent work [3] with neural networks, the authors reported that the final results did not reach the quality of a Q-table in which no approximation ability was used. In this paper, we continue this research with grid based function approximators. In addition, we consider the required number of state transitions and apply ideas from the field of active learning to reduce this number. We expect the learning process of a similar problem in a real world system to be significantly shorter because state transitions, which represent an object's actual movements, require much more time than basic computational processes.

1 Introduction

On Reinforcement Learning (RL) problems [13] in which it is impracticable to investigate all states and actions of interest explicitly, function approximators such as neural networks (NNs) are commonly used [10]. In fact, the NN is a popular tool in various domains of machine learning. However, convergence criteria for RL algorithms are not guaranteed when approximators are in use [1], and in practice poor results may possibly arise [2]. A main drawback of NNs is the interference problem [15] which means that new learning samples in one area of the state space may cause unlearning in another area. The obvious development of a RL process is to learn the reinforcement values near the goal first, before continuing with states further away. Therefore, the following training input may affect the already learned states in a negative way. This leads to a cleavage in the use of such approximators, while other methods can still be useful [12]. In this paper, we use grid based function approximators to investigate another possibility to approach RL solutions more deeply.

In addition, we apply methods derived from the principles of active learning [4, 9, 11] to approach the problem. Active learning becomes more and more popu-

S. Wermter et al. (Eds.): Biomimetic Neural Learning, LNAI 3575, pp. 235–244, 2005.
© Springer-Verlag Berlin Heidelberg 2005

lar, and there are already several results on this topic. The most common aspect of active learning is active data acquisition where problem-specific algorithms are designed to optimize the retrieval of costly learning data [5, 16]. Acquisition of learning data does not require a significant amount of time in a simulation, but this aspect becomes important when dealing with real world systems where robots, for example, are used. In this case, other computations of the learning algorithm are usually performed essentially faster. Several studies demonstrate the capability of active learning in the field of supervised learning. For RL tasks however, active learning applications are still scarce and also problematic, since the possibilities of actively picked training samples from the state space are still be to investigated for the RL environment [6]. In this paper, we want to make a step towards the possibilities of active learning in the field of RL. The algorithm and its general idea are described in section 2. Then, section 3 shows experimental results of the "Mountain-Car" task. The conclusions are presented in section 4 and offer starting points for further work.

2 Q-Learning with Grid Based Function Approximators

The experiments in this paper are oriented towards the work done by Carreras, Ridao, and El-Fakdi [3], and use Q-learning [14] methods as well. Where they relied on NNs, however, we use piecewise linear grid based approximators (PLGBAs) which store and retrieve data exactly at certain given grid points, and approximates linearly in between. This section describes the approximators we used, as well as our learning algorithm which applies PLGBAs on RL problems.

2.1 Piecewise Linear Grid Based Approximators

In the following we describe piecewise linear grid based approximators (PLGBA) used in the experiments of this paper. This kind of function approximator assumes a d-dimensional grid and a triangulation into simplices given. In figure 1 we see an example of a two-dimensional grid with a Kuhn triangulation. The representable functions are linear on the simplices and continuous on the simplex boundaries. They can be described as a linear combination of generalised hat functions.

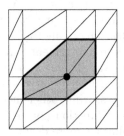

Fig. 1. Support of a hat function on a two-dimensional grid with Kuhn triangulation

In the example in figure 1 the support of a two-dimensional hat function is shaded. The black dot is the top of the hat function, where it attains the value 1. At the boundaries of the support the hat function vanishes, and on each simplex it is linear. These conditions uniquely define the hat function centred on the black dot. The set of all hat functions $\{\varphi_1, \ldots, \varphi_F\}$ corresponding to the F grid nodes is a basis of the space of representable functions, i.e. $f(x) = \sum_{i=1}^{F} \varphi_i(x) \, w_i$. The w_i are the adjustable parameters of the function approximator. They can be interpreted as the height of the different hat functions. The feature $\varphi_i(x)$, which is equivalent to the value of the i-th hat function at point x, determines the weight that is given to grid node i. The vector $\varphi(x) = (\varphi_1(x), \ldots, \varphi_F(x))^\top$ contains the barycentric coordinates of x. It satisfies $0 \le \varphi_i(x) \le 1$ and $\sum_{i=1}^{F} \varphi_i(x) = 1$.

PLGBAs are a generalization of piecewise constant grid based approximators (PCGBAs). In PCGBAs one usually uses constant indicator functions of the Voronoi cell around each grid point as basis functions. In PLGBAs this indicator function is replaced by a generalized piecewise linear and continuous hat function, which makes the whole approximator continuous, which is not the case for PCGBAs.

2.2   The Learning Algorithm

Our algorithm consists of two parts: The first part handles the acquisition of learning samples, and the second part performs the actual learning. These two parts form an *episode* of the learning process and are repeated during the learning. The following lines of code show the schema of the algorithm.

Schema of the learning algorithm

```
while true
  /* a new episode begins here */
  choose_new_starting_state;

  /* Part 1: data acquisition */
  while goal_not_reached and transition_limit_not_reached
    choose_next_greedy_action;
    perform_state_transition;
    determine_grid_point_belonging_to_last_state_transition;
    determine_database_learning_sample_belonging_to_grid_point;
    if last_state_transition is_closer_to grid_point
        than database_learning_sample
      replace_database_learning_sample_with_last_state_transition;
    end;
  end;

  /* Part 2: learning */
  for each learning_sample in database
    perform_Q_learning_update;
  end;
end;
```

Learning samples are created on *trajectories*, which refer to sequences of *state transitions* (the system's transition from state s_t to a following state s_{t+1} by performing action a_t). A trajectory ends when the goal or a given limit of state transitions is reached. After that, a new episode can be run, beginning from a new independently chosen starting state. The execution phase of a state transition on a certain trajectory is also referred as a *cycle*.

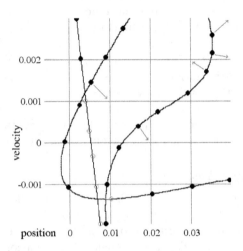

Fig. 2. An exemplary part of the database. Specific learning samples are selected from the trajectories by the algorithm for each episode, shown as black dots. The gray samples have been replaced during the learning process. For each grid point, the closest sample is preferred. The arrows indicate to which grid point the particular sample correspond

For the acquisition of training data, a state transition (s_t, a_t, s_{t+1}) is processed in each cycle of the k-th episode by evaluating the Q-values $Q_{k-1}(s_t, a_t)$ currently stored in the approximator. The current values determine the next action with a greedy policy. While a trajectory is being performed, each state transition may be inserted into a database of learning samples, replacing an already present sample if necessary. The database is organized as a grid, and each state is associated with the nearest particular grid point. Learning samples which belong to the same grid point compete against each other to remain in the database. The preference is given to those samples which have a smaller distance to their corresponding grid points. Figure 2 illustrates this method. The learning samples are stored separately for each possible action $a \in \{-1, 0, 1\}$. Since our algorithm is designed for a deterministic problem, only one learning sample is stored for each pair of state and action. For a non-deterministic problem, the database could be extended to store multiple entries per pair of state and action.

The selection and substitution of learning samples ensure an upper limit for the total number of samples in the database, although new state transition are considered in each episode. The choice is made with regards to two aspects. First, this method leads to a set of learning samples evenly distributed over the state space.

Redundant training data is kept at a minimum by this means. The second reason is explained by the nature of the PLGBA. In [8] it was shown that grid based approaches to reinforcement learning problems may lead to divergence. In contrast, the authors also showed the existence of "safe points" for state transitions at which the PLGBA can be updated without the risk of divergence. The grid points are such safe points, so we favor samples near the grid points in our algorithm.

The actual learning is done by evaluating the learning samples from the database. According to the standard Q-learning update, new target values

$$Q_k(s_t, a_t) = \alpha \left[R(s_t, a_t) + \gamma \min_{a_{t+1}} Q_{k-1}(s_{t+1}, a_{t+1}) \right] + (1 - \alpha) \, Q_{k-1}(s_t, a_t) \quad (1)$$

are computed. The learning part can be executed in parallel with the acquisition of training data, and therefore does not pose an additional delay to the costly data acquisition.

2.3 Additional Aspects of the Algorithm

Reflecting on the results, we concluded that the problem was simplified fundamentally by allowing the learning algorithm to access any state in the state space directly. Especially, in regards to the "Mountain-Car" task – which served as a benchmark in the next section – new trajectories were allowed to begin anywhere in the state space, as it was proposed in the referenced paper [3]. A state in the "Mountain-Car" thereby consists of a position p and a velocity v. With this possibility at hand however, the entire problem could be solved a lot easier. In this case, only a few learning samples of state transitions were required in an offline learning process to solve the problem. Well distributed learning samples across the state space and the approximation ability of the PLGBA are able to represent the whole state space in good quality, as the results in section 3 demonstrate.

Further experiments were therefore subjected to the restriction that new trajectories had to start with zero car velocity ($v = 0$). States with non-zero velocity had to be explored by running episodes based on repeated car movement. This restriction was added in regards to practical RL tasks. It is to be expected that direct access to any state will not be possible in general.

In a second step, the database of learning samples also serves for decisions about where new training trajectories will begin next. Each grid point (corresponding to zero starting velocity, since only these grid points are valid as new starting positions) has a counter, and for each state transition, the counter of the corresponding grid point is increased. New trajectories start near grid points whose neighbourhoods were visited least. This decision is orientated towards the key idea of active learning to spend more effort in those areas where detailed knowledge about the state space is still lacking.

3 Experimental Results

To validate our results, we used a benchmark called the "Mountain-Car" problem – a common benchmark to evaluate the quality of RL algorithms. The behaviour of the system is detailed in [7]. Please note that there exist various variants of the "Mountain-Car" task, and most of them specify the task with the car placed specifically into the middle of the valley [7, 12], while our paper wants to cover all starting states in general according to the SONQL paper [3]. Comparisons in general are therefore difficult.

For the results in this section, the quality criterion of the experiments was the averaged number of cycles required for a trajectory to reach the goal. The number of cycles was measured on an evaluation set with 1000 independently and randomly chosen starting states which remained fixed for all experiments in this section. An experiment was considered finished when the quality criterion was stable. This number of cycles on average was then seen as the final quality of the learning process. A stable quality means that the averaged number of cycles stayed fixed (with a tolerance of 0.05) for the following 10^6 Q-table updates. New to [3], the total number of state transitions performed to reach the final quality was calculated as our main performance criterion. This criterion is important in real world systems where movement of objects such as robots occur, in contrast to simulation only. Computation times for updates to the approximator, for example, are then negligible.

In the next subsection, the performance baseline presented in the SONQL paper [3] is reproduced. Standard Q-learning is used with a PCGBA to show that the evaluation set of their work is comparable to ours. Experiments with various PLGBAs in the following subsection then demonstrate the approximation capability of PLGBAs. The learning restriction of zero starting velocity is introduced in the third subsection and is also valid for the last one. The actual application of our learning algorithm and its results are presented in the final subsection.

3.1   The Original Performance Baseline

In our first experiment, we used the standard Q-learning algorithm to get a performance baseline. This baseline was also presented in [3]. The state space was discretized into 181 and 141 points for position and velocity, respectively. Since there are three actions available in each state, a corresponding PCGBA, with $181 \cdot 141 \cdot 3 = 76563$ grid points, was used as an approximator. With the parameter configuration given in the SONQL paper, the performance baseline of 50 cycles per trajectory on average could be confirmed. In our experiment however, $2.13 \cdot 10^7$ learning cycles (which correspond to 250000 episodes approximately) instead of $1 \cdot 10^7$ learning cycles were necessary. Since we were not interested in analysing the standard Q-learning algorithm, we accepted this result. Note that for the standard on-policy Q-learning algorithm, the number of cycles are equal to the number of performed state transitions, our main performance criterion.

3.2   Experiments with PLGBAs

Further experiments used PLGBAs to store and retrieve the Q-values. In order to motivate our approaches towards actively chosen learning samples, the next experiments demonstrate the capability of the PLGBA to learn the given task

Table 1. Results of different PLGBA dimensions

PLGBA dimensions	number of learning samples	cycles per trajectory on average	number of iterations	updates to Q-table
$181 \cdot 141 \cdot 3$	76563	49.8	21	1607823
$91 \cdot 71 \cdot 3$	19383	50.5	15	290745
$37 \cdot 29 \cdot 3$	3219	51.3	15	48285
$19 \cdot 15 \cdot 3$	855	53.0	11	9405

The learning samples are placed on the grid points and thus are evenly distributed across the state space. The final learning quality is given by the averaged number of cycles per trajectory, measured on the evaluation set. The number of iterations denotes how many times the entire set of learning samples has been updated.

in an offline learning process. The state space was discretized as before, but this time a single learning sample (s_t, a_t, s_{t+1}) was drawn at every grid point. The offline learning process consisted of updating the PLGBA's values repeatedly according to (1). No additional training samples were used during the learning phase, so the set of learning samples remained fixed for every experiment in this subsection. The schema of the algorithm in this subsection is summarized in the following lines of code.

Schema of the algorithm in this subsection

```
/* Part 1: data acquisition */
for each grid_point in PLGBA
  create_learning_sample_at_grid_point;
end;

/* Part 2: learning */
while true
  /* a new iteration begins here */
  for each learning_sample in database
    perform_Q_learning_update;
  end;
end;
```

The first grid had $181 \cdot 141 \cdot 3 = 76563$ grid points, and the final result was 49.8 cycles per trajectory on average, measured on the same evaluation set as for the PCGBA. Additional PLGBAs further showed the efficiency of this proceeding, whereby smaller grids got slightly worse results as shown in Table 1. Increasing the grid density did not further reduce the average trajectory length.

Please keep in mind that according to the discussion from section 2, these results were for analytical use only since training samples placed on the grid points meant that the starting velocity of the car could be chosen freely. Therefore, the next step was to apply the knowledge from this subsection to learning based on actual car movement.

3.3   Adding a Realistic Learning Restriction

With regards to the discussion about the starting velocity $v = 0$ from section 2, the standard Q-learning was used again to give a new performance baseline. Both a PCGBA and a PLGBA were tested hereby, and the dimensions were corresponding to each other resulting in $181 \cdot 141 \cdot 3$ grid points. While using a PCGBA, $15.0 \cdot 10^6$ state transitions and 162000 episodes were executed to reach the final result of 50.8 cycles per trajectory on average. Please note that the number of cycles has increased in these experiments. The reason is that some states are not accessible anymore due to the restriction of the starting velocity, so the state space has changed. The Q-learning algorithm itself did not perform worse in these experiments.

$6.5 \cdot 10^6$ state transitions in 75500 episodes were necessary for the same result of 50.8 cycles per trajectory on average while using the PLGBA. This shows that the use of a linear approximation can surpass the learning performance of the PCGBA while retaining the learning quality at the same time. The learning process still required much time however, and further learning methods are expected to improve the performance beyond these results.

3.4   Actively Chosen Learning Samples

This subsection applies the learning algorithm from section 2 onto the "Mountain-Car" task. The following experiments were done with a database storing the learning samples, and the selection of state transitions near the grid points was applied, as it has been detailed in before. In the end, the same averaged number of 50.8 cycles per trajectory was reached, showing the same learning quality as the standard Q-learning algorithm. However, while the standard algorithm needed a big amount of state transitions, our learning algorithm could reduce this number by performing only 470000 state transitions approximately. The number of episodes could be reduced as well since only 5200 episodes on average were necessary for these results.

The final step includes the selection of starting positions for new learning episodes, while other aspects of the learning algorithm remained the same. Figure 3 shows the learning progression of both methods in comparison with the standard Q-learning algorithm. With actively chosen starting states, the number of required state transitions to achieve the final result of 50.8 cycles per trajectory could be reduced further on to 427000, and 4720 episodes were run during this learning process. Since our algorithm updated the Q-table at all learning samples in the database in every episode, it performed much more of these updates than the standard Q-learning. We regard this drawback as negligible for learning processes with real world systems, because an object's movements usually require several dimensions more of time than updates to the approximator. Table 2 summarizes the results of the different methods.

The learning progressions, however, reveal that the selection of starting states did not improve the learning process in general. In fact, as long as the state space is not completely explored yet, randomly chosen starting states perform equally well, since the visited states during the learning process mainly depend on the

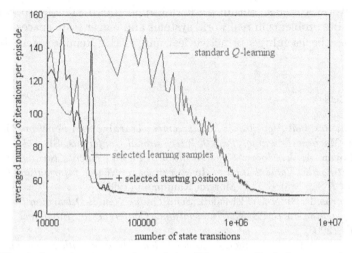

Fig. 3. Learning progressions of the different methods in comparison

inaccuracy of the incompletely learned Q-values. Specifically chosen starting position quicken the search for the final results not before the optimal quality is approached.

Table 2. Results of the different learning methods with learning on trajectories

learning method	number of state transitions	cycles per trajectory on average	number of episodes	updates to Q-table
1	$15.0 \cdot 10^6$	50.8	162000	$15.0 \cdot 10^6$
2	$6.5 \cdot 10^6$	50.8	75500	$6.5 \cdot 10^6$
3	470000	50.8	5200	$51.0 \cdot 10^6$
4	427000	50.8	4720	$46.3 \cdot 10^6$

Method 1: Standard Q-learning with a PCGBA. Method 2: Standard Q-learning with a PLGBA. Method 3: Selection of learning samples near grid points of a PLGBA. Method 4: Third method with additional selection of starting states.

4 Conclusions

In this paper, we showed that it is possible to use piecewise linear grid based approximators to attain the quality of an approximator with constant approximation, and moreover reduce the required effort during the learning process at the same time. This approach led to the application of our learning algorithm based on the ideas of active learning which was also able to improve the learning performance on the given task. The algorithm and its underlying idea have been detailed and tested on the well-known "Mountain-Car" task. Specifically, this was demonstrated by reducing the amount of required state transitions during the learning process, a criterion which plays the essential role when working with real world systems. Future work will apply this technique to Reinforcement Learning problems in real world systems and bigger state spaces to further demonstrate the usefulness of active learning in the Reinforcement Learning domain.

References

1. L. Baird: *Residual Algorithms: Reinforcement Learning with Function Approximation.* In *Machine Learning: Twelfth International Conference*, SF, USA (1995)
2. J. A. Boyan, A. W. Moore: *Generalization in Reinforcement Learning: Safely Approximating the Value Function.* In *Advances in Neural Information Processing Systems 7*, San Mateo, CA. Morgan Kaufmann (1995)
3. M. Carreras, P. Ridao, A. El-Fakdi: *Semi-Online Neural-Q_learning for Real-time Robot Learning.* In *IEEE/RSJ International Conference on Intelligent Robots and Systems* (2003)
4. D. Cohn, L. Atlas, R. Ladner: *Improving Generalization with Active Learning.* In *Machine Learning*, 15:201-221 (1994)
5. D. A. Cohn, Z. Ghahramani, M. I. Jordan: *Active Learning with Statistical Models.* In *Journal of Artificial Intelligence Research*, 4:129-145, Cambridge, MA (1996)
6. M. Hasenjäger, H. Ritter: *Active Learning in Neural Networks.* In *New learning paradigms in soft computing*, p. 137 - 169, Physica-Verlag Studies In Fuzziness And Soft Computing Series (2002)
7. R. M. Kretchmar, C. W. Anderson: *Comparison of CMACs and Radial Basis Functions for Local Function Approximators in Reinforcement Learning.* In *International Conference on Neural Networks*, Houston, Texas, USA (1997)
8. A. Merke, R. Schoknecht: *A Necessary Condition of Convergence for Reinforcement Learning with Function Approximation.* In *ICML*, p. 411 - 418 (2002)
9. J. Poland, A. Zell: *Different Criteria for Active Learning in Neural Networks: A Comparative Study.* In M. Verleysen, editor, *Proceedings of the 10th European Symposium on Artificial Neural Networks*, p. 119 - 124 (2002)
10. M. Riedmiller: *Concepts and Facilities of a neural reinforcement learning control architecture for technical process control (draft version).* In *Neural Computation and Application Journal*, 8:323-338, Springer Verlag London (1999)
11. K. K. Sung, P. Niyogi: *Active learning for function approximation.* In G. Tesauro, D. Touretzky, T. K. Leen, editors, *Advances in Neural Processing Systems*, 7:593-600, MIT Press, Cambridge, MA (1998)
12. R. S. Sutton: *Generalization in Reinforcement Learning: Successful Examples Using Sparse Coarse Coding.* In M. E. Hasselmo, D. S. Touretzky, M. C. Mozer, editor, *Advances in Neural Information Processing Systems*, 8. MIT Press (1996)
13. R. S. Sutton, A. G. Barto: *Reinforcement Learning: An Introduction.* MIT Press, Cambridge, MA (1998)
14. C. J. C. H. Watkins, P. Dayan: *Q-learning.* In *Machine learning*, 8:279-292 (1992)
15. S. Weaver, L. Baird, M. Polycarpou: *An Analytical Framework for Local Feedforward Networks.* In *IEEE Transactions on Neural Networks*, 9(3) (1999)
16. Z. Zheng, B. Padmanabhan: *On Active Learning for Data Acquisition.* In *2002 IEEE International Conference on Data Mining*, p. 562ff, Maebashi City, Japan

Spatial Representation and Navigation in a Bio-inspired Robot

Denis Sheynikhovich, Ricardo Chavarriaga, Thomas Strösslin,
and Wulfram Gerstner

Swiss Federal Institute of Technology,
Laboratory of Computational Neuroscience,
CH-1015 Lausanne, Switzerland
{denis.sheynikhovich, ricardo.chavarriaga,
thomas.stroesslin, wulfram.gerstner}@epfl.ch

Abstract. A biologically inspired computational model of rodent representation–based (locale) navigation is presented. The model combines visual input in the form of realistic two dimensional grey-scale images and odometer signals to drive the firing of simulated place and head direction cells via Hebbian synapses. The space representation is built incrementally and on-line without any prior information about the environment and consists of a large population of location-sensitive units (place cells) with overlapping receptive fields. Goal navigation is performed using reinforcement learning in continuous state and action spaces, where the state space is represented by population activity of the place cells. The model is able to reproduce a number of behavioral and neurophysiological data on rodents. Performance of the model was tested on both simulated and real mobile Khepera robots in a set of behavioral tasks and is comparable to the performance of animals in similar tasks.

1 Introduction

The task of self-localization and navigation to desired target locations is of crucial importance for both animals and autonomous robots. While robots often use specific sensors (e.g. distance meters or compasses), or some kind of prior information about the environment in order to develop knowledge about their location (see [1] for review), animals and humans can quickly localize themselves using incomplete information about the environment coming from their senses and without any prior knowledge. Discovery of location and direction sensitive cells in the rat's brain (see Sect. 2) gave some insight into the problem of how this self-localization process might happen in animals. It appears that using external input and self-motion information various neural structures develop activity profiles that correlate with current gaze direction and current location of the animal. Experimental evidence suggests that in many cases activity of the place and direction sensitive neurons underlies behavioral decisions, although some results are controversial (see [2], Part II for review).

S. Wermter et al. (Eds.): Biomimetic Neural Learning, LNAI 3575, pp. 245–264, 2005.
© Springer-Verlag Berlin Heidelberg 2005

The first question that we try to answer in this work is what type of sensory information processing could cause an emergence of such a location and direction sensitivity. Particular constraints on the possible mechanism that we focus on are *(i)* the absence of any prior information about the environment, *(ii)* the requirement of on-line learning from interactions with the environment and *(iii)* possibility to deploy and test the model in a real setup. We propose a neural architecture in which visual and self-motion inputs are used to achieve location and direction coding in artificial place and direction sensitive neurons. During agent-environment interactions correlations between visually– and self-motion–driven cells are discovered by means of unsupervised Hebbian learning. Such a learning process results in a robust space representation consisting of a large number of localized overlapping place fields in accordance with neuro-physiological data.

The second question is related to the use of such a representation for goal oriented behavior. A navigational task consists of finding relationships between any location in the environment and a hidden goal location identified by a reward signal received at that location in the past. These relationships can then be used to drive goal–oriented locomotor actions which represent the navigational behavior. The reinforcement learning paradigm [3] proposed a suitable framework for solving such a task. In the terms of reinforcement learning the states of the navigating system are represented by locations encoded in the population activity of the place sensitive units whereas possible actions are represented by population activity of locomotor action units. The relationships between the location and the goal are given by a state-action value function that is stored in the connections between the place and action units and learned online during a goal search phase. During a goal navigation phase at each location an action with the highest state-action value is performed resulting in movements towards the goal location. The application of the reinforcement learning paradigm is biologically justified by the existence of neurons whose activity is related to the difference between predicted and actual reward (see Sect. 2) which is at the heart of the reinforcement learning paradigm.

The text below is organized as follows. The next section describes neurophysiological and behavioral experimental data that serve as a biological motivation for our model. Section 3 reviews previous efforts in modeling spatial behavior and presents a bio-inspired model of spatial representation and navigation. Section 4 describes properties of the model and its performance in navigational tasks. A short discussion in Sect. 5 concludes the paper.

2   Biological Background

Experimental findings suggest that neural activity in several areas of the rat's brain can be related to the self-localization and navigational abilities of the animals. Cells in the hippocampus of freely moving rats termed *place cells* tend to fire only when the rat is in a particular portion of the testing environment, independently of gaze direction [4]. Different place cells are active in different parts of the environment and activity of the population of such cells encode the current location of the rat in an allocentric frame of reference [5]. Other cells found in the hippocampal formation [6], as well as in other parts of the brain, called *head direction cells*, are active only when the rat's head is oriented towards a specific direction independently of the location (see [2], Chap. 9 for review). Different head direction cells have different preferred orientations and the population of such cells acts as an internal neural compass. Place cells and head-direction cells interact with each other and form a neural circuit for spatial representation [7].

The hippocampal formation receives inputs from many cortical associative areas and can therefore operate with highly processed information from different sensory modalities, but it appears that visual information tends to exert a dominant influence on the activity of the cells compared to other sensory inputs. For instance, rotation of salient visual stimuli in the periphery of a rat's environment causes a corresponding rotation in place [8] and head direction [6] cell representations. On the other hand, both place and head direction cells continue their location or direction specific firing even in the absence of visual landmarks (e.g. in the dark). This can be explained by taking into account integration over time of vestibular and self-movement information (that is present even in the absence of visual input), which is usually referred to as the ability to perform 'path integration'. There is an extensive experimental evidence for such 'integration' abilities of place and head direction cell populations (reviewed in [9] and [2], Chap. 9, respectively).

One of the existing hypotheses of how the place cells can be used for navigation employs a reinforcement learning paradigm in order to associate place information with goal information. In the reinforcement learning theory [3] a state space (e.g. location-specific firing) is associated with an action space (e.g. goal-oriented movements) via a state-action value function, where the value is represented by an expected future reward. This state-action value function can be learned on-line based on the information about a current location and a difference between the predicted and an actual reward. It was found that activity of dopaminergic neurons in the ventral tegmental area (VTA) of the brain (a part of the basal ganglia) is related to the errors in reward prediction [10, 11]. Furthermore these neurons project to the brain area called nucleus accumbens (NA) which has the hippocampus as the main input structure and is related to motor actions [12, 13, 14, 15]. In other words neurons in the NA receive spatial information from the hippocampus and reward prediction error information from the VTA. As mentioned before, these two types of information are the necessary prerequisites for reinforcement learning. This data supports the hypothesis that the neural substrate for goal learning could be the synapses between the hippocampus and NA. The NA further projects to the thalamus which is in turn interconnected with the primary motor cortex, thus providing a possibility that the goal information could be used to control actions. The model of navigation described in this paper is consistent with these experimental findings.

On the behavioral level, several experimental paradigms can be used to test navigational abilities of animals in the tasks in which an internal space representation is necessary for the navigation (so called *locale navigation*, see [16] for a

review of navigational strategies). Probably the most frequently used paradigm
is the hidden platform water maze [17]. The experimental setup consists of a
circular water pool filled with an opaque liquid and a small platform located
inside the pool, but submerged below the surface of the liquid. At the beginning
of each trial a rat is placed into the pool at a random location and its task is to
find the platform. Since no single visual cue directly identifies the platform and
the starting locations are random, animals have to remember the location of the
hidden platform based on the extra-pool visual features. After several trials rats
are able to swim directly to the hidden platform from any location in the pool,
which indicates that they have acquired some type of spatial representation and
use it to locate the platform.

Extensive lesion studies show that damage to brain areas containing place or
direction sensitive cells, as well as lesions of the fornix (nerve fibers containing
projections from the hippocampus to the NA) or the NA itself selectively impair
navigational abilities of rats in tasks where an internal representation of space
is necessary [16, 15, 18].

This and other experimental data suggest that the hippocampal formation can
serve as the neural basis for spatial representation underlying navigational behavior. This hypothesized strong relation between behavior and neuro-physio-logical
activity can be elaborated by means of computational models, that can in turn
generate predictions testable on the level of both neurophysiology and behavior.

3   Modeling Spatial Behavior

The ability of animals to navigate in complex task-environment contexts has
been the subject of a large body of research over the last decades. Because of
its prominent role in memory and its spatial representation properties described
above the hippocampus has been studied and modeled intensively. In the next
section we review several models of the mechanisms yielding place cell activity
and its role in locale navigation. In Sect. 3.2 we describe our own model in detail.

3.1   Previous Models

In this section we focus on those models which were tested in navigational tasks
in real environments using mobile robots. Readers interested in theoretical and
simulation models as well as in models of different types of navigation are referred
to reviews of Trullier et al. [19] and Franz and Mallot [1].

Recce and Harris [20] modeled the hippocampus as an auto-associative memory which stored a scene representation consisting of the bearings and distances
of the surrounding landmarks and of a goal location. The landmark bearings
and distances were extracted from omnidirectional sonar scans. During a first
exploration phase the scenes were stored in the memory and each stored scene
was associated with a place cell. During a second goal navigation phase the currently perceived scene was compared to the scenes stored in the memory. When
the scenes matched, the stored scene was activated (i.e. the place cell fired)

together with the goal location information. Once the scene was recalled, the robot moved directly to the goal. Information about the landmark positions and orientations were updated using integrated odometer signals, but the place cell activity depended only on the visual input.

Burgess et al. [21, 22] described a robotic implementation of an earlier neuro-physiological model of the rat hippocampus [23]. Some place cells were shown to fire at a relatively fixed distance from the walls of a testing environment [24]. This property inspired the place recognition mechanism of the robot of Burgess et al. which visually estimated distances to the surrounding walls by detecting the position of a horizontal line at the junction of the walls and the floor in the input image. During a first exploration phase, the robot rotated on the spot at all locations of the arena to face all walls and to estimate their distances. The robot's orientation with respect to a reference direction was derived from path integration which was periodically reset by using a dedicated visual marker. A competitive learning mechanism selected a number of place cells to represent the specific wall distances for each place. In a second goal search phase, once the goal was found the robot associated four goal cells with the place cells representing four locations from which the direction towards the goal was known. During goal navigation, the goal direction could be computed from the relative activity of all goal cells using population vector technique [25].

In the model by Gaussier et al. [26, 27, 28], at each time step during exploration, a visual processing module extracted landmark information from a panoramic visual image. For each detected landmark in turn its type (e.g. vertical line in the image) and its compass bearing (their robot had a built-in magnetic compass) were merged into a single "what" (landmark type) and "where" (landmark bearing) matrix. When a place cell was recruited the active units from the "what-where" matrix were connected to it. The activity of the place cells was calculated in two steps: first, the initial activation of a place cell was determined as a product of the recognition level of a given feature and its bearing. Second, a winner-take-all mechanism reset the activities of all but the winning cell to zero. A delay in activation of the place cells between successive time steps allowed the next layer to learn place transitions: an active cell from the previous time step and an active cell from the current time step were connected to a transition cell using Hebbian learning rule. This way when a place cell was active (i.e. a place was recognized), it activated the associated transition cells thus "predicting" all possible (i.e. experienced in the past) transitions from that place. A transition selection mechanism was trained in the goal search phase: after the goal was found, the transition cells leading to the goal were activated more than others. This goal-oriented bias in the competition among possible transitions allowed the agent to find the goal.

The model of Arleo et al. [29, 30, 31] is an earlier version of the model presented in the next section. In this model position and direction information extracted from the visual input were combined with information extracted from the self-motion signals and merged into a single space representation which was then used for goal navigation. The visual processing pathway transformed a two-dimensional camera image into a filter-based representation by sampling it with a set of orientation-sensitive filters. At each time step during an exploration phase, the agent took four snapshots, one in each cardinal direction. For each orientation, the filter activities were stored in a so called view cell. A downstream population of visual place cells combined the information from all simultaneously active view cells using a Hebbian learning rule. In the parallel self-motion processing pathway an estimation of position was performed by integrating signals from odometers. The self-motion position estimation was calibrated using the visual position estimation. Similarly, the direction estimation was performed by integrating rotations, but calibrated using a dedicated landmark (a lamp). The self-motion and visual estimations of position were then combined in "hippocampal" place cells population using Hebbian learning. The authors proposed a locale navigation system using reinforcement learning where the population of the hippocampal place cells served as a state space. Each place cell projected to four action cells, that coded for a movement in directions north, south, east and west respectively. The projection weights stored an approximated state-action value function and were modified using a reward-based learning method during a goal search phase. During navigation the action cells population vector encoded the direction of movement to the goal from any location in the environment.

3.2   A Model of Space Representation and Navigation

The computational model of the rat spatial behavior presented in this paper is an extension of the previous model by Arleo et al. (Sect. 3.1) and is able to learn a representation of the environment by exploration. Starting with no prior knowledge, the system grows incrementally based on agent–environment interaction. Information about locations visited for the first time is stored in a population of place cells. This information is subsequently used for self-localization and navigation to desired targets.

In the neural model of place cells, *allothetic* (visual) information is correlated with *idiothetic* information (rotation and displacement signals from the robot's odometers) using Hebbian learning. This yields a stable space representation where ambiguities in the visual input are resolved by the use of the idiothetic information, and a cumulative error of path integration is accounted for by using unbiased visual input.

Figure 1 presents a functional diagram of the model. Visual stimuli are encoded in the population of View Cells (VCs), which project to the population of Allothetic Place Cells (APCs) where a vision-based position estimation is acquired and to the population of Allothetic Heading Cells (AHCs) where current gaze direction is estimated from the visual input. The transformation of the visual input to the vision-based representation is a part of the allothetic pathway leading to the population of Hippocampal Place Cells (HPCs). In the second, idiothetic, pathway, displacement and rotation signals from the odometers are integrated over time to build an internal estimation of position in the Path Integration Cells (PIC) and gaze direction in the Heading Integration Cells (HIC). The path and heading integration systems allow the rat to navigate in darkness

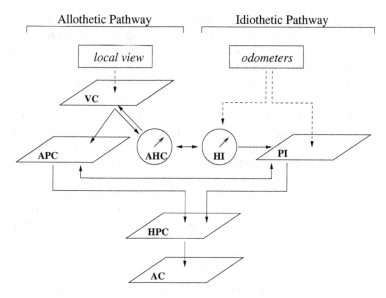

Fig. 1. Functional diagram of the model. Dashed lines denote neural transformation of a sensory input, solid lines denote projections between populations. See explanations in the text

or in the absence of visual cues. Both allothetic (APC) and idiothetic (PIC) populations project onto the HPC population where the final space representation is constructed in the form of location sensitive cells with overlapping receptive fields. Once the space representation has been learned, it can be used for navigational tasks. A direction of movement to the goal from any location in the environment is learned in the Action Cells population using temporal-difference learning technique.

The model is tested in navigational tasks using a computer simulation as well as a real Khepera robot which we refer to as 'agent' in the text below. We now discuss our model in detail.

Idiothetic Input. The idiothetic input in the model consists of rotation and displacement signals from the agent's odometers. In order to track current gaze direction we employ a population of 360 Heading Integration Cells (HIC), where each cell is assigned a preferred heading $\psi_i \in [0°, 359°]$. If $\hat{\phi}$ is the estimate of a current gaze direction, the activity of cell i from the HIC population is given by

$$r_{\psi_i}^{\text{HIC}} = \exp(-(\psi_i - \hat{\phi})^2/2\sigma_{\text{HIC}}^2) \ , \tag{1}$$

enforcing a Gaussian activity profile around $\hat{\phi}$, where σ_{HIC} defines the width of the profile. A more biologically plausible implementation of the neural network with similar properties can be realized by introducing lateral connections between the cells where each cell is positively connected to the cells with similar preferred directions and negatively connected the other cells. The attractor dynamics of such an implementation accounts for several properties of real head direction cells [32]. Here we employ the simpler algorithmic approach (1) that preserves network properties relevant for our model. When the agent enters a new environment arbitrary direction ψ_0 is taken as a reference direction. Whenever the agent performs a rotation, the rotational signal from the odometers is used to shift the activity blob of the HIC population. Here again a simple algorithmic approach is used where the new direction is explicitly calculated by integrating wheel rotations and a Gaussian profile is enforced around it, although more biologically plausible solutions exist [33, 34].

Having a current gaze direction encoded by the HIC population, standard trigonometric formulas can be used to calculate a new position with respect to the old position in an external Cartesian coordinate frame whenever the agent performs a linear displacement. We define Path Integration Cells (PIC) population as a two–dimensional grid of cells with predefined metric relationships, each having its preferred position $\boldsymbol{p}_i = (x_i, y_i)$ and activity

$$r_i^{\mathrm{PIC}} = \exp(-(\boldsymbol{p}_i - \hat{\boldsymbol{p}})^2/2\sigma_{\mathrm{PIC}}^2) \ , \qquad (2)$$

where $\hat{\boldsymbol{p}} = (\hat{x}, \hat{y})$ is the estimate of position based on idiothetic information only. Origin $\boldsymbol{p}_0 = (0,0)$ is set at the entry point whenever the agent enters a new environment. The PIC population exhibits a two-dimensional Gaussian profile with width σ_{PIC} around the current position estimation.

While the agent moves through an environment the activities of HICs (1) and PICs (2) encode estimates of its position and heading with respect to the origin and the reference direction based only on the idiothetic input. They enable the agent to navigate in darkness or return to the nest location in the absence of visual cues, properties that are well known in animals [35]. The estimation of direction and position will drift over time due to accumulating errors in the odometers. Another problem is that the abstract Cartesian frame is mapped onto the physical space in a way that depends on the entry point. Both problems are addressed by combining the idiothetic input with a visual (allothetic) input and merging the two information streams into a single allocentric map.

Allothetic Input. The task of the allothetic pathway is to extract position and heading information from the external (visual) input. Based on the visually driven representation the agent should be able to recognize previously visited locations from a current local view[1]. Such a localization property can be implemented by comparing a current local view to all previously seen local views and using some similarity measure to recognize visited places (with a natural assumption that similar local views signal for spatially close locations). This comparison of the local views should take into account information about current heading that can be estimated from the relative angles between the current and all stored local views where the relative angles can in turn be computed from the amount of overlap between the local view representations.

---

[1] The term "local view" is used to denote information extracted from the visual input at a given time step.

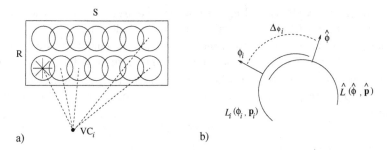

Fig. 2. Visual input and heading estimation. a: Two-dimensional panoramic image is processed by a grid of $S \times R$ points with 8 Gabor filters of different orientations at each point (filters are shown as overlapping circles, different orientations are shown only in the lower-left circle). Responses of the filters are stored in a View Cell. b: Current heading $\hat{\phi}$ can be estimated from the maximal overlap C_i between the current and a stored local views corresponding to the angular difference Δ_{ϕ_i} (3,5)

The raw visual input in the model is a two-dimensional grey-level image, received by merging several snapshots captured by the video camera of the robot into a single panoramic $(320° - 340°)$ picture imitating the rat's wide view field. Note that the individual directions and the number of the snapshots are not important as long as the final image is of the required angular width (the model would perfectly suit for a single panoramic image as an input). In order to neurally represent visual input the image is sampled with a uniform rectangular grid of $S \times R$ points (see Fig. 2(a), for the results presented here we used $S = 96$ columns and $R = 12$ rows). At each point of the grid we place a set of 8 two-dimensional Gabor filters with 8 different orientations and a spatial wavelength matched to the resolution of the sampling grid. Gabor filters are sensitive to edge-like structures in the image and have been largely used to model orientation-sensitive simple cells in the visual cortex [36]. Responses F_k of $K = S \times R \times 8 = 9216$ visual filters constitute the local view information $L(\phi, \boldsymbol{p}) = \{F_k(\phi, \boldsymbol{p})\}_1^K$ that depends on the heading direction ϕ and the position $\boldsymbol{p}$ where the local view was taken.

Local views perceived by the agent are stored in the population of View Cells (VCs). At each time step a new VC i is recruited that stores the amplitudes of all current filter responses $L(\phi_i, \boldsymbol{p}_i)$. As the agent explores an environment the population of View Cells grows incrementally memorizing all local views seen so far. The place is considered to be well memorized if there is sufficient number of highly active View Cells. The information stored by the VC population can be used at each time step to estimate current heading and position as follows.

Allothetic Heading Estimation. In order to estimate current heading based on the visual input we employ a population of 360 Allothetic Heading Cells (AHC) with preferred directions uniformly distributed in $[0°, 359°]$. Suppose that a local view $L_i(\phi_i, \boldsymbol{p}_i)$ taken at position $\boldsymbol{p}_i$ in direction ϕ_i is stored by a View Cell i at time step t and the agent perceives a new local view $\hat{L}(\hat{\phi}, \hat{\boldsymbol{p}})$ at a later time step t', where $\hat{\phi}$ and $\hat{\boldsymbol{p}}$ are unknown (the local view information is time independent).

This is schematically shown in Fig. 2(b), where the arcs illustrate panoramic images (that elicit filter responses constituting the local views) with arrows showing corresponding gaze directions. In order to estimate the current heading $\hat{\phi}$ based on the stored local view L_i we first calculate the angular difference Δ_{ϕ_i} (measured in columns of the filter grid) between $\hat{L}$ and L_i that maximizes the sum of products C_i of corresponding filter values (i.e. gives maximum of the correlation function)

$$\Delta_{\phi_i} = \max_{\Delta_\phi} C_i(\Delta_\phi) \ , \tag{3}$$

$$C_i(\Delta_\phi) = \sum_s \boldsymbol{f}_i(s) \cdot \hat{\boldsymbol{f}}(s + \Delta_\phi) \ . \tag{4}$$

Here $\boldsymbol{f}_i(s)$ and $\hat{\boldsymbol{f}}(s)$ are the sets of all filter responses in vertical column s of the stored L_i and current $\hat{L}$ local views respectively, s runs over columns of the filter grid $s \in [0, S-1]$.

The estimation of the current heading $\hat{\phi}$ performed using information stored by a *single* View Cell i is now given by

$$\hat{\phi} = \phi_i + \delta_{\phi_i} \ , \quad \text{where} \tag{5}$$

$$\delta_{\phi_i} = \Delta_{\phi_i} \cdot V/S \tag{6}$$

is the angular difference measured in degrees corresponding to the angular difference measured in filter columns Δ_{ϕ_i}, V is the agent's view field in degrees.

Let us transform the algorithmic procedure (3)-(5) into a neuronal implementation. $C_i(\Delta_\phi)$ is calculated as a sum of scalar products and can hence be regarded as the output of a linear neuron with synaptic weights given by the elements of $\boldsymbol{f}_i$ applied to a shifted version of the input $\hat{\boldsymbol{f}}$. We now assume that this neuron is connected to an allothetic head direction cell with preferred direction $\psi_j = \phi_i + \delta_\phi$ and the firing rate

$$r_{\psi_j}^{\text{AHC}} = C_i(\Delta_\phi) \ , \tag{7}$$

taking into account (6). The maximally active AHC would then code for the estimation $\hat{\phi}$ of the current heading based on the information stored in VC i.

Since we have a population of View Cells (incrementally growing as the environment exploration proceeds) we can combine the estimates of *all* View Cells in order to get a more reliable estimate. Taking into account the whole VC population the activity of a single AHC will be

$$r_{\psi_j}^{\text{AHC}} = \sum_{i \in VC} C_i(\Delta_\phi) \ , \tag{8}$$

where for each AHC we sum correlations $C_i(\Delta_\phi)$ with Δ_ϕ chosen such that $\phi_i + \delta_\phi = \psi_j$. The activations (8) result in the activity profile in the AHC population. The decoding of the estimated value is done by taking a preferred direction of the maximally active cell.

Allothetic Position Estimation. As mentioned before, the idea behind the allo-thetic position estimation is that similar local views should signal for spatially close locations. A natural way to compare the local views is to calculate their difference

$$\Delta\tilde{L}(\phi_i, \boldsymbol{p}_i, \hat{\phi}, \hat{\boldsymbol{p}}) = \left| L_i(\phi_i, \boldsymbol{p}_i) - \hat{L}(\hat{\phi}, \hat{\boldsymbol{p}}) \right| = \sum_s \left| \boldsymbol{f}_i(s) - \hat{\boldsymbol{f}}(s) \right|_1 , \qquad (9)$$

where $\boldsymbol{f}_i(s)$ and $\hat{\boldsymbol{f}}(s)$ are defined as in (4).

While exploring an environment the agent makes random movements and turns in the azimuthal plane, hence stored local views correspond to different al-locentric directions. For the difference (9) to be small for spatially close locations the local views must be aligned before measuring the difference. It means that (9) should be changed to take into account the angular difference $\delta_{\phi_i} = \hat{\phi} - \phi_i$ where $\hat{\phi}$ is provided by the AHC population:

$$\Delta L(\phi_i, \boldsymbol{p}_i, \hat{\phi}, \hat{\boldsymbol{p}}) = \sum_s \left| \boldsymbol{f}_i(s) - \hat{\boldsymbol{f}}(s + \Delta_{\phi_i}) \right|_1 . \qquad (10)$$

In (10) Δ_{ϕ_i} is the angular difference δ_{ϕ_i} measured in columns of the filter grid (i.e. a whole number closest to $\delta_{\phi_i} \cdot S/V$).

We set the activity of a VC to be a similarity measure between the local views:

$$r_i^{VC} = \exp\left(\frac{-\Delta L^2(\phi_i, \boldsymbol{p}_i, \hat{\phi}, \hat{\boldsymbol{p}})}{2\sigma_{VC}^2 N_\Omega} \right) , \qquad (11)$$

where N_Ω is the size of the overlap between the local views measured in filter columns and σ_{VC} is the sensitivity of the View Cell (the bigger σ_{VC} the larger is the receptive field of the cell). Each VC "votes" with its activity for the estimation of the current position. The activity is highest when a current local view is identical to the local view stored by the VC, meaning by our assumption that $\hat{\boldsymbol{p}} \approx \boldsymbol{p}_i$.

Each VC estimates current position based only on a single local view. In order to combine information from several local views, all simultaneously active VCs are connected to an Allothetic Place Cell (APC). Unsupervised Hebbian learning is applied to the connection weights between VC and APC populations. Specifically, connection weights from VC j to APC i are updated according to

$$\Delta w_{ij} = \eta \, r_i^{APC}(r_j^{VC} - w_{ij}) , \qquad (12)$$

where η is a learning rate. Activity of an APC i is calculated as a weighted average of the activity of its afferent signals.

$$r_i^{APC} = \frac{\sum_j r_j^{VC} w_{ij}}{\sum_j w_{ij}} . \qquad (13)$$

The APC population grows incrementally at each time step. Hebbian learning in the synapses between APCs and VCs extracts correlations between the View Cells so as to achieve a more reliable position estimate in the APC population.

Combined Place Code. The two different representations of space driven by visual and proprioceptive inputs are located in the APC and PIC populations respectively. At each time step the activity of PICs (2) encode current position estimation based on the odometer signals, whereas the activity of APCs (13) encode the position estimation based on local view information.

Since the position information from the two sources represent the same physical position we can construct a more reliable combined representation by using Hebbian learning.

At each time step a new Hippocampal Place Cell (HPC) is recruited and connected to all simultaneously active APCs and PICs. These connections are modified by Hebbian learning rule analogous to (12). The activity of an HPC cells is a weighed average of its APC and PIC inputs analogous to (13).

For visualization purposes the position represented by the ensemble of HPCs can be interpreted by population vector decoding [37]:

$$\hat{\boldsymbol{p}}^{\text{HPC}} = \frac{\sum_j r_j^{\text{HPC}} \boldsymbol{p}_j^{\text{HPC}}}{\sum_j r_j^{\text{HPC}}} , \tag{14}$$

where $\boldsymbol{p}_j^{\text{HPC}}$ is the center of the place field of an HPC j.

Such a combined activity at the level of HPC population allows the system to rely on the visual information during the self-localization process at the same time resolving consistency problems inherent in a purely idiothetic system.

Goal Navigation Using Reinforcement Learning. In order to use the position estimation encoded by the HPC population for navigation, we employ a Q-learning algorithm in continuous state and action space [38, 39, 40, 3]. Values of the HPC population vector (14) represent a continuous state space. The HPC population projects to the population of N^{AC} Action Cells (AC) that code for the agent's motor commands. Each AC i represents a particular direction $\theta_i \in [0°, 359°]$ in an allocentric coordinate frame. The continuous angle θ^{AC} encoded by the AC population vector

$$\theta^{\text{AC}} = \arctan\left(\frac{\sum_i r_i^{\text{AC}} \cdot \sin(2\pi i/N^{\text{AC}})}{\sum_i r_i^{\text{AC}} \cdot \cos(2\pi i/N^{\text{AC}})}\right) \tag{15}$$

determines the direction of the next movement in the allocentric frame of reference. The activity $r_i^{\text{AC}} = Q(\hat{\boldsymbol{p}}^{\text{HPC}}, a_i) = \sum_j w_{ij}^a r_j^{\text{HPC}}$ of an Action Cell i represents a state-action value $Q(\hat{\boldsymbol{p}}^{\text{HPC}}, a_i)$ of performing action a_i (i.e. movement in direction θ_i) if the current state is defined by $\hat{\boldsymbol{p}}^{\text{HPC}}$. The state-action value is parameterized by the weights w_{ij}^a of the connections between HPCs and ACs.

The state-action value function in the connection values w_{ij}^a is learned according to the Q-learning algorithm using the following procedure [38]:

1. At each time step t the state-action values are computed for each action $Q(\hat{\boldsymbol{p}}^{\text{HPC}}(t), a_i) = r_i^{\text{AC}}(t)$.

2. Action $a = a^*$ (i.e. movement in the direction θ^{AC} defined by (15)) is chosen with probability $1 - \epsilon$ (exploitation) or a random action $a = a^r$ (i.e. movement in a random direction) is chosen with probability ϵ (exploration).

3. A Gaussian profile around the chosen action a is enforced in the action cells population activity resulting in $\tilde{r}_i^{AC} = \exp(-(\theta - \theta_i)/2\sigma_{AC}^2)$, where θ and θ_i are the directions of movement coded by the actions a and a_i respectively. This step is necessary for generalization purposes and can also be performed by adding lateral connectivity between the action cells [38].

4. The eligibility trace is updated according to
   $e_{ij}(t) = \alpha \cdot e_{ij}(t-1) + \tilde{r}_i^{AC}(t) \cdot r_j^{PC}(t)$ with $\alpha \in [0,1]$ being the decay rate of the eligibility trace.

5. Action a is executed (along with time step update $t = t + 1$).

6. Reward prediction error is calculated as
   $\delta(t) = R(t) + \gamma \cdot Q(\hat{\boldsymbol{p}}^{HPC}(t), a^*(t)) - Q(\hat{\boldsymbol{p}}^{HPC}(t-1), a(t-1))$,
   where $R(t)$ is a reward received at step t.

7. Connection weights between HPC and AC populations are updated according to $\Delta w_{ij}^a(t) = \eta \cdot \delta(t) \cdot e_{ij}(t-1)$ with $\eta \in [0,1]$ being the learning rate.

Such an algorithm enables fast learning of the optimal movements from any state, in other words given the location encoded by the HPC population it learns the direction of movement towards the goal from that location. The generalization ability of the algorithm permits calculation of the optimal movement from a location even if that location was not visited during learning. Due to the usage of population vectors the system has continuous state and action spaces allowing the model to use continua of possible locations and movement directions using a finite number of place or action cells.

4   Experimental Results

In this section we are interested in the abilities of the model to *(i)* build a representation of a novel environment and *(ii)* use the representation to learn and subsequently find a goal location. The rationale behind this distinction relates to the so called latent learning (i.e. ability of animals to establish a spatial representation even in the absence of explicit rewards [41]). It is shown that having a target–independent space representation (like the HPC place fields) enables the agent to learn target–oriented navigation very quickly.

For the experiments discussed in the next sections we used a simulated as well as a real Kephera robots. In the simulated version the odometer signals and visual input are generated by a computer. The simulated odometers error is taken to be 10% of the distance moved (or angle rotated) at each time step. Simulated visual input is generated by a panoramic camera placed into a virtual environment.

4.1   Development and Accuracy of the Place Field Representation

To test the ability of the model to build a representation of space we place the robot in a novel environment (square box 100 cm.×100 cm.) and let it move in random directions incrementally building a spatial map.

Figure 3(a) shows an example of the robot's trajectory at the beginning of the exploration (after 44 time steps). During this period 44 HPCs were recruited as shown in Fig. 3(b). The cells are shown in a topological arrangement for visualization purposes only (the cells that code for close positions are not necessarily neighbors in their physical storage). After the environment is sufficiently explored (e.g. as in Fig. 3(d) after 1000 time steps), the HPC population encodes estimation of a real robot's position (Fig. 3(c)).

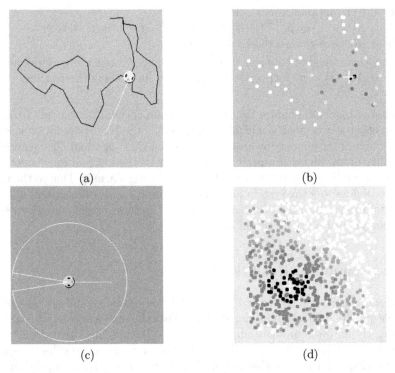

Fig. 3. Exploration of the environment and development of place cells. The grey square is the test environment. a: Exploratory trajectory of the robot after 44 time steps. Light grey circle with three dots is the robot, black line is its trajectory. The white line shows its gaze direction. b: HPCs recruited during 44 steps of exploration shown in (a). Small circles are the place cells (the darker the cell the higher its activity). c: The robot is located in the SW quadrant of the square arena heading west, white arc shows its view field (340°). d: Population activity of the HPC population after exploration while the robot's real location is shown in (c). The white cross in (b) and (d) denotes the position of the HPC population vector (14)

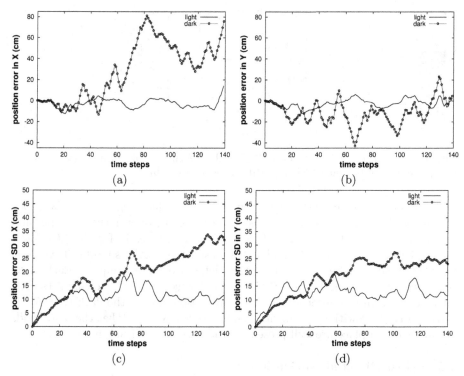

Fig. 4. a,b: Position estimation error in a single trial in X (a) and Y (b) directions with (light conditions) and without (dark conditions) taking into account the visual input. c,d: Position estimation error SD over 50 trials in X (c) and Y (d) directions in the light and dark conditions

To investigate self-localization accuracy in a familiar environment we let the robot run for 140 steps in the previously explored environment and note the error of position estimation (i.e. difference between the real position and a value of the HPC population vector (14)) at each time step in the directions defined by the walls of the box.

Figures 4(a),(b) show the error in vertical (Y) and horizontal (X) directions versus time steps ('light' conditions, solid line) in a single trial. For comparison we also plot the position estimation error in the same trials computed only by integrating the idiothetic input, i.e. without taking into account visual input ('dark' conditions, line with circles). A purely idiothetic estimate is affected by a cumulative drift over time. Taking into account visual information keeps the position error bounded.

Figure 4(c),(d) show the standard deviation (SD) of the position estimation error in light and dark conditions over 50 trials in X and Y directions (the mean error over 50 trials is approximately zero for both conditions). The error SD in light conditions is about 12 cm (that corresponds to 12% of the length of the wall).

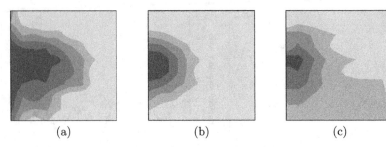

(a)   (b)   (c)

Fig. 5. a. Receptive field of a typical APC. b. Receptive field of a typical HPC. c. Receptive field of the same cell as in (b) but in dark conditions

In order to inspect the receptive fields of the place cells we let the robot systematically visit 100 locations distributed uniformly over the box area and noted the activity of a cell at each step. Contour graphs in Fig. 5(a),(b) show the activity maps for an APC and a HPC respectively. APCs tend to have large receptive fields, whereas HPC receptive fields are more compact. Each HPC combines simultaneously active PICs and APCs (see Sect. 1) allowing it to code for place even in the absence of visual stimulation, e.g. in the dark (Fig. 5(c)). This is consistent with experimental data where they found that the place fields of hippocampal place are still present in the absence of visual input [42].

4.2   Goal Directed Navigation

A standard experimental paradigm for navigational tasks that require internal representation of space is the hidden platform water maze [17]. In this task the robot has to learn how to reach a hidden goal location from any position in the environment.

The task consists of several trials. In the beginning of each trial the robot is placed at a random location in the test environment (already familiar to the robot) and is allowed to find a goal location. The position of the robot at each time step is encoded by the HPC population. During movements the connection weights between HPC and AC populations are changed according to the algorithm outlined in Sect. 1. The robot is positively rewarded each time it reaches the goal and negatively rewarded for a wall hit. The measure of performance in each trial is the number of time steps required to reach the goal (that corresponds to the amount of time required for a rat to reach the hidden platform).

After a number of trials the AC population vector (15) encodes learned direction of movement to the goal from any location $\hat{p}^{\mathrm{HPC}}$. A navigation map after 20 trials is shown in Fig. 6(a). The vector field representation of Fig. 6(a) was obtained by rastering uniformly over the whole environment: the ensemble responses of the action cells were recorded at 100 locations distributed over 10×10 grid of points. At each point (black dots in Fig. 6(a)) the population vector (15) was calculated and is shown as a black line where the orientation of the line corresponds to ϕ^{AC} and the length corresponds to the action value $Q(\hat{p}^{\mathrm{HPC}}, a^*)$.

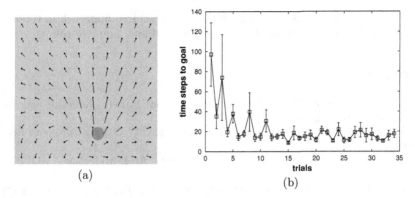

(a)

(b)

Fig. 6. a: Navigation map learned after 20 trials, dark grey circle denotes the goal location, black points denote sample locations, lines denote a learned direction of movement. b: Time to find a goal versus the number of trials

As the number of learning trials increase, the number of time steps to reach the goal decreases (Fig. 6(b)) in accordance with the experimental data with real animals [43].

5   Conclusion

The work presents a bio-inspired model of a representation–based navigation which incrementally builds a space representation from interactions with the environment and subsequently uses it to find hidden goal locations.

This model is different from the models mentioned in Sect. 3.1 in several important aspects. First, it uses realistic two–dimensional visual input which is neurally represented as a set of responses of orientation–sensitive filter distributed uniformly over the artificial retina (the visual system is similar to the one used by Arleo et al. [31], but in contrast it is not foveal in accordance with the data about the rat's visual system [44]). Second, the direction information is available in the model from the combination of visual and self-motion input, no specific compass or dedicated orientational landmark are used. Third, as in the model by Arleo et al. the integration of the idiothetic information (i.e. path integration) is an integrative part of the system that permits navigation in the dark and supports place and head direction cells firing in the absence of visual input.

The model captures some aspects of related biological systems on both behavioral (goal navigation) and neuronal (place cells) levels. In experimental neuroscience the issue of relating neuro-physiological properties of neurons to behavior is an important task. It is one of the advantages of modeling that potential connections between neuronal activity and behavior can be explored systematically. The fact that neuro-mimetic robots are simpler and more experimentally transparent than biological organisms makes them a useful tool to check new hypotheses and make predictions concerning the underlying mechanisms of spatial behavior in animals. On the other hand, a bio-inspired approach in robotics may help to discover new ways of building powerful and adaptive robots.

References

1. Franz, M.O., Mallot, H.A.: Biomimetic robot navigation. Robotics and Autonomous Systems **30** (2000) 133–153
2. Jeffery, K.J., ed.: The neurobiology of spatial behavior. Oxford University Press (2003)
3. Sutton, R., Barto, A.G.: Reinforcement Learning - An Introduction. MIT Press (1998)
4. O'Keefe, J., Dostrovsky, J.: The hippocampus as a spatial map. preliminary evidence from unit activity in the freely-moving rat. Brain Research **34** (1971) 171–175
5. Wilson, M.A., McNaughton, B.L.: Dynamics of the hippocampal ensemble code for space. Science **261** (1993) 1055–1058
6. Taube, J.S., Muller, R.I., Ranck Jr., J.B.: Head direction cells recorded from the postsubiculum in freely moving rats. I. Description and quantitative analysis. Journal of Neuroscience **10** (1990) 420–435
7. Knierim, J.J., Kudrimoti, H.S., McNaughton, B.L.: Place cells, head direction cells, and the learning of landmark stability. Journal of Neuroscience **15** (1995) 1648–1659
8. Muller, R.U., Kubie, J.L.: The effects of changes in the environment on the spatial firing of hippocampal complex-spike cells. Journal of Neuroscience **7** (1987) 1951–1968
9. McNaughton, B.L., Barnes, C.A., Gerrard, J.L., Gothard, K., Jung, M.W., Knierim, J.J., Kudrimoti, H., Qin, Y., Skaggs, W.E., Suster, M., Weaver, K.L.: Deciphering the hippocampal polyglot: the hippocampus as a path integration system. J Exp Biol **199** (1996) 173–85
10. Schultz, W., Dayan, P., Montague, P.R.: A neural substrate of prediction and reward. Science **275** (1997) 1593–1599
11. Schultz, W.: Predictive Reward Signal of Dopamine Neurons. Journal of Neurophysiology **80** (1998) 1–27
12. Freund, T.F., Powell, J.F., Smith, A.D.: Tyrosine hydroxylase-immunoreactive boutons in synaptic contact with identified striatonigral neurons, with particular reference to dendritic spines. Neuroscience **13** (1984) 1189–215
13. Sesack, S.R., Pickel, V.M.: In the rat medial nucleus accumbens, hippocampal and catecholaminergic terminals converge on spiny neurons and are in apposition to each other. Brain Res **527** (1990) 266–79
14. Eichenbaum, H., Stewart, C., Morris, R.G.M.: Hippocampal representation in place learning. Journal of Neuroscience **10(11)** (1990) 3531–3542
15. Sutherland, R.J., Rodriguez, A.J.: The role of the fornix/fimbria and some related subcortical structures in place learning and memory. Behavioral and Brain Research **32** (1990) 265–277
16. Redish, A.D.: Beyond the Cognitive Map, From Place Cells to Episodic Memory. MIT Press-Bradford Books, London (1999)
17. Morris, R.G.M.: Spatial localization does not require the presence of local cues. Learning and Motivation **12** (1981) 239–260
18. Packard, M.G., McGaugh, J.L.: Double dissociation of fornix and caudate nucleus lesions on acquisition of two water maze tasks: Further evidence for multiple memory systems. Behavioral Neuroscience **106(3)** (1992) 439–446

19. Trullier, O., Wiener, S.I., Berthoz, A., Meyer, J.A.: Biologically-based artificial navigation systems: Review and prospects. Progress in Neurobiology **51** (1997) 483–544
20. Recce, M., Harris, K.D.: Memory for places: A navigational model in support of Marr's theory of hippocampal function. Hippocampus **6** (1996) 85–123
21. Burgess, N., Donnett, J.G., Jeffery, K.J., O'Keefe, J.: Robotic and neuronal simulation of the hippocampus and rat navigation. Phil. Trans. R. Soc. Lond. B **352** (1997) 1535–1543
22. Burgess, N., Jackson, A., Hartley, T., O'Keefe, J.: Predictions derived from modelling the hippocampal role in navigation. Biol. Cybern. **83** (2000) 301–312
23. Burgess, N., Recce, M., O'Keefe, J.: A model of hippocampal function. Neural Networks **7** (1994) 1065–1081
24. O'Keefe, J., Burgess, N.: Geometric determinants of the place fields of hippocampal neurons. Nature **381** (1996) 425–428
25. Georgopoulos, A.P., Kettner, R.E., Schwartz, A.: Primate motor cortex and free arm movements to visual targets in three-dimensional space. II. Coding of the direction of movement by a neuronal population. Neuroscience **8** (1988) 2928–2937
26. Gaussier, P., Leprêtre, S., Joulain, C., Revel, A., Quoy, M., Banquet, J.P.: Animal and robot learning: Experiments and models about visual navigation. In: 7th European Workshop on Learning Robots, Edinburgh, UK (1998)
27. Gaussier, P., Joulain, C., Banquet, J.P., Leprêtre, S., Revel, A.: The visual homing problem: An example of robotics/biology cross fertilization. Robotics and Autonomous Systems **30** (2000) 155–180
28. Gaussier, P., Revel, A., Banquet, J.P., Babeau, V.: From view cells and place cells to cognitive map learning: processing stages of the hippocampal system. Biol Cybern **86** (2002) 15–28
29. Arleo, A., Gerstner, W.: Spatial cognition and neuro-mimetic navigation: A model of hippocampal place cell activity. Biological Cybernetics, Special Issue on Navigation in Biological and Artificial Systems **83** (2000) 287–299
30. Arleo, A., Smeraldi, F., Hug, S., Gerstner, W.: Place cells and spatial navigation based on 2d visual feature extraction, path integration, and reinforcement learning. In Leen, T.K., Dietterich, T.G., Tresp, V., eds.: Advances in Neural Information Processing Systems 13, MIT Press (2001) 89–95
31. Arleo, A., Smeraldi, F., Gerstner, W.: Cognitive navigation based on nonuniform gabor space sampling, unsupervised growing networks, and reinforcement learning. IEEE Transactions on Neural Networks **15** (2004) 639–652
32. Zhang, K.: Representation of spatial orientation by the intrinsic dynamics of the head-direction cell ensemble: A theory. Journal of Neuroscience **16(6)** (1996) 2112–2126
33. Arleo, A., Gerstner, W.: Spatial orientation in navigating agents: Modeling head-direction cells. Neurocomputing **38–40** (2001) 1059–1065
34. Skaggs, W.E., Knierim, J.J., Kudrimoti, H.S., McNaughton, B.L.: A model of the neural basis of the rat's sense of direction. In Tesauro, G., Touretzky, D.S., Leen, T.K., eds.: Advances in Neural Information Processing Systems 7, Cambridge, MA, MIT Press (1995) 173–180
35. Etienne, A.S., Jeffery, K.J.: Path integration in mammals. HIPPOCAMPUS (2004)
36. Daugman, J.G.: Two-dimensional spectral analysis of cortical receptive field profiles. Vision Research **20** (1980) 847–856
37. Pouget, A., Dayan, P., Zemel, R.S.: Inference and computation with population codes. Annu. Rev. Neurosci. **26** (2003) 381–410

38. Strösslin, T., Gerstner, W.: Reinforcement learning in continuous state and action space. In: Artificial Neural Networks - ICANN 2003. (2003)
39. Foster, D.J., Morris, R.G.M., Dayan, P.: A model of hippocampally dependent navigation, using the temporal difference learning rule. Hippocampus **10(1)** (2000) 1–16
40. Doya, K.: Reinforcement learning in continuous time and space. Neural Computation **12** (2000) 219–245
41. Tolman, E.C.: Cognitive maps in rats and men. Psychological Review **55** (1948) 189–208
42. Quirk, G.J., Muller, R.U., Kubie, J.L.: The firing of hippocampal place cells in the dark depends on the rat's recent experience. Journal of Neuroscience **10** (1990) 2008–2017
43. Morris, R.G.M., Garrud, P., Rawlins, J.N.P., O'Keefe, J.: Place navigation impaired in rats with hippocampal lesions. Nature **297** (1982) 681–683
44. Hughes, A.: The topography of vision in mammals of contrasting life style: Comparative optics and retinal organisation. In Crescitelli, F., ed.: The Visual System in Vertebrates. Volume 7/5 of Handbook of Sensory Physiology. Springer-Verlag, Berlin (1977) 613–756

Representations for a Complex World: Combining Distributed and Localist Representations for Learning and Planning

Joscha Bach

University of Osnabrück,
Department of Cognitive Science
jbach@uos.de

Abstract. To have agents autonomously model a complex environment, it is desirable to use distributed representations that lend themselves to neural learning. Yet developing and executing plans acting on the environment calls for abstract, localist representations of events, objects and categories. To combine these requirements, a formalism that can express neural networks, action sequences and symbolic abstractions with the same means may be considered advantageous. We are currently exploring the use of compositional hierarchies that we treat both as *Knowledge Based Artificial Neural Networks* and as localist representations for plans and control structures. These hierarchies are implemented using *MicroPsi node nets* and used in the control of agents situated in a complex simulated environment.

1 Introduction

Plan based control of agents typically requires the localist representation of objects and events within the agent's world model: to formulate a plan, individual steps have to be identified, arranged into sequences and evaluated. The ordering of the plan components asks for some kind of pointer structure that is usually expressed as a symbolic formalism. On the other hand, to have an agent act in a complex dynamic environment with properties and structure unknown to the agent, it is often desirable to use sub-symbolic representations of the environment that lend themselves to autonomous reinforcement learning. These demands result in hybrid architectures [11, 5], which combine symbolic and sub-symbolic layers of description. Usually, these layers are implemented with different techniques, for instance by defining a number of learnable low-level behaviors implemented in neural networks, which are switched and parameterized by a symbolic, non-neural layer.

In our approach, we make use of a different setting: we are using a kind of executable semantic network, called *MicroPsi node nets* [2] that can act both as feed-forward networks suitable for back-propagation learning and as symbolic plan representations. Even the control structures of our agents are implemented within the same networks as are their plans and their representations of the environment. This has a number of advantages: it is not necessary, for example, to draw a sharp

S. Wermter et al. (Eds.): Biomimetic Neural Learning, LNAI 3575, pp. 265–280, 2005.
© Springer-Verlag Berlin Heidelberg 2005

boundary between categorical abstractions and sensory-motor behavior. Rather, we express rules and abstractions as instances of localist neural network structures that may even be used to facilitate neural learning. We may thus mix distributed representations at all descriptional levels with rules, and we can also use rules at the lowest sensory-motor levels, if this is appropriate for a given task.

2   MicroPsi Agents

We are currently developing a cognitive architecture that is called *MicroPsi* [1] and focuses on the autonomous acquisition of grounded representations by agents, based on motivation. MicroPsi is partially derived from ideas of the "Psi"-theory of psychologist Dietrich Dörner [6, 7, 8]. Here, agents do not possess predefined knowledge of the world, but a set of predefined modes of access to it (i.e. sensors and actuators and some low-level processing for sensor data). Additionally, MicroPsi agents have a fixed set of motivational parameters (*urges*), which measure demands (like food or uncertainty reduction). In the pursuit of the demands, the agents' action control establishes consumptive goals (which consist in fulfilling of these demands) and directs the agents' behavior.

While some of the urges allude to physical needs (like food, water and integrity), there are also urges that are directed on cognitive aspects (like *competence*, i.e. the effectiveness in attaining goals, and *uncertainty reduction*, which measures the degree

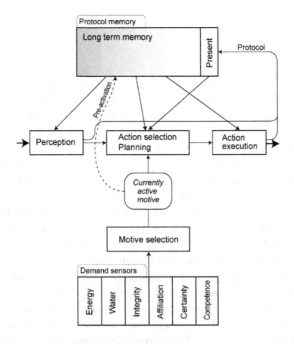

Fig. 1. Agent architecture as suggested by Dörner 2002 [8]

of exploration of accessible environmental features). Dörner also suggests a social urge, called *affiliation*, which is directed at receiving positive socially interpreted signals from other agents. Each urge may give rise to a respective *active motive*, where the motive strength depends on the deviation of the demand from its target value, and the chance of selecting it is proportional to both the estimate of reaching a related goal and the intensity of the urge. Initially, the agents do not know which operations on the environment are effective in addressing the demands, so they have to resort to a try-and-error strategy. If actions have an effect (positive or negative) on the demands, a connection between the action, the surrounding perceived situation and the demand is established. If the respective urge signal gives rise to the related motive later on, it will pre-activate the associated situation and action context in the agent's memory. Thus, the urges may govern the behavior of the agent, based on its previous experiences.

Perception, action control, memory retrieval and planning may furthermore be modified by a set of modulating parameters: the *selection threshold* determines motive stability, the *resolution level* controls the accuracy and speed of perception and planning by affecting the number of features that are tested, the *activation* controls the action readiness and is basically inverse to the resolution level, and the *securing level* affects the ratio of orientation behavior. Together with the current urge changes (which are interpreted as pleasure or displeasure signals) and the competence

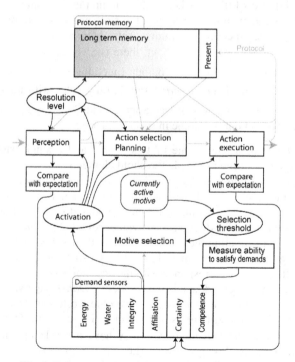

Fig. 2. Influences of modulators in the Dörner model

and uncertainty levels, the state of the modulators might be interpreted as an *emotional configuration* of the agent.

2.1 MicroPsi Node Nets: Executable Spreading Activation Networks

Internally, MicroPsi agents are made up of a spreading activation *node net*. This section gives a (somewhat simplified) description:

$$NN = \langle U, V, DataSources, DataTargets, Act, f_{net} \rangle \qquad (1)$$

$$U = \{(id, type, I, O, f_{node})\}, f_{node} : NN \rightarrow NN \qquad (2)$$

$$I = \{(slotType, in)\}, in_{i_j^u} = \sum_{n=1}^{k} \omega_{v_n} out_n \qquad (3)$$

$$O = \{(gateType, \alpha, out, \theta, min, max, amp, f_{act}, f_{out})\} \qquad (4)$$

$$f_{act}^{u,o} : in_{u,i} \times \theta \rightarrow \alpha \qquad (5)$$

$$f_{out} : \alpha \times Act \times amp \times min \times max \rightarrow out \qquad (6)$$

$$V = \{(o_i^{u_1}, i_j^{u_2}, \omega, st)\}, st \in \mathbb{R}^4; st = (x, y, z, t) \qquad (7)$$

Its building blocks, the net-entities U, are connected via weighted, directional links V. Net entities possess slots I (this is where links sum up their transmitted activation) and gates O, which take activation values from the slots and calculate output activations with respect to gate specific activators Act. Activators allow controlling the directional spread of activation throughout: there is an activator $act_{gateType} \in Act$ for each gate type, and since the node output function (6) is usually computed as $out = act_{gateType_o} amp \cdot min(max(\alpha, min), max)$, only gates with non-zero activators may propagate activation. Vectors of DataSources and DataTargets connect the net to its environment. The net entities come in several flavors:

- *Register nodes* have a single gate of type "*gen*" and a single slot, also of type "*gen*"; they are often used as simple threshold elements.
- *Sensor nodes* and *actuator nodes* provide the connection to the environment. Their activation values are received from and sent to the agent world (which can be a simulation or a robotic environment). Sensor nodes do not need slots and have a single gate of type "*gen*", which takes its value from a *DataSource*. Actuator nodes transmit the input activation they receive through their single slot (also type "*gen*") to a *DataTarget*. At the same time, they act as sensors and receive a value from a *DataSource* that usually corresponds with the actuator's *DataTarget*: The technical layer of the agent framework sends the respective *DataTarget* value to the agent's world-server which maps it to an operation on the world and sends back a success or failure message, which in turn is mapped onto the actuator's *DataSource*.
- *Concept nodes* are like register nodes; they have a single incoming slot, but in addition several kinds of outgoing links (i.e. types of gates). Each kind of links can be turned on or off to allow for directional spreading activation throughout the network, using the corresponding activator. Concept nodes allow the construction

of partonomic hierarchies: the vertical direction is made of by the link type "sub", which encodes a part-whole relationship of two nodes, and the link type "sur", which encodes the reciprocal relationship. Horizontally, concept nodes may be connected with "por" links, which may encode a cause-effect relationship, or simply an ordering of nodes. The opposite of "por" links are "ret" links. Additionally, there are link types for encoding categories ("cat" and "exp") and labeling ("sym" and "ref"). (Note that *link type* translates into a link originating from a gate of the respective type.)

- *Activator* and *associator nodes* are special entities that can be used to manage link weights, control directional spreading of activation and the values of activation of individual or groups of nodes in the network.

It is possible to write arbitrary control structures for agents using these node types.

We have implemented a graphical editor/simulator to do this. However, we found that the manual design and debugging of large control programs using individual linked nodes is practically not feasible (trying to do this literally gives the notion of spaghetti code a new meaning). If a structure within the agent is not meant to be accessed as a collection of nodes, it is more straightforward to implement it using a programming language such as Java. This purpose is served by

- *Native modules* – this are net entities encapsulating arbitrary functions f_{node} acting on the net, implemented in a native programming language, and executed whenever the module receives activation.

- *Node spaces* are net entities that encapsulate a collection of nodes:

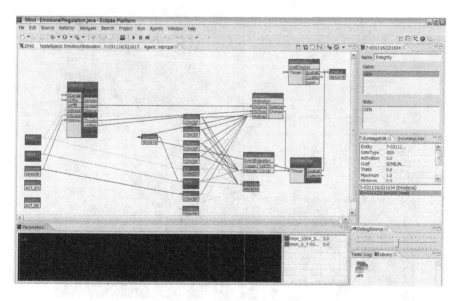

Fig. 3. MicroPsi node net editor

$$S = \left\{ U^S, DataSources^S, DataTargets^S, f_{net}^S \right\} \qquad (8)$$

Activator and associator nodes act only within the level of their node spaces. Because node spaces can contain other node spaces, agents may be split into modules, which make their innards a lot more accessible to human readers.

A slightly more detailed description of MicroPsi node nets is given in [2].

2.2 Situatedness of the Agents

Most of our experiments take place in a simulated environment that provides necessary resources and some hazards to the agents. Objects within the environment appear as co-located collections of features to the agents, where features correspond to sensory modalities of the agents. The simulation world is a plane consisting of different terrain types (the terrain has an effect on locomotion and may also provide a hazard, i.e. certain areas may cause damage to the agent). Different modes of locomotion are available to the agents, such as simple grid-based movement or a simulation of a pair of stepper motor driven wheels.

In the past, we have presented agents with identifiers for each object, along with a spatial position relative to the agent. Currently, we start using abstract basic modalities, such as *Gestalt* identifiers, spatial extensions, relative positions, color

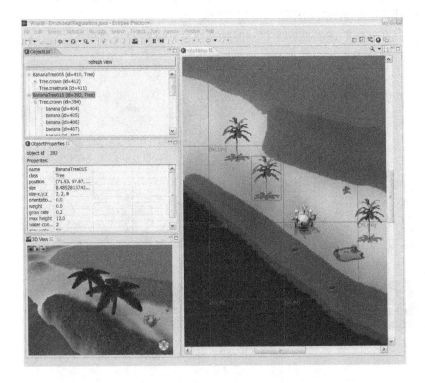

Fig. 4. World editor/simulator

values and weights, which might be replaced by a more low-level interface (like bitmaps and surface textures) in the future. Besides sensing, agents may probe their environment by acting upon it and examining the outcome. Actually, sensing may be seen as a form of action, and consequently, the definition of the sensory appearance of an object may amount to a script that encodes a sequence of actions necessary to recognize it.

Based on the affordances of the agents [9, 10], which are constrained by the sensory modalities and the needs of the agents, internal representations are derived from interactions with the environment.

Internally, world objects may be composed of sub-objects in spatial arrangements; the world maintains interactions between these objects by performing a discrete simulation, typically asynchronous to the agents. Agents, environment and computationally expensive world components (such as a simulation of plant growth) may run in a distributed network.

Using a constructed virtual environment for learning and classification experiments is not without difficulty: in many cases, agents may do nothing but rediscover a portion of the ordering that has been carefully engineered into their environment before, which limits the complexity of what is learned to what has been pre-programmed elsewhere. Additionally, the bias introduced by the artificial world may make it difficult to perform meaningful evaluations of the learning and classification results. On the other hand, because of shortcomings in perceptual and motor abilities, robots tend to be confined to a highly restricted and artificial environment as well. To put it a little provocatively: contemporary robots are often almost deaf, functionally blind, have restricted locomotion and only the simplest of push/grasp interactions available. It seems that many robotic testbeds (such as robotic soccer) can be quite satisfactorily simulated, and even significantly enhanced by incorporating additional modes of interaction and perception.

The simulation world is part of an integral experimental framework, along with the network simulator and a number of tools that aid in performing experiments. [3, 4]. (We are also using the toolkit for other agent designs, for instance in Artificial Life experiments, and to control Khepera robots.)

3   Representation Using Compositional Hierarchies

3.1   Partonomies

In MicroPsi agents, there is no strict distinction between symbolic and sub-symbolic representations. The difference is a gradual one, whereby representations may be more localist or more distributed. For many higher-level cognitive tasks, such as planning and language, strictly localist structures are deemed essential; in these procedures, individual objects of reference have to be explicitly addressed to bring them into a particular arrangement. However, a node representing an individual concept (such as an object, a situation or an event) refers to sub-concepts (using the "sub"-linkage) that define it. These sub-concepts in turn are made up of more basic sub-concepts and so on, until the lowest level is given by sensor nodes and actuator nodes. Thus, every concept acts as a reference point to a structured interaction context; symbols are grounded in the agent's interface to its outer and inner environment.

There are several requirements to such a representation:

- *Hierarchies:* abstract concepts are made up of more basic concepts. These are referenced using "sub"-links (i.e. these sub-concepts are "part-of" a concept). Because these sub-concepts are in turn made up of sub-concepts as well, the result is a *compositional hierarchy* (in this case, a *partonomy*). For example, a hierarchy representing a face can be made out of a concept of a face, "sub"-linked to concepts for the eyes, the nose, the mouth etc. The concept for the eye points to concepts for eyelids, iris etc. until the lowest level is made up of primitive image sensors like local contrasts and directions. Note that this representation is devoid of categories, we are only representing individual instances of objects. However, if similar instances are encountered later on, the representation may act as a classifier for that object structure.

- *Sequences:* to encode protocols of events or action sequences, sequences of concepts need to be expressed. This is done by linking nodes using "por"-connections. "por" acts as an ordering relation and is interpreted as a subjunction in many contexts. The first element of such a "por"-linked chain is called the head of a chain and marks the beginning of execution on that level. In our face-example, the "sub"-linked parts of the face concept could be connected using spatially annotated "por"-links that define a plan to first recognize the left eye, then the right eye, then the nose, then the mouth. These sequences may occur on all levels of the hierarchy. The mouth-concept for instance might be made up of a sequence looking for the upper lip and the lower lip etc.

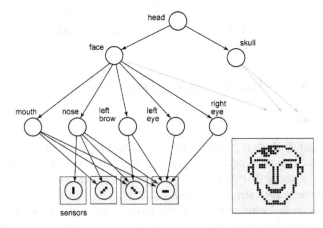

Fig. 5. Hierarchy of nodes, only "sub"-links are shown, reciprocal "sur"-links omitted

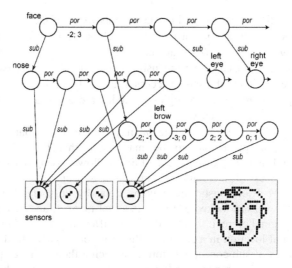

Fig. 6. Sequences in a hierarchy, reciprocal "sur" links and "ret" links omitted

- *Disjunctions:* Since there might be more than one way to reach a goal or to recognize an object, it should be possible to express alternatives. Currently this is done by using "sub"-linked concepts that are *not* "por"-linked, that is, if two concepts share a common "sur/sub" linked parent concept without being members of a "por"-chain, they are considered to be alternatives. This allows to link alternative sub-plans into a plan, or to specify alternative sensory descriptions of an object concept.
- *Conjunctions:* in most cases, conjunctions can be expressed using sequences ("por"-linked chains), or alternatives of the same concepts in different sequence (multiple alternative "por"-linked chains that permute over the possible sequential orderings). However, such an approach fails if two sub-concepts need to be activated in parallel, because the parts of the conjunction might not be activated at the same time. Currently we cope with this in several ways: by using weights and threshold values to express conjunctions (fig. 7a), with branching chains (fig. 7b) or with reciprocal "por"-connections (fig. 7c). In the first case, we encode the

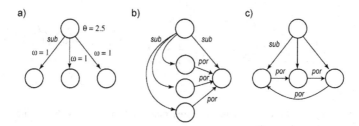

Fig. 7. Expressing conjunctions, reciprocal link directions ("ret" and "sur") have been omitted

relationship to the parent by setting the weights $\omega_{1..n,i}$ of the "sur/sub"-links from the alternatives $u_{1..n}$ to the parent u_i and a threshold value θ_i of u_i such that $\omega_{1..n,i} > \theta_i$ and $\omega_{1..n,i} - \omega_{j,i} < \theta_i$ for all individual weights $\omega_{j,i}$ of an alternative $u_j \in \{u_{1..n}\}$. In the second case, we are using two "por"-links (i.e. two "por"-linked chains) converging onto the same successor node, and in the third, we are defining that fully "por"-connected topologies of nodes are given a special treatment by interpreting them as conjunctive.

- *Temporary binding:* because a concept may contain more than one of a certain kind of sub-concept, it has to be made sure that these instances can be distinguished. Linking a concept several times allows having macros in scripts and multiple instances of the same feature in a sensor schema. In some cases, distinguishing between instances may be done by ensuring that the respective portions of the net are looked at in a sequential manner, and activation has faded from the portion before it is re-used in a different context (for instance, at a different spatial location in a scene). If this can not be guaranteed, we may create actual instances of sub-concepts before referencing them. This can be signaled by combining partonomies with an additional link-type: "cat/ref", which is explained below. Note that sensors and actuators are never instantiated, i.e. if two portions of the hierarchy are competing for the same sensor, they will either have to go through a sequence of actions that gives them exclusive access, or they will have to put up with the same sensory value.

3.2 Taxonomic Relationships

If two different "por"-linked chains share neighboring nodes, and the relationship between these node is meant to be different in each chain (for instance, there is a different weight on the "por" and "ret" links, or the direction of the linkage differs, if they have different orderings in the respective chains), the specific relationship can not be inferred, because "por"-links are not relative to the context given by the parent. This can be overcome by making the chain structure itself specific to the parent, and linking the nodes to the chain structure via "cat/exp"-links (fig. 8).

Thus, the structural intermediate node may hold activation values of the "exp"-linked actual concept, which itself may be used in other contexts as well. Of course, an intermediate node may have more than one "exp"-link. In this case, the linked

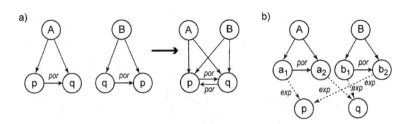

Fig. 8. a) Sharing of differently related features may lead to conflicts. b) Separating features and relationship with respect to parent

concepts become interchangeable (element abstraction). The intermediate node may be interpreted as a category of the "exp"-linked concepts. Using "cat" and "exp" links, it is possible to build taxonomic hierarchies. In conjunction with "sub" and "sur", MicroPsi node nets may be used to express hybrid or *parse structures* [12].

Within MicroPsi agents, "cat/exp" links are also used to reference different instances of the same concept, for instance in plans and in the local perceptual space. Here, "cat" links may act as pointers to the actual concepts in long term memory. "cat" may usually be interpreted as an "is-a" relationship.

3.3 Execution

Behavior programs of MicroPsi agents could all be implemented as chains of nodes. The most simple and straightforward way probably consists in using linked concept nodes or register nodes that are activated using a spreading activation mechanism. Conditional execution can be implemented using sensor nodes that activate or inhibit other nodes. Portions of the script may affect other portions of the script by sending activation to associator nodes or activator nodes. However, for complex scripts, backtracking and re-using portions of the script as macros become desirable.

For our purposes, a hierarchical script consists of a graph of options O, actions A and conditions C. Options might follow each other or might contain other options, so they can be in the relationships $succ(o_1, o_2)$, $pred(o_1, o_2)$ iff $succ(o_2, o_1)$, $contains(o_1, o_2)$ and $part\text{-}of(o_1, o_2)$ iff $contains(o_2, o_1)$. They might also be conjunctive: $and(o_1, o_2)$ iff $and(o_2, o_1)$, or disjunctive: $or(o_1, o_2)$ iff $or(o_2, o_1)$. The following restriction applies: $and(o_1, o_2) \vee or(o_1, o_2) \vee succ(o_1, o_2) \rightarrow \exists o_3: part\text{-}of(o_1, o_3) \wedge part\text{-}of(o_2, o_3)$.

Options always have one of the states *inactive, intended, active, accomplished* or *failed*. To conditions, they may stand in the relationship $is\text{-}activated\text{-}by(c, o)$, and to actions in $is\text{-}activated\text{-}by(o, a)$ and $is\text{-}activated\text{-}by(a, o)$. Options become *intended* if they are part of an *active* option and were *inactive*. They become *active*, if they are *intended* and have no *predecessors* that are not *accomplished*. From the state *active* they may switch to *accomplished* if all conditions they are *activated by* become *true* and for options that are *part of* them holds either, that if they are member of a conjunction, all their conjunction partners are *accomplished*, or that at least one of them is not part of a conjunction and is *accomplished* and has no predecessors that are not *accomplished*. Conversely, they become *failed* if they are *active*, one of the conditions they are *activated by* becomes *failed* or if all options that are *part of* them and are neither in *conjunctions* nor *successor* or *predecessor* relationships turn *failed*, or if they contain no options that are not in *conjunctions* or *successions* and one of the *contained* options becomes *failed*. And finally, if an option is *part of* another option that turns from *active* into any other state, and it is not *part of* another *active* option, it becomes *inactive*.

The mapping of a hierarchical script as defined above onto a MicroPsi node net is straightforward: options may be represented by concept nodes, the part-of relationship using "sub" links, the successor relationship with "por" links etc. (In order to use macros, "exp"-links have to be employed as discussed in section 3.2.)

Conditions can be expressed with sensor nodes, and actions with actuator nodes, whereby the activation relationship is expressed using "gen" links. Disjunctions simply consist in nodes that share the same "sur" relationship, but are not connected to each other. This way, there is no difference between sensory schemas that are used to describe the appearance of an object, and behavior programs: a sensory schema is simply a plan that can be executed in order to try to recognize an object.

Even though the notation of a script is simple, to execute hierarchical scripts, some additional measures need to be taken. One way consists in employing a specific script execution mechanism that controls the spread of activation through the script. We have implemented this as a script execution module that will "climb" through a hierarchical script when linked to it (fig. 9).

Here, the currently active option is marked with a link and receives activation through it. "sub"-linked options get their *intended* status by a small amount spreading activation. By preventing this pre-activation from spreading (for instance by using inhibitory connections from outside the script), it is possible to block portions of the script from execution.

Actions are handled by sending activation into an actuator node and waiting for a specified amount of time for its response. If the actuator node does not respond with a success signal, the script will fail at the respective level and backtrack; backtracking positions are held in a stack that is stored within the script execution module.

The drawbacks of this approach are obvious:

-      There is no parallel processing. Only one option is being activated at a time. In the case of conjunctive nodes, the activation focus is given to the one with the highest pre-activation first. If all conjunctive options have the same activation, one is randomly chosen

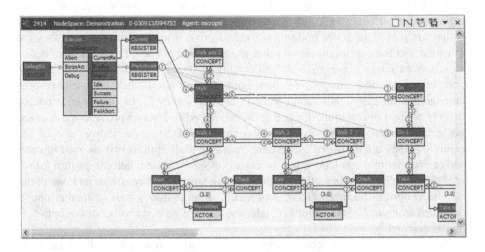

Fig. 9. Using a native module for script execution

- The activation of the individual nodes poorly reflects the execution state, which is detrimental to some learning methods (like decaying of rarely used links).
- The approach does not seamlessly integrate with distributed representations, i.e. it is for example not advisable to perform back-propagation learning on the node hierarchy. (It is still possible to add lower, distributed layers that will be interpreted just like sensor and actuator nodes, though.)

On the other hand, it is also possible to devise a specific node type that spreads activation in the following manner: each node has two activation values, the *request activation* a_r, determining whether a node attempts to get confirmed by "asking" its sub-nodes, and a *confirm activation* a_c that states whether a node confirms to its parent concepts, where for each node: $0 \le a_c \le a_r$ (or $a_c < 0$ to signal failure). When a node gets first activated, it switches its state from *inactive* to *requested*. It then checks for "por"-linking neighbors (i.e. the corresponding slot): if it has no unconfirmed predecessors (i.e. nodes that possess a "por"-link ending at the current node), it becomes *requesting* and starts propagating its request activation to its "sub"-linked sub-concepts. In the next step, it switches to the state *wait for confirmation*, which is kept until its "sub"-linked children signal either confirmation or failure, or until their "sub"-linking parent stops sending a request signal. After confirmation, the node checks if it has "por"-linked unconfirmed successors. If this is not the case, a_c gets propagated to the "sub"-linking parent node, otherwise a_c is propagated to the successor node only. The node then remains in the state *confirmed* until its parent node stops requesting, then goes back to *inactive*. (Failures are propagated immediately.)

With this mechanism, we can describe conjunctions and disjunctions using weighted links. Since the execution of a script is now tantamount to pre-activating a hypothesis (the portion of the script we want to try) and its failure or success translates into a match with a sensor configuration, we may use the data structure for back-propagation and other neural learning methods. The distributed nature of execution makes supervision of the execution more difficult, but enables parallel distributed processing. (It should be mentioned that we can not use simple chains of "por"-linked nodes with this approach, without also "sub"-linking each of them to the same parent node. This is less of an issue for the script execution module, because it can determine the parent of each element of a sequence by parsing backwards along the "ret" links to the first element. But because this might take additional time in the case of backtracking, it seems always a good idea to declare the part-of relationship of each sequence element explicitly.)

4 Perception

The perceptual mechanism of the agents follows a bottom-up/top-down approach. If a sensor is triggered by some prominent environmental feature, activation spreads upward ("sur") and activates all concepts this sensor is part of. These concepts are then marked as perceptual hypothesis by linking them to an activation source. From now on, script execution is performed as described above: On each level of each of these concepts, activation is spreading down again to test the other parts deemed necessary to identify them. If the sensors connected to these other parts report success (i.e. find a corresponding feature in the environment), then the concept itself becomes confirmed, otherwise it will be marked as "failed". Thus, activation spreads upward, "suggesting" theories of what is there to be perceived in the environment, then down, "testing" those theories, then up again, confirming or disconfirming them. The remaining confirmed concepts are considered the immediate percepts of the agent and are made available in a distinct node-space for further processing. (In Dörner's work, this mechanism is called "hypothesis based perception".)

4.1   Building Structured Representations of the Environment

Before partonomies can be employed to recognize objects, the agent has to construct them. There are several ways to do this. The first step, called *accommodation*, has been suggested by Dörner [6] and consists in using basic perceptual levels to arrive at a simple division of compositional layers: for instance by putting direction sensitive detectors for line elements at the lowest level. These make up contour segments, which are in turn parts of gestalts, and these may be combined into object schemas, which finally make up more complex objects and situations. However, the main problem turns out not to be the initial construction of such a simple partonomy, but its extension and modification, whenever new examples are encountered by the agent. In realistic environments, perception tends to be partial and somewhat uncertain, and different instances of a perceptual class may have somewhat different appearances. We combine two different approaches:

Whenever we come about a partonomic graph describing a percept that has not been encountered before in exactly the same way, we perform a graph matching procedure with existing object representations to obtain a similarity measure. For this, we are using the MatchBox algorithm [13] that estimates the best match according to node topology and link weights with a Hopfield network.

If we encounter a prototype with sufficient similarity, the new percept is merged with it by adding new nodes where necessary, and by adjusting link weights. If the percept does not exhibit much similarity to anything encountered before, a new prototype is created. (However, if the new object lends itself to similar interactions as another already known object, the perceptual descriptions might be merged as disjunctions in the future.)

The other method makes use of *Knowledge Based Artificial Neural Networks* (KBANN) [14, 15]. It works by using a partial, possibly incorrect domain theory that is given to the agent in the form of propositional logical clauses. These are converted into a hierarchical graph, where the lower level (sensor nodes and actuator nodes) are the precedents which are "sur"-linked to their antecedents. Conjunctions, disjunctions and negations are expressed by using weighted links and threshold values.

When activation is given to the sensor nodes according to the logical values of the antecedents, it spreads along the links and activates the compositional nodes higher up in the hierarchy in such a way as to evaluate the clauses.

The graph is then extended by a few randomly inserted nodes and additional sensor nodes to lend it more flexibility. These new nodes are given link weights close to zero, so they do not affect the outcome of the spreading activation. Furthermore, weak links are added across the hierarchy. The network is then used to classify examples encountered by the agent and is adapted using standard back-propagation. The additional nodes allow the insertion of additional abstract features, and the additional links make generalization and specialization possible.

4.2 Learning Action Sequences and Planning

The planning capabilities of our agents are still very basic. To achieve at a repertoire of action sequences, they maintain a protocol memory made up of "por"-linked situations in the environment, whereby a situation consists of a spatial arrangement of recognized objects, along with actions performed on them. Whenever a situation is correlated with a positive or negative effect upon the urges of the agent (for example, the agent has reduced its need for water by collecting some), the links to the preceding situations are strengthened according to the impact of this event. Because in protocol memory, links below a certain strength decay over time, event chains leading to situations of importance to the agent (where it satisfied a need or suffered damage) tend to persist.

When the agent establishes a goal from an urgent need, it first looks for a direct chain leading from the current situation to the goal situation. If it can not find such an automatism, it attempts to construct such a chain by combining situations and actions from previous experiences, using a limited breadth-first search. If it does not succeed in constructing a plan, it resorts to either a different goal or to try-and-error behavior to create more knowledge.

5 Outlook

In MicroPsi agents, we combine neural network methods with compositional hierarchies that lead to localist, abstract representations of the environment suitable for planning. However, much work remains to be done. Currently, we are concerned especially with the representation of space and time and the implementation of an inheritance mechanism that allows distributing information along taxonomical hierarchies, as these seem to be a prerequisite for more generalized approaches to perception and memory.

While we are busy extending the simulation environment with more operators, more complex objects and relationships between them, we are planning to use the architecture more extensively with real-world sensor data and for the control of robots in the near future.

Acknowledgements

This work is the result of the efforts of many people, especially Ronnie Vuine, who supervised the implementation of large parts of the technical framework, the planning and the perception mechanism, Colin Bauer, who has implemented support for KBANNs and has designed the graph matching algorithm. Matthias Füssel has implemented the better part of the simulation environment, and David Salz is the creator of a 3D display component for the agent world.

References

1. Bach, J. (2003). The MicroPsi Agent Architecture Proceedings of ICCM-5, International Conference on Cognitive Modeling, Bamberg, Germany (pp. 15-20)
2. Bach, J., Vuine, R. (2003). Designing Agents with MicroPsi Node Nets. Proceedings of KI 2003, Annual German Conference on AI. LNAI 2821, Springer, Berlin, Heidelberg. (pp. 164-178)
3. Bach, J. (2003). Connecting MicroPsi Agents to Virtual and Physical Environments. Workshops and Tutorials, 7th European Conference on Artificial Life, Dortmund, Germany. (pp. 128-132)
4. Bach, J., Vuine, R. (2003). The AEP Toolkit for Agent Design and Simulation. M. Schillo et al. (eds.): MATES 2003, LNAI 2831, Springer Berlin, Heidelberg. (pp. 38-49)
5. Burkhard, H.-D., Bach, J., Berger, R., Gollin, M. (2001). Mental Models for Robot Control, Proceedings of Dagstuhl Workshop on Plan Based Robotics 2001
6. Dörner, D. (1999). *Bauplan für eine Seele*. Reinbeck
7. Dörner, D. (2003). *The Mathematics of Emotion*. International Conference on Cognitive Modeling, Bamberg
8. Dörner, D., Bartl, C., Detje, F., Gerdes, J., Halcour, (2002). *Die Mechanik des Seelenwagens. Handlungsregulation*. Verlag Hans Huber, Bern
9. Gibson, J.J. (1977). The theory of affordances. In R. E. Shaw & J. Bransford (Eds.), Perceiving, Acting, and Knowing. Hillsdale, NJ: Lawrence Erlbaum Associates
10. Gibson, J.J. (1979). The Ecological Approach to Visual Perception. Boston: Houghton Mifflin
11. Maes, P. (1990), Situated Agents Can Have Goals, Robotics and Autonomous Systems, 6 (p. 49-70)
12. Pfleger, K. (2002). On-line learning of predictive compositional hierarchies. PhD thesis, Stanford University
13. Schädler, K.,Wysotzki, F. (1998). Application of a Neural Net in Classification and Knowledge Discovery. In: Michael Verleysen (ed.): Proc. ESANN'98, D-Facto, Brussels
14. Towell, G. Shavlik, J. (1992). Using symbolic learning to improve knowledge-based neural networks. In Proceedings of the Tenth National Conference on Artificial Intelligence, 177-182, San Jose, CA. AAAI/MIT Press
15. Towell, G. Shavlik, J. 1994. Knowledge-based artificial neural networks. Artificial Intelligence, 70 (p. 119-165)

MaximumOne: An Anthropomorphic Arm with Bio-inspired Control System

Michele Folgheraiter and Giuseppina Gini

DEI Department, Politecnico di Milano,
piazza Leonardo da Vinci 32, Italy

Abstract. In this paper we present our bio-mimetic artificial arm and the simulation results on its low level control system. In accordance with the general view of the Biorobotics field we try to replicate the structure and the functionalities of the natural limb. The control system is organized in a hierarchical way, the low level control reproduces the human spinal reflexes and the high level control the circuits present in the cerebral motor cortex and the cerebellum. Simulation show how the system controls the single joint position reducing the stiffness during the movement.

1 Introduction

Developing an artificial arm that mimics the morphology and the functionalities of a human limb is the principal goal of our work. In designing the arm we adopted a biomimetic approach; this means that at the beginning we dedicated a big amount of time in studying the natural limb from the anatomical, physiological and neurological point of view. After these studies we tried to integrate knowledge from different scientific and technological fields in order to synthesize the robot. We can state that the approach we adopted is in accordance with the general view of the Biorobotics field.

People involved in this robotics branch ,[1],[2],[3],[4],[5],[6] believe that studying and mimicking a biological organism allows us to design a robot with more powerful characteristics and functionalities than a classical robot, as well as to better understand the organism itself. Indeed, if we think of the history of technology, often humans were inspired by nature. Famous are the studies conducted by Leonardo da Vinci between 1487 and 1497 on the flying machines, that were inspired by the birds. This does not mean that observing and studying nature we can find out the best solution for a specific problem. In fact, for example, our technology can synthesize flying machines that are much faster than any biological organism.

An important question, that will arise if you are involved in this research field, is: why emulate the human body? Many scientists are convinced that for a robot whose purpose is to work with people, human morphology is necessary. In millions of years the human species has adapted the environment to its needs, developing tools and things that are suitable for its morphology. So, if we want

S. Wermter et al. (Eds.): Biomimetic Neural Learning, LNAI 3575, pp. 281–298, 2005.
© Springer-Verlag Berlin Heidelberg 2005

a robot to collaborate with a human being in a unstructured environment, it must have human shape and human-like manipulation capabilities. It is clear that, from a technical point of view, it is not possible, and at the same time not necessary, to reproduce in detail the human body's functionalities and morphology. Instead what is desirable for a futuristic humanoid robot is the same human mobility, manipulation capability and adaptability. Another aspect that justifies research in this field, as well as in the biorobotics field in general, is the utilization of biomimetic robotic systems as a new tool to investigate cognitive and biological questions. Collaboration between neurologists, psychologists and roboticians can be a useful way to improve in each specific field. Engineers can be inspired by neurological and psychological studies in the synthesis of the artificial system, and at the same time, neurologists and psychologists can better understand the biological system analyzing results coming from the experimentation on the artificial system.

In section 2 we will introduce the state of the art in this robotics field. Section 3 describes our experimental prototype from the kinematical and mechanical point of view. In Section 4 the control system architecture is outlined with particular attention to the reflex module. Finally the last section brings the conclusions to this work.

2   State of the Art

Robotics since its origin was involved in replicating human manipulation capabilities. In order to better understand the motivation pushing researchers toward humanoid robotics it is useful to look at the robot arms' history. One of the first robots for research purposes was the Stanford arm, designed in the Stanford Artificial Intelligence Lab. This robot has 6 DOFs (Degrees Of Freedom) , five revolute joints and one prismatic, therefore it can not be classified as anthropomorphic, nevertheless it was one of the first attempt to reproduce human arm manipulation capabilities.

In the sixties General Motor (the first to apply a robot in industry) financed a research program at MIT that developed another famous robot: the PUMA (Programmable Universal Manipulator for assembly).

This manipulator has 6 rotational DOF's and therefore it is classified as anthropomorphic; we can say that this robot was clearly inspired by biology. Indeed it is possible to compare this robot to a human arm; we can divide the mechanical structure in three principal blocks: the shoulder with two DOF, the elbow with 1 DOF and the wrist with another three DOF. The Puma has a dexterity that is quite near to that of a human arm, even though the human shoulder has more than two DOF. The analogy between the human arm and the PUMA manipulator is true only from a kinematic point of view, because the two systems have completely different performances. We can assert that this robot is more precise than the human arm, but at the same time the human arm can exhibit a compliant behavior that is indispensable to perform certain tasks like use a screwdriver or clean a complex surface.

It is clear that for industrial applications, a classical manipulator is better than a human arm. For example a manipulator is stronger than a human limb. The load for a medium size robot is about 10 Kg, but a human being finds it difficult to move, in every position of the workspace, such a weight. Manipulators are more precise and accurate in positioning the end-effector and furthermore they are free from fatigue problems that affect the human arm during intense activities.

Nevertheless from another point of view, the human arm is superior to robots. It is lighter and therefore it has a big force to weight ratio (100N/20N=5) with respect to an artificial manipulator(100N/3000N=0.03). Right now, with present technology, we are far away from the possibility to emulate human arm efficiency and functionality. What is lacking today is a system that presents the same flexibility and the same compliant behavior as the human limb. In this context, the applicability of industrial robots remains confined in the factories. Therefore, at the moment, a lot of research in order to bring robot systems also in the household and in the public environments is still needed.

Right now there are many research groups involved in developing humanoid artificial arms; usually the simple robot structure comprises one or two arms, a torso and a head equipped with a vision system. Because light-weight and a compliant behavior is needed for the robot, a lot of research was done on novel actuators able to mimic, at least from the macroscopic point of view, the human muscle.

At the Center for Intelligent Systems (Vanderbilt University) Prof. Kawamura and its group are working on the ISAC humanoid robot.

This robot consists of a human-like trunk equipped with two six-DOF arms moved by McKibben artificial muscles[4]. The system has also a four-DOF stereo vision head with voice recognition that permits interaction between the robot and humans. Each joint is actuated by two antagonistic actuators that are controlled by a system able to emulate the electromyogram patterns (EMG) of a human muscle. In particular the pressure inside the actuator is governed by a control signal analogous to the tonic and phasic activation of the muscle; it consists in three phases (agonist-antagonist-agonist) that permits the single joint to reach a precise position. The sensorial information are used to correct for misperceived loading conditions and to compensate eventually variations of the physical characteristics of the robot's actuators. The arm, during a fast reaching movement, can avoid an obstacle performing a reflex behavior [7], furthermore the phasic pattern is autonomously adjusted when a reach trajectory doesn't closely match a desired response. The main advantage of this bio-mimetic control architecture is the possibility to reduce the joint stiffness during a movement execution; this permits at the same time to save energy and to perform movements that are not dangerous for human beings.

Another project in the same direction is that one at the Biorobotics Laboratory in Washington University. Here Prof. Hannaford and his team have worked intensely on the emulation of the human arm [8] [3]. The goal of this research is to transfer knowledge from human neuro-musculo-skeletal motion control to robotics in order to design an "anthroform" robotic arm system. They introduce the new word "anthroform" to describe a robotic arm in which all aspects of its design are specified in terms of emulation of the corresponding functions of the human arm. They tested the elastic property of the McKibben actuators [9] [10] and proposed a more accurate dynamic model. In comparison with experiments conducted on human and animals muscle[11] they show how these type of actuators are, actually, the best choice to implement an anthropomorphic robot arm. Following the bio-mimetic approach they also developed a new kind of sensor [12] [13], whose purpose is to replicate a mammalian muscle spindle cell, that measures the contraction and the muscle velocity.

Since they maintain that it is very hard to create a realistic model of the human arm, they prefer to make experiments directly on the robotic arm and subsequently compare the data with that of a human limb. They are interested not only in the emulation of the human arm actuation system but also in the emulation of the spinal cord reflexes to control the artefact. Here, in comparison with the Kawamura et al. approach, they based the control system on studies conducted by neurophysiologists on the neural circuits delegated to generate the basic arm reflexes. In order to build a real time controller they implemented the neural circuit in a DSP (Digital Signal Processor) and acquired data coming from the force and position sensor with a dedicated computer.

The principal experiment conducted on this system was the cyclic application of a noise force on the forearm and the measurements of the joint angle deviation. This was made in many conditions and changing the neural network parameters. After a large amount of experiments they calculate the covariance between the more important variables in order to better understand their correlation. This analysis shows which are the variables and the sub-networks involved in a certain behavior, and allows formalizing hypothesis also on the human limb. The results show that muscle co-contraction and other circuit parameters can regulate the joint stiffness and damping.

3   The ARM Prototype

The arm developed in our laboratory (Figure 1), is intended to be the ideal test-bed for testing the control system architecture proposed in this work and for improving new technologies applicable to humanoid robotics. The arm, without considering the wrist and hand, that are still under development, has two joints for a total of four degrees of freedom. The shoulder consists in a spherical joint with 3 DOF, and the elbow is a rotational joint with 1 DOF. Joints are moved by tendons connected with McKibben artificial muscles, which in turn are bonded with the support structure and the upper arm. Each "muscle" is equipped with a force sensor mounted in series to the actuator (comparable, from a functional point of view, with the Golgi tendon organ in the human arm) and of a position sensor located in parallel to the external shell that covers the artificial muscle (comparable, from a functional point of view, with the muscle spindle in the human arm). The elbow joint has also an angular sensor (Figure 1)that

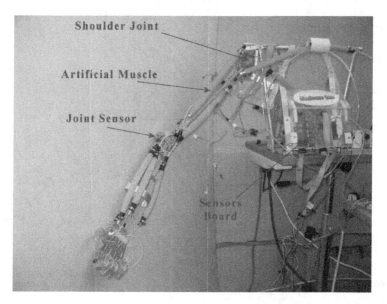

Fig. 1. The Arm Prototype, MaximumOne, Artificial Intelligence and Robotics Laboratory, Politecnico di Milano

measures the joint position and velocity with more precision. Sensor signals are conditioned and gathered by dedicated boards and sent to a PC A/D card. The control system runs in real time on a target PC, and its output are converted in appropriate signals that feed the actuation system.

As it is possible to see in the prototype picture (Figure 1), this arm has an anthropomorphic design. In particular, during the design, we have tried to reproduce the human arm dimensions and proportions, the articulation mobilities, the muscle structure, and the same sensorial capabilities. The actuation system is composed of seven muscles: five actuate the shoulder joint and two the elbow. This permits us to fully actuate the joints but at the same time to have a minimal architecture. The five shoulder actuators emulate the function of: pectoralis major, dorsal major, deltoid, supraspinatus and subscapularis muscles. The two elbow actuators emulate the function of biceps and triceps muscles. In comparison with the human arm musculature, the actuation system of our prototype is quite different, for example the biceps and triceps artificial muscles are mono-articular in the sense that they are dedicated only for the elbow actuation.

4   Architecture of the Control System

In designing the control system for the arm we tried to mimic the human nervous system; the architecture is organized in a modular-hierarchical fashion. At the bottom level (Figure 2) there are the artificial reflex modules that govern the ac-

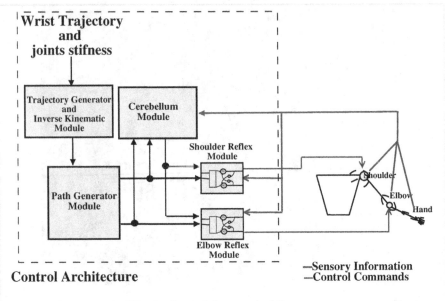

Fig. 2. Control System Architecture

tuator's contraction and force. These modules receive inputs from the joint path generator, which in turn is fed by the inverse kinematic module that computes the target actuators lengths. The reflex modules also receive inputs from the cerebellar module whose function is to regulate the path generator outputs. The cerebellum module, as inputs, receives signals from the path generator modules and the error signals from the reflex modules. The inputs of the entire control system are: the final hand position in the cartesian space, the GO signal that scale the speed of movement and the P signal that scales the level of artificial muscles co-activation (simultaneously activation of the muscle that govern the same joint).

From a hierarchical point of view, we can distinguish three principal levels:

High level controller: composed of the Inverse Kinematic and the cerebellum modules that cooperate in parallel to control the path generator activity

Medium level controller: composed of the path generator module. This is capable of generating desired arm movement trajectories by smoothly interpolating between the initial and the final length commands for the synergetic muscles that contribute to a multi-joint movement.

Low level controller: composed of the reflex modules that control the artificial muscles activities

The signals transmitted from one module to another are expressed in a vectorial form, where each vector component corresponds to one of the seven artificial muscles that compose the actuation system. Therefore L_T represents the target lengths vector for the actuators, V_T represents the target velocity vector for the actuators, E_L represents the length vector error of the actuators, C_S is the signal vector generated by the cerebellum module, and P is the stiffness command vector. At the level of each single module these signals are decomposed in their components and sent to the appropriate submodules.

In this document, for brevity, we do not deal with the cerebellum module that operate in feedforward manner in order to compensate Coriolis and inertia force during rapid movements [14],[15]. Instead we will concentrate our attention on the modules that govern the arm during normal conditions.

4.1   Inverse Kinematic Module

Given the final position for the wrist, in order to calculate the muscles lengths, it is necessary to solve the inverse kinematics problem. A necessary, but not sufficient, condition for the existence of the solution, is that the point that we want to reach is inside the arm's workspace. In robotics terminology the manipulator workspace is the portion of the space that is reachable by the robot's hand. If we take into account only the target point and we do not consider the arm orientation when the target is reached, the inverse kinematic problem, in our case, has an infinite number of solutions. This is due to the fact that to reach a point in a three dimensional space only three degrees of freedom (3DOFs) are needed,

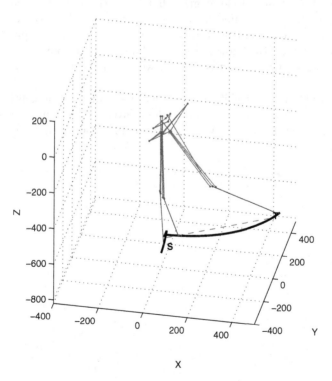

Fig. 3. Frontal View of the Arm in three different configurations, each line depicts a link or an artificial muscle

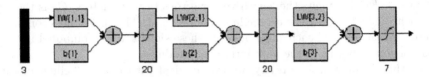

Fig. 4. Architecture of the Multilayer Perceptron

but our arm has four DOFs. To find a single solution we impose, in the system equations, a constraint on the orientation of the plane formed by the upper and forearm with the robot's sagittal plane. Normally this angle in a human being is about 20 degrees and remains fixed during the arm movement (Figure 3).

It is possible to approach the inverse kinematic problem in two ways: using the direct kinematics to generate input-output pairs that can be used to train a neural network, or solving other systems of equations where the wrist position is imposed. We followed the first approach which seems to be more appropriate to this situation. In order to obtain the input-output pairs necessary to train a neural network, we executed the direct kinematic algorithm on a sufficient set of inputs. Each input was an admissible vector of actuator lengths (as admissible we intend a set of muscle lengths that bring the wrist in a position inside the workspace). To determine the correct intervals for the actuator lengths, we performed some measurements directly on the real arm prototype. In a second step, when this data was known, we created a data set of points each representing a vector of actuators lengths. Finally we calculated the corresponding wrist position.

As a neural network architecture we chose a multilayer perceptron [16] with an input layer of three neurons, two hidden layers of 20 neurons each, and an output layer of seven neurons. As an activation function for the neurons we chose

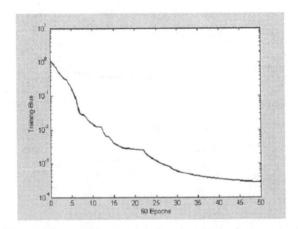

Fig. 5. Trend of the medium square error after 50 epochs

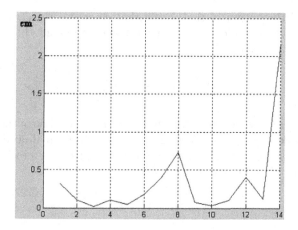

Fig. 6. The Generalization Error on the wrist position (calculated on 28 points)

a sigmoid, and for the network training the back propagation algorithm. Before training, the input-output pairs were normalized in order to obtain values in the interval (-1,1). We have used a set of 1081 training data, that was obtained using points in the workspace at distances of 5cm. The trend for the medium square error is reported in figure 5.

As we see from the graph the error, after 50 epochs, decreases under the value $2 \cdot 10^{-3}$. After 1000 epochs the error reached the value of $1.6 \cdot 10^{-4}$. This required 16 hours of computation on a Pentium-4 (2GHz) equipped with 500 Mb of memory. After the net was trained we conducted a series of tests to understand if the neural network exhibits a generalization behavior. Therefore we gave to the net positions for the wrist that were different from the positions used for training, and we calculated the error of generalization. In figure 6 we can see the error for 28 wrist positions.

The median value in positioning the wrist is about 0.8 cm, which is a good value for our purposes.

The main advantage in using a neural network to compute the inverse kinematics is that we can train the network on values acquired directly on the real arm. This overcame the limitations in using an approximate arm model, and is suitable especially for complex kinematic chains as found in a humanoid robot. The other advantage is that the time required for the network to compute the inverse kinematic, for a given point in the workspace, is low in comparison with other algorithms. Indeed, when the network is trained, the operations required to calculate the outputs are simple additions and multiplications.

4.2   Path Generator Module

The path generator module is capable of generating desired arm movement trajectories by smoothly interpolating between the initial and the final length commands for the synergetic muscles that contribute to a multi-joint movement. The rate of the interpolation is controlled by the product of two signals: the start signal GO and the output of the V_i cell, that computes the error in the length of the muscle i^{th}. The Go signal is a volitional command that in our case is formalized by the equation 1:

$$GO(t) = Go\frac{(t-\tau_i)^2}{k+(t-\tau_i)^2}u[t-\tau_i] \tag{1}$$

where parameter Go scales the GO signal, τ_i is the onset time of the ith volitional command, k takes into account the time that the GO signal needs to reach the maximum value and $u[t]$ is a step function that jumps from 0 to 1 to initiate the movement. The V_i cell dynamics is defined

$$\frac{d}{dt}V_i = K(-V_i + Lt_i - A_i) \tag{2}$$

where Lt_i is the target length for the ith muscle, the constant K defines the dynamic for the cell and A_i defines the present length command for the ith muscle. The model for the neuron A_i is defined in equation 3.

$$\frac{d}{dt}A_i = GO \cdot Th(V_i) - GO \sum_{k=1,k\neq i}^{n} \cdot Th(V_k) \tag{3}$$

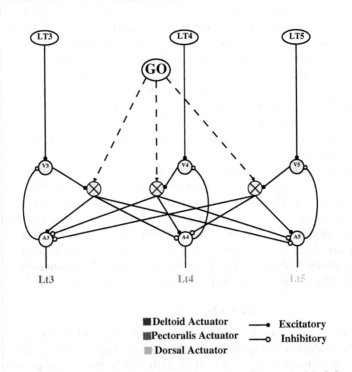

Fig. 7. Architecture of the Path Trajectory Generator Module

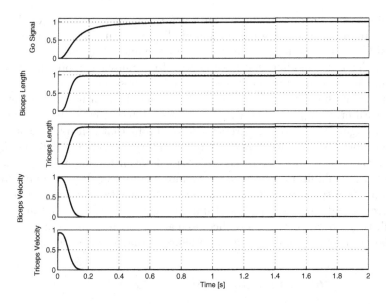

Fig. 8. Signals generated by the Path Generator Module that governs the elbow

Where again the Th is the same threshold function used for all the cells of our model. Study of Bullock and Grossberg [17] have demonstrated that this path generator model can be used to explain a large number of robust kinematic features of voluntary point to point movements with a bell-shaped velocity profile. The architecture for the path generator is presented in figure 7. The circuit is suitable for the three muscles of the shoulder joint that permit the upper arm flexion-extension and adduction-abduction. The trajectory generators for the arm's other muscles are quite similar to that one presented. The inputs for the system are the target lengths Lt_i for each muscle and the outputs are the signals A_i that will feed the inputs of the reflex module of each joint. An example of signals generated by the elbow path generator is presented in figure 8.

4.3   Joint Reflex Control Module

In order to perform the Reflex behaviors we implement a simplified model of the natural circuits present in the human spinal cord. With respect to other models in literature [18],[19],[20],[21], or to hardware solutions [22] we decided to neglect the spike behavior of the neuron for all the artificial cells, instead we concentrated our attention on modelling its membrane potential. From an information point of view the spiking behavior in the neuron is not so crucial. In a living organism the action potential mechanism permits to convert information, represented by the neuron potential (analog signal), into an impulsive signals. In such a manner the information is transmitted modulating the frequency of the impulsive signal. This is particularly useful when the signal (of few mV) is transmitted over a long distance, for example from the arm receptors (peripheral nervous system) to the central nervous system. In our system (arm prototype) the entity of the sensor signals are in the order of some volts, and all the information are processed in a normal CPU, so it is not efficient to convert the analog signals into a impulsive signals. The reflex module that governs the elbow muscle is represented in figure 9. It implements an opponent force controller whose purposes are to attempt to implement the path generator module commands, measure movements error and return error signals when the execution is different from the desired movement. In figure 9 M_6 and M_7 are the motoneurons that control the contraction rate and force of the triceps and biceps actuators respectively. I_a6 and I_a7 are the interneurons that receive the error signals from the artificial spindles and project, with inhibitory synapses, to the motoneurons of the antagonist muscles M_7 and M_6 respectively. R_6 and R_7 represent the Renshaw cells that receive the error signals from spindles and inhibit the corresponding motoneuron and I_a cell, they are important to reduce oscillations of the joint around the target angular position. I_b6 and I_b7 are interneurons that receive the signals coming

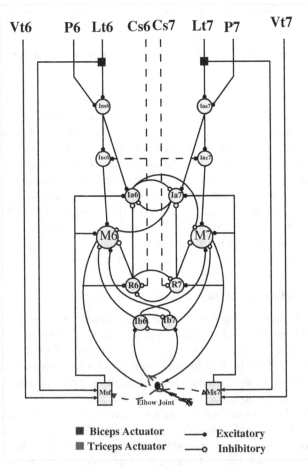

Fig. 9. Architecture of the Elbow Reflex Module

from the artificial Golgi tendon organs (that in this system are represented by a normalized force measurements). $I_{nc}6$ and $I_{nc}7$ are interneurons whose purpose is to integrate information coming from the cerebellum (signals C_s6 and C_s7) and from the $I_{ns}6$ and $I_{ns}7$ interneurons, thanks to these cells the cerebellum module can apply its influence on the overall joint movement. $I_{ns}6$ and $I_{ns}7$ are the interneurons that integrate information of stiffness and target length commands. Finally M_s6 and M_s7 represent the artificial muscle spindle receptors. As inputs they receive the muscle velocity command, the muscle target length command and the actual muscle length and in turn excite the corresponding motoneuron and I_a interneurons.

Neurons Model. It is possible to model the artificial neuron using a not linear differential equation. Different neurons, in the circuits, differ only for the constant values and the inputs, therefore here we will describe only the motoneuron equations.

The motoneuron receives its inputs from almost all the cells that compose the neural circuit. In equation 4 M_i represents the potential (membrane potential) of the motoneuron **i**.

$$\frac{d}{dt}M_i = (1 - M_i)(exc_i) - M_i(inh_i) \tag{4}$$

where the terms exc_i and inh_i are expressed by equations 5

$$exc_i = w_1 \cdot E_i + w_2 \cdot Inc_i + \sum_{k=1, k \neq i}^{n}(z_k \cdot Ib_k)$$
$$inh_i = K + w_3 \cdot R_i + w_4 \cdot Ib_i + \sum_{k=1, k \neq i}^{n}(v_k \cdot Ia_i) \tag{5}$$

the motoneuron output is

$$Mo_i = Th(M_i) \tag{6}$$

where the threshold function is defined by equations 7 :

$$Th(x) = \begin{cases} x & \text{if } 0 \leq x \leq 1 \\ 0 & \text{if } x \leq 0 \\ 1 & \text{if } x \geq 1 \end{cases} \tag{7}$$

The first term in the right side of the equation 4 is the gain for the excitatory part (**exc**); this gain is a function of the motoneuron potential. Therefore, the more the neuron is active the smaller the gain will became. This avoids the neuron's potential saturating rapidly when the excitatory synapses are strongly stimulated. The second part of the equation 4 gathers the inhibitory signals that feed the motoneuron (**inh**). In the (**inh**) term the inhibitory signals are multiplied by the corresponding synapse's gain w_i and v_k, and added together. It is clear, that the gain for the excitatory part (Equation 4) will decrease when the motoneuron potential increases. This contributes to maintain the neuron activation confined under the maximum value. The summation in the (**inh**) part, takes into account of the inhibitory action of the antagonistic Ia_i, the summation is extended to **n**, the number of motoneurons that constitute the reflex circuit (n=2 for the elbow reflex circuit and 3 for the shoulder reflex circuit).

The term K represents the leaky current of the neuron membrane. When the neuron is not excited its potential will decrease thanks to this term. Finally E_i is the error signal coming from the spindle cell Ms_i.

5   Test on the Reflex Module

In the first simulation we tested the capability of the reflex module to govern the actuator pressures in order to regulate the joint position. In this simulation the biceps and triceps length commands were manually set, therefore the path generator module, and the inverse kinematic module are not yet connected to the reflex circuit. In figure 10 the elbow angular position during the entire motion is reported. We see that the elbow position in the first movement reaches 0.4 radians (24.2°), with the second movement that starts at the fifth second it reaches 1.15 radians (70°), and finally the joint is restored to the first position.

Note that in the first movement there is a big over-elongation, partially due to the fact that when the first movement starts all the neurons potentials are set at the minimum value, and it take a certain time for the neurons to reach the operative value. In the arm prototype a minimum motoneuron activity is needed in order to maintain a sufficient pressure inside the artificial muscles. This is to avoid the detachment of the inner tube from the external braided shell.

It is possible to note from the graph, that also in the second joint movement there is a certain over-elongation, but this is reduced in comparison with the first one.

From figure 10 it is possible to see how the elbow's velocity follows a human bell shape profile, thanks to the smooth control behavior of the motoneurons.

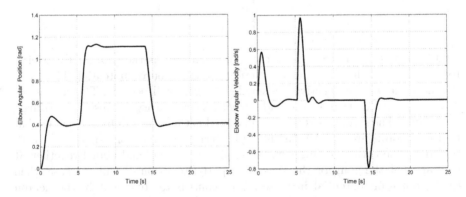

Fig. 10. The Angular position and velocity of the Elbow

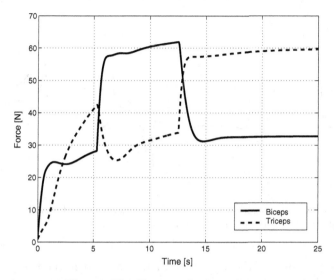

Fig. 11. The biceps and triceps forces

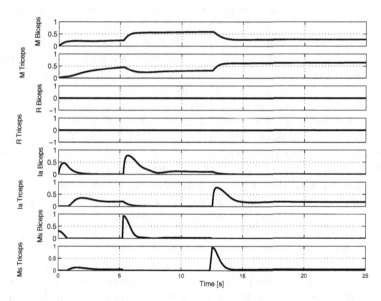

Fig. 12. Motoneuron and interneurons activities during the Elbow flexion

In figure 12 are reported the motoneuron and interneuron signals during the elbow flexion. Starting from the bottom we can see the activities of the artificial spindles Ms_i that measure the length and velocity errors in the biceps and triceps actuators. When the first elbow movement starts, the biceps's spindle increases its activity rapidly this is because, in comparison with the length command, the actuator should be shorter. After 0.8 seconds the biceps's Ms decreases its

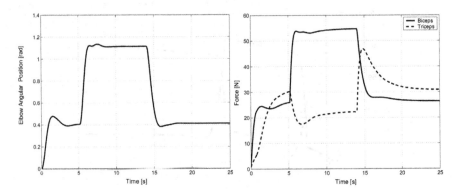

Fig. 13. The Angular position and the forces generated by the Biceps and Triceps actuators of the Elbow in the second experiment

activity to zero , but at the same time there is a burst in the triceps's Ms, due to the fact that the elbow has overshoot the target position and therefore the triceps should be contracted. Looking at the axes that report the Ia interneuron outputs, it is possible to note that the activity of these neurons are strictly correlated with those of the Ms. Nevertheless their effect, now, is transmitted on the antagonistic motoneuron. This action is very important for the elbow joint control. Indeed thanks to this cross inhibition a big length or velocity error on an artificial muscle, not only increases its pressure, but decreases at the same time the pressure in the antagonistic artificial muscle. We can see this influence in the motoneurons activities or directly on the actuator force.

In this first simulation we prevented the action of the R_i (Renshaw cells) interneurons, as it is possible to see in the graph of figure 12. They are important to maintain the motoneuron activity under control when the elbow has reached a stable position. From the graph that depicts the actuator force (Figure 11) it is possible to note that when each movement is ended the force increases autonomously in both the motoneurons, this causes a stiffness increasing in the elbow joint. In humans this disease is called hypertonia.

In the following simulation we enabled the R_i interneurons and performed the same movements as the first experiment (Figure 13).

This time, even thought the elbow performed the same movements, the actuators force changed. Indeed from the second graph of figure 13 it is possible to note that after each movement the forces don't increase like in the first experiment. This behavior is due to the R_i interneurons that limit the motoneurons potential when the elbow doesn't move. It is possible also to note that for the same elbow movement the forces applied by the two actuators are smaller in comparison with the other experiment. This behavior is very important because it permits the saving of energy especially during operations that do not require the movement of objects.

6   Conclusions

The main aim of this work was the development of a human-like artificial arm for application in the field of humanoid robotics. Because mimicking the human arm from the mechanical and functional point of view was one of our principal research aims, we conducted an intensive study of the natural limb. We concentrated our attention to the design and the implementation of a real human-like robotic arm, and at the same time, to developing a possible control model based on the actual knowledge that neurophysiologists have of the human nervous system.

Our arm differs from other analogous systems [23], [3], [24], by the presence of a full 3DOF shoulder joint moved by five artificial muscles. Furthermore, thanks to the employment of light materials, the system can be integrated with a whole humanoid robot. Results, on the low level control system, show how it can set the single joint position and stiffness in a efficiently way.

References

1. Beer, R., Chiel, H., Quinn, R., Ritzmann, R.: Biorobotic approaches to the study of motor systems. Current Opinion in Neurobiology **8** (1998) 777–782
2. Metta, G.: Babyrobot A Study on Sensori-motor Development. PhD thesis, University of Genoa (1999)
3. Hannaford, B., andChing Ping Chou, J.M.W., Marbot, P.H.: The anthroform biorobotic arm: A system for the study of spinal circuits. Annals of Biomedical Engineering **23** (1995) 399–408
4. Kawamura, K., II, R.P., Wilkes, D., Alford, W., Rogers, T.: Isac: Foundations in human-humanoid interaction. IEEE Intelligent Systems **15** (2000) 38–45
5. Webb, B.: Can robots make good models of biological behaviour? Behavioural and Brain Sciences **24** (2001) 1033–1094
6. Dario, P., Guglielmelli, E., Genovese, V., Toro, M.: Robot assistant: Application and evolution. Robotic and Autonomous Systems **18** (1996) 225–234
7. Gamal, M.E., Kara, A., Kawamura, K., Fashoro, M.: Reflex control for an intelligent robotics system. Proceeding of the 1992 IEEE/RSJ Intelligent Robots and System **2** (1992) 1347–1354
8. Hannaford, B., Chou, C.P.: Study of human forearm posture maintenance with a physiologically based robotic arm and spinal level neural controller. Biological Cybernetics **76** (1997) 285–298
9. Klute, G.K., Hannaford, B.: Accounting for elastic energy storage in mckibben artificial muscle actuators. ASME Journal of Dynamic Systems, Measurement, and Control, **122** (2000) 386–388,
10. Chou, C.P., Hannaford, B.: Measurement and modeling of mckibben pneumatic artificial muscles. IEEE Transactions on Robotics and Automation **12** (1996) 90–102
11. Klute, G.K., Czerniecki, J.M., Hannaford, B.: Mckibben artificial muscles: Pneumatic actuators with biomechanical intelligence. IEEE/ASME 1999 International Conference on Advanced Intelligent Mechatronics (1999)
12. Hannaford, B., Jaax, K., Klute, G.: Bio-inspired actuation and sensing. Robotics **11** (2001) 267–272

13. Jaax, K.N.: A Robotic muscle spindle: neuromechanics of individual and ensemble response. Phd thesis, University of Washington (2001)
14. Diener, H.C., Hore, J., Dichgans, J.: Cerebellar dysfunction of movement and perception. The Canadian Journal of Neurological Sciences (**20**) 62–69
15. Kawato, M., Gomi, H.: A computational model of 4 regions of the cerebellum based on feedback-error learning. Biological Cybernetics **68** (1992) 95
16. Haykin, S.: Neural Network A Comprensive Foundation. Macmillan College Publishing Company (1994)
17. Bullock, D., Grossberg, S.: Neural dynamics of planned arm movements:emergent invariants and speed-accuracy properties during trajectory formation. Psychological Review **95** (1988) 49–90
18. Folgheraiter, M., Gini, G.: Simulation of reflex control in an anthropomorphic artificial hand. Proc. VIII ISCSB 2001 (Int Symposiumon Computer Simulation in Biomechanics), Milan, Italy, (2001)
19. Folgheraiter, M., Gini, G.: Human-like hierarchical reflex control for an artificial hand. Proc. IEEE Humanoids 2001, Waseda University, Tokyo (2001)
20. Sterratt, D.C.: Locust olfaction synchronous oscillations in excitatory and inhibitory groups of spiking neurons. Emergent Neural Computational Architectures **2036** (2001) 270–284
21. Kuntimad, G., Ranganath, H.S.: Perfect image segmentation using pulse coupled neural networks. IEEE Transaction on Neural Network **10** (1999) 591–598
22. Omura, Y.: Neuron firing operations by a new basic logic element. IEEE Electron Device Letters **20** (1999) 226–228
23. Kawamura, K., II, R.P., Wilkes, D., Alford, W., Rogers, T.: Isac: Foundations in human-humanoid interaction. IEEE Intelligent Systems **15** (2000) 38–45
24. Brooks, R., Breazeal, C., Marjanovic, M., Scassellati, B., Williamson, M.: The cog project: Building a humanoid robot. Computation for Metaphors, Analogy and Agents **LNCS 1562** (1999) 52–87

LARP, Biped Robotics Conceived as Human Modelling

Umberto Scarfogliero, Michele Folgheraiter, and Giuseppina Gini

DEI Department, Politecnico di Milano,
piazza Leonardo da Vinci 32, Italy
scarfogl@elet.polimi.it

Abstract. This paper presents a human-like control of an innovative biped robot. The robot presents a total of twelve degrees of freedom; each joint resemble the functionalities of the human articulation and is moved by tendons connected with an elastic actuator located in the robot's pelvis. We implemented and tested an innovative control architecture (called elastic-reactive control) that permits to vary the joint stiffness in real time maintaining a simple position-control paradigm. The controller is able to estimate the external load measuring the spring deflection and demonstrated to be particularly robust respect to system uncertainties, such as inertia value changes. Comparing the resulting control law with existing models we found several similarities with the Equilibrium Point Theory.

1 Introduction: The Biped Robot

The development of a humanoid robot usually requires relevant investments, comprehensive design and complex mathematical models. With LARP (Light Adaptive-Reactive biPed) we designed a simple and easy-to-reproduce biped, which could be at the same time cheap and efficient. The goal of this research is to create a system that could be used as a model of human lower limbs. Having such a model can help understanding how the natural walking motion is achieved and how it can be implemented in a humanoid robot [1]. An interesting research in this direction is the functional electrical stimulation in paraplegics [2],[3]; to make this people walk using external input signals to the muscles, it is fundamental to have a reliable walking pattern and a good muscle model. Indeed, the development of prosthesis, orthosis and rehabilitation devices for human lower limbs requires the knowledge of kinematics and dynamics of human walking.

For this reasons, we implemented an anthropomorphic biped robot, with feet, knees and a mass-distribution similar to the human limbs (fig.1). The prototype has twelve active degrees of freedom, disposed as follows: three in the hip, one in the knee and two in the ankle (pitch and roll) of each leg.

In particular, the foot is provided with two passive degrees of freedom, representing the heel and the toe (fig. 2.a); in this way, we have a planar base on

S. Wermter et al. (Eds.): Biomimetic Neural Learning, LNAI 3575, pp. 299–314, 2005.
© Springer-Verlag Berlin Heidelberg 2005

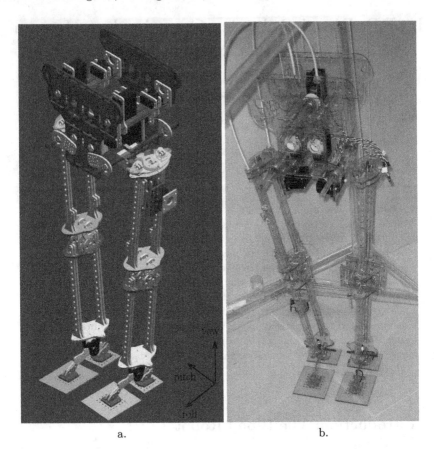

a.   b.

Fig. 1. (a) The 3D cad assembly of the robot. We can notice that the motors are collected in the upper part of the robot. The transmission to the joints is performed by tendons. (b) The prototype itself, here with only one actuated leg

which lean during the whole step. In addition, a pseudo Achilles tendon is used to absorb the impact with the ground, as it is performed in humans [4],[5]. Another innovation of the mechanical design is the knee. This joint was studied to reproduce the human articulation: made up with two circular supports, the joint is fastened by five sinews, as shown in figure 2.b [6].

The robot has a total of twelve active degrees of freedom disposed as in most of the modern biped robots: two in the ankle, one in the knee and three in the hip of each leg. The range of motion of each joint is similar to that of humans during normal walking. Joint torques are provided by servo motors disposed in the upper part of the robot. In this way, we can obtain a very light leg, even with 6 actuated degrees of freedom. The transmission is performed by a simple system of cables and levers. The servo motors are equipped with a spring and a damper to permit the joint stiffness control, which is described in the following paragraphs. The biped robot built with these features is 90

a.   b.

Fig. 2. (a) The foot under development. (b) The knee articulation, with five tendons and two circular supports

cm high, has twelve active and four passive degrees of freedom and weight less then 5 kg. We managed to make it that light thanks to the particular design, which characterizes this robot; as a matter of facts, the prototype is made with pieces cut out from a polycarbonate sheet, linked by plastic and carbon-fibre parts.

In the following paragraph we present the elastic actuator and control algorithm developed for the biped. First, we describe the control basic laws, then a more complete algorithm with acceleration feedback. The fourth paragraph introduce the simulations made on a computer model of the biped, showing the advantages of the elastic regulator respect to a classical position control.

2   The Elastic Actuator

The actuator is composed by a servo motor (we used big servos with 24 kg cm torque), a torsional spring and a damper. The resulting assembly is small, lightweight and simple, as we use a single torsional spring.

Using a spring between the motor and the joint let us have a precise force feedback simply measuring the deflection of the spring. Also, the resulting actuator has a good shock tolerance thanks to the damper. Similar actuators, with a DC motor and a spring, have been successfully used in biped robotics by Pratt et al. [7] and Yamaguchi and Takanishi [8].

The choice of the servos and the materials was made basically on cheap and off-the-shelf components. The main characteristic of this actuator is that the joint stiffness is not infinite, as it is in servo motors, and it can be changed in real time despite the constant stiffness of the spring. This has been achieved through a right choice of spring-damper characteristics and thanks to an intuitive control algorithm. We must underline here that as joint stiffness we consider k_g

$$k_g = \frac{M_e}{\varepsilon}$$

where M_e is the external load and ε is the position error. A first prototype of our actuator was composed by two motors and two springs, working as agonist and antagonist muscles in humans. This let us vary the joint stiffness even when no external load is acting, pre-tensioning the joint. With only one motor and one spring, the initial stiffness of the joint is fixed by the spring constant. This because the motor needs some time to tension the spring and counteract the external torque. Also, in this conditions, the presence of the damper avoid high initial errors due to rapidly varying loads.

The damping factor can be chosen constant, at its critical value ($\xi = 1$)

$$\begin{cases} w_n = \sqrt{k_g/I} \\ d = 2\xi w_n I; \end{cases} \qquad (1)$$

or can be varied during motion, in order to save motor torque and make the system faster. In the following paragraph we present the first alternative.

3   The Control Architecture

Regarding the low-level control of the motor, the spring-damper actuator can be used in a torque control loop: the controller assigns the torque to be delivered and, measuring the spring deflection, the actuator performs the task. A way to assign joint torques is the Virtual Model Control developed by J. Pratt et al. [9],[7]. In this approach, the controller set the actuator torques using the simulation results of a virtual mechanical component: like a spring, damper or any other mechanical device. In this way, the input torque can be computed intuitively shifting from virtual to real actuators.

In other classical approaches [10] the calculation of the joint torques is based on the dynamic model of the robot, that in many cases is complicated and imprecise. Indeed the biped robot can be formalized with a multi input multi output (MIMO) non linear system, that sometimes presents also time variant dynamical behavior. In these conditions a classical PID (Proportional Integral Derivative) controller is not suitable and more complex control strategies are needed. On the other hand, if we apply only a simple position controller it remains to solve how to control the joint stiffness.

To solve these issues we developed a simple algorithm that can control the joint stiffness and position providing the worth torque with a simple position control paradigm; no inverse-dynamic problem has to be solved for the positioning task. Actually, this kind of control is an implementation of the Equilibrium Point (EP) hypothesis as described by Feldman [11] [12] [13] regarding the control of the movements in humans. This theory suggests that the segmental reflexes together, with the muscolo-skeletal system, behave like a spring. Movement is achieved just by moving the equilibrium position of that spring. Virtually it is like pushing, or pulling, on one end and the other end will follow. And this is actually what happens with our actuators and the control law we adopted. Also, we developed a more articulated algorithm, with acceleration and velocity feedback. This can provide an estimation of the external torque acting on the link, and modify the joint stiffness accordingly.

These algorithms are described in detail in the next two sections. In the first of these, the basic control law is presented, while in the second, the more complex algorithm with acceleration feedback is illustrated.

3.1   The Simplest Control Model: An Implementation of the EP Hypothesis

The basic control algorithm is very simple, and very close to a classical model of the Equilibrium Point hypothesis; it needs the reference position $\bar{\varphi}$ and the joint stiffness k_g as inputs, and gives in output the motor position α_0. The only state information needed is the actual joint position, that must be measured and feedback to the regulator. We may remind that the difference between the actual position and the motor one is covered by the spring deflection. The control law is expressed by equation (2):

$$\alpha_0 = \frac{k_g}{k}(\bar{\varphi} - \varphi) + \varphi \tag{2}$$

where k represent the spring stiffness, φ and $\bar{\varphi}$ the actual and desired angular position respectively. The result is that a virtual spring with k_g stiffness is acting between the reference angle and the actual position. As a matter of facts, the finite joint stiffness betokens the presence of an error and one may define the time by which the desired position must be reached, accordingly with the joint stiffness. If this is very high, the error will be small, and the actual trajectory very close to the assigned one; this means that in presence of a step in $\bar{\varphi}$, high acceleration peaks can be generated. If the joint stiffness is small, one may expect relevant differences in the reference and actual trajectories, as the inertia and the damping oppose to fast movements. The static error ϵ depends anyway on the external load (T_{ext}), as

$$\epsilon = \frac{T_{ext}}{k_g} \tag{3}$$

Equation (3) represents also a way to determine a proper joint stiffness, deciding the maximum error tolerance and estimating the external maximum load. Note that k_g can be changed in real time, accordingly to the precision needed in critical phases of the motion.

To define the reference trajectory we used a step function filtered by a second order filter defined by a suited time constant T. In this way we can characterize the reference pattern with a single parameter.

To maintain the controller and the mechanical structure simple, the damping factor is set to a constant value that keep the system at the critical damping, as in equation (1).

We simulated the control of a simple 1-dof pendulum, and the results confirm the theoretical approach. In the simulation, gravity and varying external loads were included. Also friction was included to test the robustness of the algorithm.

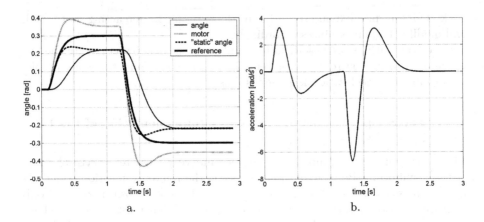

Fig. 3. (a) The link rotation and the motor position referred to the commanded angle. We can see that the actual angle approaches the reference accordingly to the set stiffness and external load ("static" angle). (b) The acceleration pattern presents two peaks, characteristic of damped systems. The change at about t=1.5 s is due to the limit on servo maximum torque

The system parameters are:

$$m = 1.2 \ kg; l = 0.3 \ m; I_g = 7.35 \cdot 10^{-2} \ kg \ m^2; k = 6 \ Nm/rad; kg = 10 \ Nm/rad$$

where l is the distance between the center of mass and the joint axis.

Fig. 3.a shows the behavior of the system: the commanded angle goes from zero to 0.3 rad at 0.1 sec and from 0.3 rad to -0.3 rad at 1.2 sec with a constant time T=0.08 s. Here, only gravity is acting. The actual joint angle and the motor position are also showed. With "static angle", we denote the position that the joint would have if the link inertia was zero and the damper was not present. To keep the figure clear the chosen stiffness is quite weak the error is about 0.1 rad only due to gravity. Looking at the motor position, we can notice that it is always opposite to the angle respect to the reference. This because here the spring stiffness is chosen lower than the joint stiffness. In this way the motor has to rotate more, but the system is less sensitive to motor position error. At about 1.4 sec., the motor rotation changes velocity due to servo maximum torque limit. In every simulation, also servo speed limitations were included.

Considering the resulting rotational acceleration, we can notice in fig.3.b that we have only two peaks, acceleration and deceleration with no oscillation. This pattern, typical of damped systems, is particularly useful when it is needed to exploit the natural dynamics of multi-link systems. For instance, when starting a step, the acceleration of the thigh can be used to bend the knee, as in passive dynamic walkers [14] [15], or, before foot-fall, the deceleration of the swing motion can be exploited to straight the leg, as in passive lower-limb prosthesis.

To figure out the influence of rapidly external loads on the system behavior, we simulated a positioning task under step-varying external torque. Figure 4

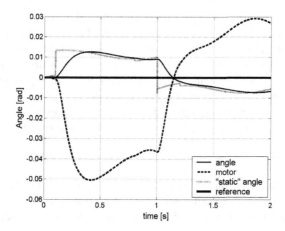

Fig. 4. The system behavior under rapidly-varying external torques. These can be seen in the "static angle" changing accordingly to the sinusoidal and step components of the load

shows the system under the action of an external load composed by a sinusoidal and constant action: at 0.1 s there is a positive step; at 1 s a negative one. Here the stiffness was highly increased, as a keep-position task was to be performed:

$$k = 10Nm/rad; kg = 50Nm/rad$$

Similar simulations have been run including a variable reference angle and friction at the joint.

Thanks to this simple control law, we do not need to solve any inverse dynamic problem, but just decide the joint stiffness - using for example equation (3) - and define the suited reference pattern.

The following section describes a more complete algorithm that can automatically adapts joint stiffness to the external load in case that this dimensioning is not accurate. Regarding to the system, the only information needed is its inertia, or its average value for a multi-link system. In the next section, it will be also shown that the controller behaves robustly respect to inertia misestimation.

3.2   The Control Law with Acceleration Feedback

Generally, in trajectory planning, not only the position is constrained, but also the velocity and acceleration must respect some limitations. This is especially important when we want to exploit the natural dynamic of the multi-body system; as we sketched above , the acceleration of the thigh can be used to bend the knee when starting the step [14] or to straight it before the foot-fall, as in passive leg prosthesis. Also velocity and acceleration limitations are needed where the inertial load due to the movement of one part can interfere with the motion of the rest of the robot. This is particularly relevant in bipedal walking.

To consider acceleration constrains, we included in our controller a sort of impedance control. By this term, we refer to the fact that the algorithm tracks the delivered torque and studies the resulting acceleration, creating a function relating these two quantities. In this way, we can create a simple dynamic model of a multi-body system without solving any inverse dynamic problem. The model can also get a good estimate of the external load acting on the joint; this can include the sole gravity or the interaction force with another links.

That can be obtained using, in the control loop, the equations:

$$T_{ext}^{i-1} = -k \cdot (\alpha_0^{i-1} - \varphi^{i-1}) + I \cdot \ddot{\varphi}^{i-1} + d \cdot \dot{\varphi}^{i-1} \qquad (4)$$

where d is the damping factor (see equation (1)), α_0 is obtained from eq. (2), I is the inertia and k the spring stiffness. The approximation we make is that between the instants *i-1* and *i* of the control loop the external load remains constant

$$T_{ext}^{i-1} = T_{ext}^i$$

Given the values of k,d,I, the position of the motor α_0 and the estimation of T_{ext}, the acceleration can be foreseen as:

$$A^i = \frac{k \cdot (\alpha_0^i - \varphi^i) + T_{ext}^{i-1} - d \cdot \dot{\varphi}^i}{I} \qquad (5)$$

This is the way in which we implement a kind of impedance control: if the acceleration (system output) in the next step is different from the foreseen one, given the calculated α_0 (system input), we infer that a different load is acting (system model has changed) and thus the motor position α_0 is corrected accordingly. In some way this is also how we sample object properties in real word; for instance, to understand if a bin is empty or not we lift it and according to the resulting motion, we estimate the mass. We do the same to evaluate a spring stiffness, for example. In a positioning task, we make this sample-evaluation-correction every instant. The controller we developed perform this simple check at any iteration, modifying in real time the system model.

The simulations on a single joint brought to interesting results; with the same system as before:

$$m = 1.2kg; l = 0.3m; I_g = 7.35 \cdot 10^{-2}kgm^2; k = 10Nm/rad; kg = 50Nm/rad$$

we could perform the motion evaluating the acceleration and the external load. In fig. 5 the results are shown with and without motor torque limitation. Here the external load is only the gravitational one. We can notice the effect of torque limit on the acceleration pattern.

As it is possible to see in fig. 5.c the characteristic is similar to the human electro-myographic activity, composed by there phases: acceleration-pause-deceleration [1], [16], and suitable for exploiting the natural dynamic of the links, i.e. in leg swinging as pointed out before.

From figures 5.e and .f we can also notice that the system perform a pretty good estimation of the external load acting on the link.

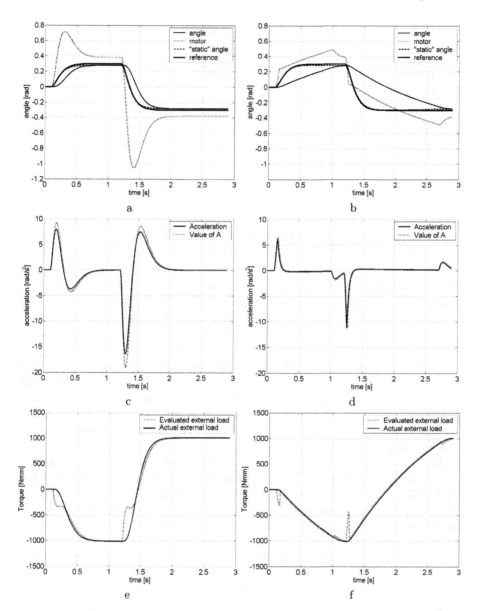

Fig. 5. (a),(c),(e) show respectively the angles, the acceleration and its evaluation, T_{ext} and its estimation when no motor torque limitation is considered. As we can see, the estimate is in good accordance with the real value. (b),(d),(f) show the same graph when a torque limitation is considered

The controller can also perform a path monitoring on the acceleration; as a matter of facts, if the joint stiffness we imposed is, for example, too high for the load applied or the reference angle changes too quickly, the controller

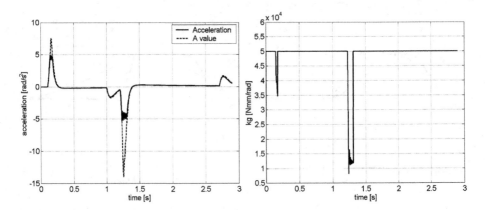

Fig. 6. The algorithm can limit the acceleration acting on the joint stiffness without compromising the final positioning. This within few lines of calculations

decrease the joint stiffness during the motion to prevent too high accelerations. This is done simply using the calculated acceleration value for the incoming iteration (eq. 5). If with the imposed stiffness the acceleration A^i is too high, k_g is changed in order to have $A^i = A_{limit}$. In this very simple way, the real value of the acceleration is kept under its maximum value.

Setting the right joint stiffness can be guided by equation (3) or with a trial-and-error procedure. For example, a learning algorithm could be used, not only to determine the k_g value, but also the time constant of the reference trajectory. The choice of this two parameters as inputs is quite relevant: as a matter of facts, these two quantities can greatly influence the joint behavior without compromising the final positioning.

The only information the controller needs about the system is its inertia; in multi-link systems it can be approximated with a constant average value computed on all the links, or it can be calculated during the motion. In any case, the controller seems to be quite robust respect to inertia uncertainties, showing no relevant changes even for errors of about 30% (see fig. 7). As a matter of facts, the difference in inertia load is considered by the controller as an additional external torque. Regarding the damping, equation 1 can be rewritten as:

$$d = 2\xi \cdot \sqrt{k_g I} \tag{6}$$

This means that the damping factor is also proportional to the square root of inertia errors: while a too high inertia make the system over-damped, an underestimation can let the system have some oscillations. Anyway, the error in the inertia must be very high (such as 50%) to see noticeable effect on the damping.

In the external torque estimation, we can notice the effect of wrong inertia input in the controller: for instance, if the real inertia value is lower, the controller acts as an additional external load is helping rotation during positive accelerations (see fig.7). In this way, the system is "automatically compensated".

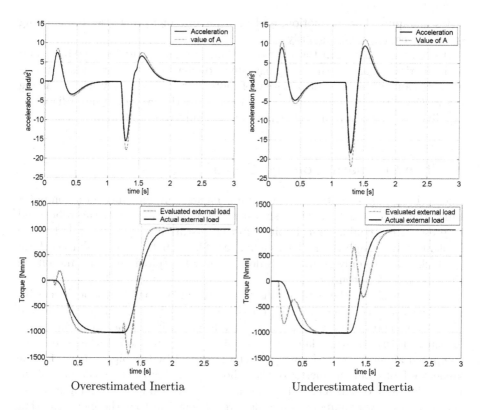

Overestimated Inertia   Underestimated Inertia

Fig. 7. As we can see, an error of 30% in inertia value does not compromise the positioning; it is considered as an external additional load. If the computed inertia is lower than the real one, for example, when the system is accelerating, the algorithm interpret the too small acceleration (system response) as an external load that is braking the motion. On the other hand, when the computed inertia is higher than the real one, the system is over-accelerated, and a virtual additional positive torque is considered acting

3.3   The Influence of Damping on the Control

The advantage of choosing joint stiffness and reference angle curve as parameters is that we can change the behaviour of the system without the risk to hamper the final positioning. For instance, the joint stiffness can be increased as the error lessens, for a fine placement, or the reference curve can be chosen more or less sharply changing to regulate inertia forces. To avoid oscillations along the assigned angle, the damping factor has been fixed to the critical value. This means that the relative damping is kept at unity during the whole motion; the drawback of this choice is that quite a relevant part of the motor torque is adsorbed by the damper even when no external load is acting. Thus, an alternative could be to use a relative damping less then one during the motion, which must be braked when the error approaches zero. A solution is then to have two values of the damping: low for motion, and high for braking; otherwise the damping should be changed continuously.

In the algorithm both the solutions could be implemented successfully, and with the right choice of the damping values, oscillations can be avoided. In this way, we can save a relevant quantity of motor torque, especially when starting the motion. The drawback of this choice is that the system is more liable to rapidly-changing external loads when the damping value is low.

4   The Simulation on the Biped

The spring-reactive control has been implemented on our biped in a computer simulation. The robot model is shown in fig.8. As a first test, the robot had to preserve the equilibrium despite external disturbances. To run this test we implemented a simplified model; as a matter of facts, 6 dof are enough to perform the task; thus we only actuate two dof in the ankle (pitch and roll) and one in the hip (yaw) for each leg.

Figures 9 shows the external disturbances applied on the robot. The joint stiffness is set according to equation (3), where ε is the maximum error and T_{ext} is the corresponding gravitational load. The value of inertia is calculated focusing on the resulting damping more than on the real value, that should be computed along the closed kinematic chain formed by the biped. Thus, for the ankle, we figure out the inertia of the robot considering the two feet coincident. Given the value of this inertia I, we evaluate the needed total damping factor d. As in the feet two dampers in parallel are present, we split the inertia so that the sum of the two dampers equal the total damping needed. Regarding the hip,

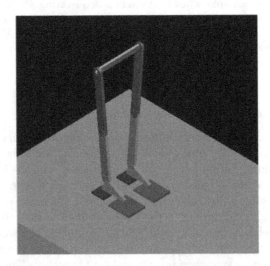

Fig. 8. The robot model in the computer simulation

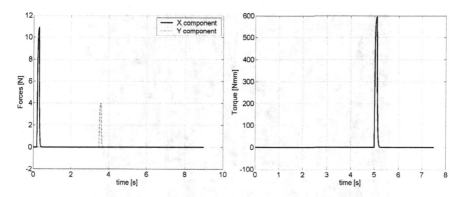

Fig. 9. The external disturbances applied to the robot, forces and torque

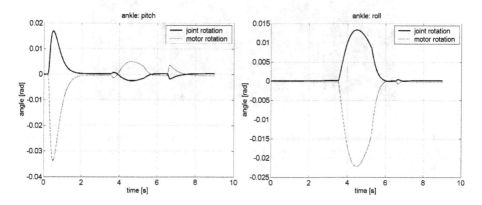

Fig. 10. The angular position in the two degrees of freedom of the ankle: the disturbances are absorbed and the robot returns in its initial position

we proceed in the same way, neglecting the leg beneath the joint for the inertia computation.

The results are shown in fig.10: we can notice that, as the disturbance is applied, a position error appears, as the actual angle differs from the reference position zero. The dotted line shows the motor rotation, that counteracts the disturbance and brings the joint back to the reference. In this way the robot is able to "react" to external loads, admitting a positioning error in order to preserve the whole balance.

Figure 11 shows the advantages of our regulator respect to a classical position control: the latter is unable to absorb and react to force disturbances, making the robot rigidly fall down. The elastic control instead, betokening the presence of a position error, can preserve the robot from falling down absorbing the disturbance.

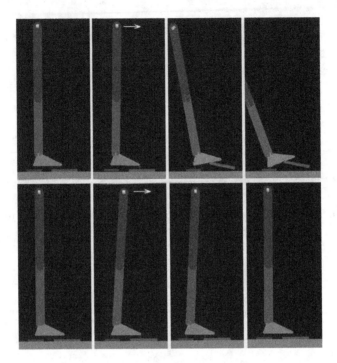

Fig. 11. The robot controlled with a classical position control (above) and with elastic control (below) under the same disturbances. The advantages of this kind of regulator are already noticeable at low level control

5   Conclusions

In this paper we presented an anthropomorphic biped robot made with innovative features in mechanical design and control architecture. Regarding the actuation system, we designed a motor device equipped with a torsional spring and a damper to mimic the elastic properties of muscles and tendons. This allows to have a good shock tolerance and, from the control point of view, to estimate the external load measuring the spring deflection. Also, a method was developed to preserve the possibility of position control even with variable joint stiffness. This aspect is fundamental in biped robotics, not only to exploit the natural dynamics of the legs, but also to face with impacts occurring at every step. With position and velocity feedback - present in musculoskeletal system through Golgi tendon organs and muscle spindles - we can perform a fine positioning task even without any a-priori knowledge of the external load. As a matter of facts, our control law is an implementation to robotics of the Equilibrium Point Hypothesis, as described by Feldman [11] [12] [13]. Indeed, our regulator perform a movement acting on the equilibrium point of the spring; using few lines of calculations, the joint stiffness can be changed in real-time, despite the constant stiffness of the spring. In addition, evaluating the acceleration, we can adopt a sort of impedance control: as the reference angle is imposed by a high level controller without any knowledge on the external load, the low level regulator can modify the joint stiffness according to the external action, avoiding, for example, too high accelerations. These arguments seem to weaken the main criticisms to Feldman lambda model; according to those criticisms it is fundamental to know the dynamics of the load to decide the equilibrium point motion. As our results showed, we can perform the movement without solving any inverse dynamic problem and, in order to face with changing external loads, we can act both on the equilibrium point and the joint stiffness, despite the a-priori defined reference angle and the constant stiffness of the system.

References

1. Kiriazov, P.: Humanoid robots: How to achieve human-like motion. Journal of Biomechanics (24) (1991) 21–35
2. H.M. Franken, P.V., Boom, H.: Restoring gait in paraplegics by functional electrical stimulation. IEEE Engineering in Medicine and Biology **13** (1994) 564–570
3. D.J. Maxwell, M.G., Baardman, G., Hermens, H.: Demand for and use of functional electrical stimulation systems and conventional orthoses in the spinal lesioned community of the uk. Artificial Organs **23** (1999) 410–412
4. Alexander, R.M.: Energetics and optimization of human walking and running: The 2000 raymond pearl memorial lecture. American Journal of Numan Biology **14** (2002) 641–648
5. Ker, R., Bennett, M., Bibby, S., R., R.K., Alexander, M.: The spring in the arc of the human foot. Nature **325** (1987) 147–149
6. U. Scarfogliero, M. Folgheraiter, G.G.: Advanced steps in biped robotics: Innovative design and intuitive control through spring-damper actuator. *In proc.* Humanoids 2004 (2004)
7. Pratt, G., Williamson, M.: Series elastic actuators. IEEE International Conferences on Intelligent Robots and Systems (1) (1995) 399–406
8. J. Yamaguchi, A.T.: Design of biped walking robot having antagonistic driven joint using nonlinear spring mechanism. IROS 97 (1997) 251–259
9. Pratt, J., Chew, C., Torres, A., Dilworth, P., Pratt, G.: Virtual model control: An intuitive approach for bipedal locomotion. The International Journal of Robotics Research **20**(2) (2001) 129–143
10. Kwek, L., Wong, E., Loo, C., Rao, M.: Application of active force control and iterative learning in a 5-link biped robot. Journal of Intelligent and Robotic Systems **37**(2) (2003) 143–162
11. Asatryan, D., Feldman, A.: Functional tuning of the nervous system with control of movement or maintenance of a steady posture - i mechanographic analysis of the work of the joint or execution of a postural task. Biofizika **10** (1965) 837–846
12. Feldman, A.: Functional tuning of the nervous system with control of movement or maintenance of a steady posture - ii controllable parameters of the muscle. Biofizika **11** (1966) 498–508

13. Feldman, A.: Functional tuning of the nervous system with control of movement or maintenance of a steady posture - iii mechanographic analysis of the work of the joint or execution of a postural task. Biofizika **11** (1966) 667–675
14. McGeer, T.: Passive walking with knees. IEEE International Conference on Robotics and Automation **2** (1990) 1640–1645
15. Collins, S., Wisse, M., Ruina, A.: A three dimensional passive-dynamic walking robot with two legs and knees. The International Journal of Robotics Research **20**(7) (2001) 607–615
16. G.L. Gottlieb, D.C., Agarwal, G.: Strategies for the control of single mechanical degree of freedom voluntary movements. Behavioral and Brain Sciences **12**(2) (1989) 189–210

Novelty and Habituation: The Driving Forces in Early Stage Learning for Developmental Robotics

Q. Meng and M.H. Lee

Department of Computer Science,
University of Wales, Aberystwyth, UK
{qqm, mhl}@aber.ac.uk

Abstract. Biologically inspired robotics offers the promise of future autonomous devices that can perform significant tasks while coping with
noisy, real-world environments. In order to survive for long periods we believe a developmental approach to learning is required and we are investigating the design of such systems inspired by results from developmental
psychology. Developmental learning takes place in the context of an epigenetic framework that allows environmental and internal constraints to
shape increasing competence and the gradual consolidation of control,
coordination and skill. In this paper we describe the use of novelty and
habituation as the motivation mechanism for a sensory-motor learning
process. In our system, a biologically plausible habituation model is utilized and the effect of parameters such as habituation rate and recovery
rate on the learning/development process is studied. We concentrate on
the very early stages of development in this work. The learning process
is based on a topological mapping structure which has several attractive
features for sensory-motor learning. The motivation model was implemented and tested through a series of experiments on a working robot
system with proprioceptive and contact sensing. Stimulated by novelty,
the robot explored its egocentric space and learned to coordinate motor
acts with sensory feedback. Experimental results and analysis are given
for different parameter configurations, proprioceptive encoding schemes,
and stimulus habituation schedules.

1   Introduction — The Developmental Framework

In recent years, there has been an increasing interest in biologically inspired
robotics. The main motivations for this trend are twofold: developing new and
more efficient learning and control algorithms for robots in unstructured and
dynamic environments on the one hand, and testing, verifying and understanding mechanisms from neuroscience, psychology and related biological areas on
the other. Developmental robotics is a multidisciplinary research area which
is greatly inspired by studies of infant cognitive development, especially the
staged growth of competence by shaping learning via the constraints in both
internal sensory-motor systems, and the external environment. Motivated by

S. Wermter et al. (Eds.): Biomimetic Neural Learning, LNAI 3575, pp. 315–332, 2005.
© Springer-Verlag Berlin Heidelberg 2005

the wealth of data from the developmental psychology literature, we are exploring a new approach strongly based on developmental learning. We believe that robotics may benefit considerably by taking account of this rich source of data and concepts. The newborn human infant faces a formidable learning task and yet advances from undirected, uncoordinated, apparently random behaviour to eventual skilled control of motor and sensory systems that support goal-directed action and increasing levels of competence. This is the kind of scenario that faces future domestic robots and we need to understand how some of the infant's learning behaviour might be reproduced.

A major inspiration for the developmental approach has been the work of the great psychologist, Jean Piaget [18] and we recognise Piaget's emphasis on the importance of sensory-motor interaction and staged competence learning. Others, such as Jerome Bruner, have developed Piaget's ideas further, suggesting mechanisms that could explain the relation of symbols to motor acts, especially concerning the manipulation of objects and interpretation of observations [7].

The first year of life is a period of enormous growth and various milestones in development are recognised. Considering just motor behaviour, newborn infants have little control over their limbs and produce uncoordinated and often ballistic actions, but over the first 12 months, control of the head and eyes is gained first, then follows visually guided reaching and grasping, and then locomotion is mastered. There are many aspects here that can be studied, but we concentrate on the sensory-motor systems that deal with local, egocentric space. In particular, our programme is investigating the developmental processes, constraints, and mechanisms needed to learn to control and coordinate a hand/eye system in order to provide mastery for reaching and grasping objects. We believe it is necessary to start from the earliest levels of development because early experiences and structures are likely to determine the path and form of subsequent growth in ways that may be crucial.

This paper focuses on novelty and habituation as motivational drivers for very early sensory-motor learning tasks. In our system, a biologically plausible habituation model is utilized and the effects of parameters of the model on the development process are studied.

2   An Experimental Developmental Learning System

In order to investigate embedded developmental learning algorithms we have built a laboratory robot system with two industrial quality manipulator arms and a motorised pan/tilt head carrying a colour CCD camera. Each arm can move within 6 degrees of freedom and each is fitted with a two-fingered gripper. The whole system is controlled by a PC running XP, and a SUN Ultra Sparc 10 workstation. The PC is responsible for controlling the two manipulators, grippers, and pan/tilt head, and processing images from the CCD camera and other sensory information. The control program is written in C++. The high level mapping and learning processes are realized in the SUN workstation and implemented in JAVA. The two machines communicate via TCP.

Fig. 1. System set up

The specific anatomy of embedded systems has important implications [17] and the configuration of our system is similar to the general spatial arrangement of an infant's arms and head. The arms are mounted, spaced apart, on a vertical backplane and operate in the horizontal plane, working a few centimetres above a work surface, while the "eye" (the colour imaging camera) is mounted above and looks down on the work area. Figure 1 shows the general configuration of the system.

We define a lateral area as a "rest" position for each arm, and define a central area between the two robot arms as the "mouth" area. Similar to newborn baby reflexes, the robot system has two given reflexes i.e. the system knows how to move its arms to the "mouth" from anywhere if it gets a suitable stimulus, and a similar reflex causes movement to return to the "rest" position. Both "mouth" and "rest" positions are small areas rather than single points, so at the very beginning, by moving between "mouth" and "rest" positions, the system already covers certain working areas to practice gaining sensory-motor skills. Some form of basic reflexes are always necessary to initiate activity. A newborn baby already has some reflexes such as sucking, arm/leg extending, and hand-to-mouth reflexes. It is believed these reflexes form a base for later skill development.

Various stimuli can be used to drive the system to explore the environment and gain hand sensory-motor mapping and eye/hand coordination skills. These include stimuli from vision (colour, shape, movement), contact sensors on arms and hands, and others such as sound, etc. We treat any newly discovered area as stimulating, thus the "mouth" and "rest" areas are initially unexplored and so become the novel stimuli that drives the robot learning process. We believe that the mapping between the hand/arm motors and proprioceptive feedback should be developed early, before introducing vision to further enrich the hand sensory-motor mapping and start eye/hand coordination development. We are investigating the optimum time for introducing vision into the learning process.

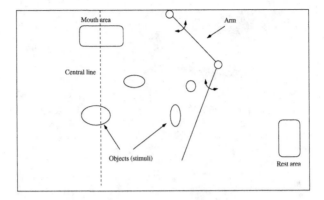

Fig. 2. Forearm and upperarm of one arm

The robot arms have proprietary controllers and can be driven by selecting commands in terms of various coordinate systems. For example, the six revolute joints can be driven directly to desired angles or the location of the end-effector can be moved to a desired location in cartesian space. The arm controllers perform any necessary kinematic computations to convert spatial transitions into motor actions. In the present experiments only two joints are used, the others being held fixed, so that the arms each operate as a two-link mechanism consisting of "forearm" and "upperarm" and sweep horizontally across the work area. The plan view of this arrangement for one arm is shown in figure 2.

Normally the arm end-points each carry an electrically driven two-finger gripper, however, for the present experiments we fitted one arm with a simple probe consisting of a 10mm rod containing a small proximity sensor. This sensor faces downwards so that, as the arm sweeps across the work surface, any objects passed underneath will be detected. Normally, small objects will not be disturbed but if an object is taller than the arm/table gap then it may be swept out of the environment during arm action.

3   Novelty and Habituation

At each stage of learning, novelty and habituation play an important role in driving the learning process. Novelty refers to new or particularly salient sensory stimuli, while habituation is defined as a decrease in the strength of a behavioural response to repeated stimulations. A habituated stimulus may be able to evoke a further response after the presentation of an intervening novel stimulus. Novelty and habituation mechanisms can help a system to explore new places/events while monitoring the current status and therefore the system can glean experience over its entire environment.

In our system we used a biologically plausible habituation model [19] which describes how excitation, y, varies with time:

$$\tau \frac{dy(t)}{dt} = \alpha[y_0 - y(t)] - S(t) \qquad (1)$$

where y_0 is the original value of y, τ and α are time constants governing the rate of habituation and recovery, and $S(t)$ represents the external stimulus.

Let $S(t)$ be a positive constant, denoted as S. Then, the solution for equation 1 is:

$$y(t) = \begin{cases} y_0 - \frac{S}{\alpha}[1 - e^{-\alpha t/\tau}], & if \ S \neq 0 \quad (a) \\ y_0 - (y_0 - y_1)e^{-\alpha t/\tau}, & if \ S = 0 \quad (b) \end{cases} \qquad (2)$$

and in recurrent form:

$$y(t+1) = y(t) + \frac{\alpha}{\tau}[y_0 - y(t)] - \frac{S}{\tau}$$

where y_1 is the value when the stimulus is withdrawn. Figure 3 shows some values for this habituation model.

There are three phases in our implementation: a habituation phase $(S \neq 0)$, a forbidden phase and a recovery phase $(S = 0)$. The implementation is based on a hierarchical mapping structure consisting of perceptual fields with different field sizes and overlaps. During the habituation phase, the strength of a field's response to repeated stimulations decreases. When the strength is below a threshold, the field is not considered in the attention selection process, i.e. the field is forbidden; After a certain period, the strength of this field's response to stimulus may be able to recover to a certain level, and the field may be selected by the attention selection process. The strength of a field's excitation is calculated using equation 2 based on whether the field is in a habituation phase or recovery phase.

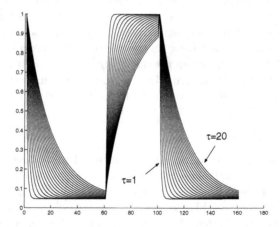

Fig. 3. The habituation model

4   The Sensory-Motor Coordination Problem

In this paper we examine a very early problem of motor control: the sensory-motor control of the limbs. We gain inspiration and insight from the infant's use of his/her arms during the first 3 months of life. Initially, it seems there is no purpose or pattern to motor acts, but actually the neonate displays very considerable learning skills in the first 3 months: from spontaneous, apparently random movements of the limbs the infant gradually gains control of the parameters and coordinates sensory and motor signals to produce purposive acts in egocentric space. Various changes in behaviour can be discerned such as (1) "blind groping" arm actions, (2) pushing objects out of the local environment, (3) touching and tracing objects. During these stages the local egocentric space becomes assimilated into the infant's awareness and forms a substrate for future cross-modal skilled behaviours. This development involves an essential correlation between sensory and motor spaces. Notice that this growth stage is prior or concurrent with visual development and so, in the experiments reported here, we do not involve the eye system. Also, for experimental simplicity, we use only one arm.

4.1   The Motor System

A two-section limb requires a motor system that can drive each section independently. A muscle pair may be required to actuate each degree of freedom, i.e. extensors and flexors, but a single motor parameter, M_i, is sufficient to define the overall applied drive strength. We assume that any underlying neuronal or engineering implementation can be abstracted into such parameters, where zero represents no actuation and increasing values represent correspondingly increasing drive strength in positive or negative directions. As we are operating in two dimensions, two motor parameters are required, one for each limb section: M_1 and M_2. It is important to recognise Bernstein's valuable contribution that there can be no one-to-one relation between the motor cortex neurons and individual muscle fibres [3]. However, we can use our abstraction to capture an overall representation of output motor activity. In this study, we operate the arms at a slow rate and so do not need to take account of effects due to gravity or dynamics.

4.2   Proprioception

The sensing possibilities for a limb include internal proprioception sensors and exterior tactile or contact sensors. Proprioception provides feedback on the sensed position of the limb in space but, in animals, the actual mechanisms and the nature of the variables sensed are not entirely known. There are two main possibilities: the **angles** of specific joints may be sensed, or the position of the limb **end-point** may be sensed. The former is technically simple, being directly derived from a local mechanical change, while the latter, being more global, requires some form of computation to derive the end-point location, as, without other external sensing such as vision, this can not be sensed directly. An attractive variation would be to relate the hands or arm end-points to the body centre line. This **body-centred** encoding would be appropriate for a "mouth-centred" space in accordance with early egocentric spatial behaviour. In this case, the limb end-points are encoded in terms of distance away and angle from the body centre.

There is one other notable spatial encoding: a Cartesian frame where the orthogonal coordinates are lateral distance (left and right) and distance from the body (near and far). The signals for this case are simply (x, y) values for the position of the end-points in a rectangular space. This encoding, referred to as **Cartesian** encoding, seems the most unlikely for a low-level biological system, however we include it due to its importance in human spatial reasoning [14].

Before vision comes into play, it is difficult to see how such useful but complex feedback as given by the three latter encodings could be generated and calibrated for local space. The dependency on trigonometrical relations and limb lengths at a time when the limbs are growing significantly makes it unlikely that these codings could be phylogenetically evolved. Only the **joint angle** scheme could be effective immediately but the others may need to develop through growth processes. Recent research [5] on the hind limbs of adult cats has discovered that *both* **joint angle** and **end-point** encodings can coexist, with some neuronal groups giving individual joint angle outputs while other neurons give foot position encodings independently of limb geometry. We investigate all four systems as candidate proprioception signals. For each encoding scheme we consider the signals, S_1 and S_2:

1. **Joint Angle** encoding.
   $S_1 = f(\theta_1)$ and $S_2 = f(\theta_2)$, where the θ are the joint angles between the limb sections and f is a near linear or at least monotonic function. θ_1 is the angle between the upperarm and the body baseline (a fixed datum in the workspace) and θ_2 is the angle between the upperarm and the axis of the forearm.

2. **Shoulder (or end-point)** encoding.
   This gives the length and angle of the limb axis from shoulder to end-point:

$$S_1 = \sqrt{l_1^2 + l_2^2 + 2l_1l_2 \cos \theta_2}$$

and

$$S_2 = \theta_1 - \arctan \frac{l_2 \sin \theta_2}{l_1 + l_2 \cos \theta_2},$$

where l_1 and l_2 are the lengths of the upperarm and forearm respectively.

We note that the length of the complete limb system, i.e. shoulder to end-point, varies only with θ_2.

We also note that if θ_2 had been measured in absolute terms, i.e. with respect to a body datum like θ_1, then the encoding formulation would be less convenient and more obscure:

$$S_1 = \sqrt{(l_1 \cos \theta_1 + l_2 \cos \theta_2)^2 + (l_1 \sin \theta_1 + l_2 \sin \theta_2)^2}$$

and

$$S_2 = \arctan \frac{l_1 \sin \theta_1 + l_2 \sin \theta_2}{l_1 \cos \theta_1 + l_2 \cos \theta_2}.$$

3. **Body-Centred** encoding.
   This is a measure of the end-point distance and angle from the body centre. Shoulder encoding effectively records the vector from shoulder to hand, while this encoding measures the vector from body to hand. Accordingly, we shift the shoulder vector given above (S_1' and S_2') by the distance *Base* which is the separation distance between the shoulder and the body centre line:

$$S_1 = \sqrt{(S_1')^2 + Base^2 - 2BaseS_1' \cos S_2'}$$

and

$$S_2 = \arctan \frac{S_1' \sin S_2'}{Base - S_1' \cos S_2'}$$

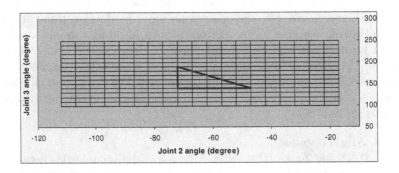

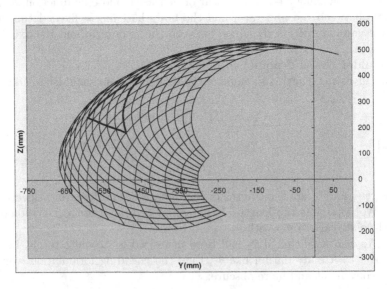

Fig. 4. Relationship between joint and Cartesian encoding

4. **Cartesian** encoding.

$S_1 = x$ and $S_2 = y$, where (x, y) is the location of the arm end-point in two-dimensional rectangular space. The actual orientation of the space is not important but for convenience the origin can be placed at the body centre point.

Figure 4 shows the relationship between the **Joint** and **Cartesian** encoding systems. We see that when the arm moves along a straight line in joint space, the arm end-path in Cartesian space is a curve, and the distortion between the maps changes considerably for different areas of the workspace.

5   Mappings as a Computational Substrate for Sensory-Motor Learning

The learning process is based on a hierarchical mapping structure which consists of fields of different sizes and overlaps at different mapping levels. Each field contains local information including sensory data and related motor movement data, statistics on activity levels and stimuli information including the time of stimuli appearance/disappearance. Fields may be merged into larger areas so that the system can become habituated to one area and then be attracted by stimuli from other unexplored areas to probe these and gain experience there.

Our maps consist of two-dimensional sheets of identical elements. Each element is a circular patch of receptive area known as a field. The fields are regularly spaced, with their centres arranged on a triangular grid. We use two variables, I, J, to reference locations on any given map; these simply define a point on the two-dimensional surface — they do not have any *intrinsic* relation with any external space.

Two parameters define a map structure: field size and inter-field spacing. These parameters determine two measures of coverage: degree of overlap (as a proportion of area covered by only one field against area covered by more than one field) and the coverage density (in terms of field area per unit surface area).

Every field can hold associated variables, as follows:

1. **Stimulus value** F_s, e.g. colour value, shape value, contact value.
2. **Excitation level** F_e, This is the strength of stimulation that a field has received.
3. **Frequency counter** F_u, This records how often the field has been used (i.e. accessed or visited).
4. **Motor values** This is a way of recording the motor parameters that were in force when this field was stimulated. It builds the cross link between sensory space and motor space, which can be reused for later movement planning in sensory space and further conversion to motor space.

The motor information within fields is effective for local areas. During the planning phase, for a target sensory field which the arm needs to move to, if the field has a cross motor link, then the motor values are utilized for driving the movement. Otherwise a local weighted regression approach is applied to generate the motor values by using any local field neighbours which have cross motor links. The motors are driven based on the estimated values which may or may not achieve the desired target field depending on how far the neighbours are from the target field and in which part of workspace the target field falls. Some areas of the workspace have more distortion than others, see figure 4 for the distortion changes across the whole workspace between Cartesian sensory encoding and joint space. After the movement, if the sensory feedback shows that the movement is not accurate enough, then another sensory-motor cross link is generated based on the motor values used and the actual sensory feedback. The details of the planning process will be discussed in another paper.

6   Experimental Procedure

The software implemented for the learning system is based on a set of four modules which operate consecutively, and in a cyclic loop. The modules are:

Action Selection: This module determines which motors should be executed. For a single limb system, this is the process of setting values for M_1 and M_2.

Motor Driver: This module executes an action based on the supplied motor values. For non-zero values of M the arm segments are set moving at constant speed and continue until either they reach their maximum extent or a sensory interrupt is raised. The ratio between the values of M_1 and M_2 determines the trajectory that the arm will take during an action.

Stimulus Processing: Upon sensory interrupt or at the completion of an action this module examines the position of the arm and returns values for proprioception, i.e. S_1 and S_2, representing the sense for the current position. If the action was terminated by a contact event then the contact value, S_C, is also returned.

Map Processing: Using S_1 and S_2 as values for I, J, this module accesses the map and identifies the set of all fields that cover the point addressed by S_1 and S_2. A **field selector** process is then used to chose a single key field, F, from the set. We use a nearest neighbour algorithm to select the key field. Any stimulus value is then entered into the field, $F_s = S_C$, and the excitation level is updated. Another field variable, the usage counter F_u, records the frequency that fields are visited, and this is then incremented and tested. If this was the first time that field had been used then an excitation level is set, otherwise, the excitation value is updated according to the function in equation 2. If this field has not yet received an entry for motor values, then the cross modal link is set up by entering the current motor values.

A motivational component is necessary to drive learning and this is implemented in the **Attention Selection** procedure. This module directs the focus of attention based on the levels of stimulation received from different sources. There is evidence from infant studies that novelty is a strong motivational stimulus. We start with a very simple novelty function to direct attention and to provide the motivational driver for sensory-motor learning. Novel stimuli get high excitation levels and are thus given high priority for attention. Thus, if a sensed value at a field remains unchanged then there is no increase in excitation, but upon change the excitation level is incremented, as described above. We use a habituation mechanism to provide decay functions and reduce overall excitation levels. The field with the highest level of excitation becomes the candidate target for the next focus of attention. Using map data from both this target field and the current location field a trajectory can be estimated and motor values are then computed and passed to **Action Selection**. In this way, motor acts are directed towards the most stimulating experiences in an attempt to learn more about them.

By investigation we can logically determine the various options available and/or appropriate for a learning agent. If no prior experience is available then there are no grounds for selecting any particular ratio between M_1 and M_2 and therefore setting both to a high positive value is the natural choice for **Action Selection**. We assume that a rest position exists (equivalent to an "origin") and the result of driving the motors "full on" from the rest position brings the hand to the body centre-line, in a position equivalent to the mouth. We also assume a reflex in the **Motor Driver** procedure that, providing there is no other activity specified, returns the arm to the rest position after any action.

The expected behaviour that we anticipate from experiments is initially spontaneous, apparently random movements of the limbs, followed by more purposive "exploratory" movements, and then directed action towards contact with objects. In particular, we expect to see the following stages:

(1) "blind groping" arm actions,
(2) unaware pushing objects out of the local environment,
(3) stopping movement upon contact with objects,
(4) repeated cycles of contact and movement,
(5) directed touching of objects and tracing out sequences of objects.

7   Experiments

In our experiments, several variables were considered: the state of the environment, the type of proprioception, the field sizes, and the habituation parameters.

The first factor is the environment and there are three possible cases:

1. No objects present
2. Some objects present, no contact sensing
3. Some objects present, contact sensing on

Noting that case 2 subsumes case 1, we can ignore case 1.

The second variable to be examined is the encoding of the proprioceptive sensing, S_1 and S_2. We can arrange that these signals are computed from one of the four schemes described above.

Another factor to be determined by experiment is the effect of different field sizes and field overlaps. We achieve this by creating three maps, each with fields of different density and overlap, and running the learning system on all three simultaneously. Each map had a different field size: small, medium and large, and the S and M signals were processed for each map separately and simultaneously. However only one map can be used for attention and action selection, because different field locations may be selected on the different maps. By running on each map in turn we can observe the behaviour and effectiveness of the mapping parameters. All maps were larger than the work space used by the arm, (any fields out of range of the arm will simply not be used).

Finally we need to experiment on the possible excitation schedules for field stimulation. In the present system this consists of the habituation parameters τ, α, and the initial excitation values set to the new stimulus fields.

Figure 5 shows the habituation and recovery phases in our experiments. We used different parameters for the mouth/rest areas and certain other fields. For the mouth/rest fields and the neighbours of any stimulated fields the following parameters were used to achieve slower habituation: $\tau = 10$, $\alpha = 1.05$. For the directly stimulated fields, another set of parameters: $\tau = 3.33$, $\alpha = 0.4$ were used to achieve faster habituation. Also, a smaller excitation value was given to neighbouring fields than for the stimulus fields. During the recovery phase, the excitation for neighbours can only reach this smaller initial value. This means that a stimulus can affect the neighbours, but the effect is less than the originating stimulus field. It should be noted that if a field is stimulated before, and even if currently it is a neighbour of the best matched field, it uses the bigger initial value (1.0 as in figure 5) during the habituation and recovery phases.

Without any contact sensing ability, the arm could not cause any stimuli during its movements, and because we did not introduce vision at the very early learning stage, the arm had just two stimuli: the mouth area and the rest area.

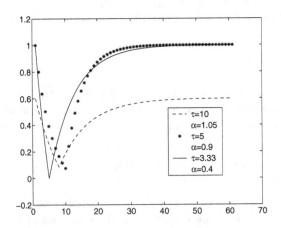

Fig. 5. Habituation and recovery in the experiments

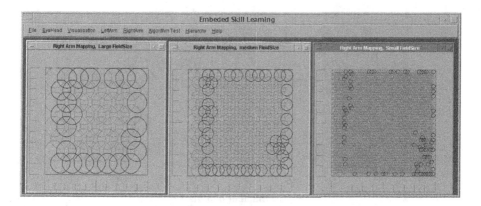

Fig. 6. Mapping with the contact sensing off

So the arm moved to its mouth and went back to the rest position and repeated the process. During this process, random values were introduced into the motor driver module if the excitation values of the mouth or rest areas fell below a threshold. This caused random arm movements that extended to the boundary of the arm space. During these movements, the arm either ignored objects on the table or swept any objects out of range. Figure 6 shows the result of this experiment. The highlighted fields in the figure shows the mouth area (near top left in each mapping), the rest area (near right bottom corner in each mapping), and the boundaries of the motor movement, respectively.

With contact sensing turned on, the arm was interrupted when it passed objects on the table, received stimuli from these interruptions, and obtained new excitation values for the best matched field based on the sensory feedback and its neighbors for the current arm position, or updated the existing ones using the habituation model shown in equation 2. The system also created/updated other

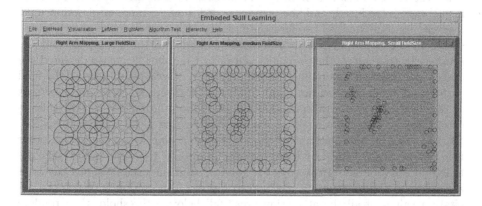

Fig. 7. Hierarchical mapping driven by novelty

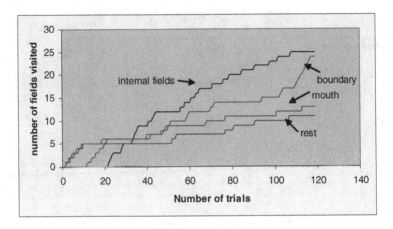

Fig. 8. Number of fields visited in the smallest field mapping

information for this field including the field access frequency, the habituation phase, motor values, and contact information. Figure 7 gives experimental results for the current experiment after 120 trials. The central highlighted area in the small scale mapping relate to the fields of the object which caused interruptions to the arm movements. It should be noted that in figure 7, we used joint angles as the arm spatial encoding system, and due to the distortion between the joint space and Cartesian space as shown in figure 4, the shape of the object displayed in figure 7 may not represent the "true" shape of the object.

Figure 8 shows the way that field visits varied with trials. The number of visits corresponds to new visits and ignores repeated visits to fields. The mapping used was the smallest field size. From the figure, we find that at the beginning, the arm moved between mouth and rest areas as there was no object in the workspace, so the only attractions were from these two areas. After a while, when the excitation levels of these two areas went down, some random movements were introduced, and the arm extended along a random direction until reaching its joint limit. Therefore the number of boundary fields increased, as shown in the figure. When an object was placed into the workspace and the arm passed the object, the proximity sensor at the robot finger tip detected the object and received a stimulus. An excitation value was computed for the field corresponding to this location, and smaller values were given to the neighbors of this field. These excitations attracted the robot's attention to visit this location and its neighbors again to explore further and glean more information. (At the moment, this means detecting the object's shape — in the future, other features may be obtained by using vision.) Gradually, the robot gains much detailed information about the object (stimulus), and the excitation level declines according to the habituation mechanism discussed in section 3. When the excitation values of all the attraction areas fall below a threshold, then random movements are allowed in order to explore novel stimuli in other newer areas. The last part of the figure (from approximately trial number 100) demonstrates this idea: the number of new internal fields visited gradually levels off and, at the same time, the number of boundary fields goes up as random movements are introduced.

In the implementation, we used the smallest field size mapping for attention selection and motor planning. The habituation function was applied to the best matched fields and their neighbouring fields. The small field mapping can habituate small areas if the fields in that area get stimulated frequently, and allow the arm to explore other nearby areas. This lets the system explore the details in some stimulus regions, which can be observed from the right mapping in figure 7. Different parameters in the habituation model can change the speed of excitation reduction, as shown in figure 5. From figure 7, we find that the area where objects were detected attracted much attention from the robot arm, and became well explored. In order to move the arm from these attractions to other areas, planning needs to shift to mappings with larger field sizes and random movements need to be introduced. By using mappings with large field size, the stimulated areas are habituated at a higher level and are explored quickly but with less detailed information obtained than when using the small field mappings. Introducing random movements at certain stages, say when the stimulated area becomes well explored, helps the system quickly jump to other workspace areas and increases the opportunity to get stimuli from those areas. Thus, there is a tradeoff between detailed exploration of small areas and a more rapid covering of the whole workspace with less detail. We may need both in the early developmental learning process in order to set up sensory-motor mapping for arms and eye/arm coordinations. The issues of local exploration versus global exploration, introducing random movements, and combining the planning across different levels are currently being investigated.

After an action is selected by the attention selection module, small levels of noise are added to the motor values. This helps the system to explore other fields nearby to the selected one and so enlarge the number of fields visited to enrich experience about the workspace. Infants also demonstrate such phenomenon in their early sensory-motor development, we find infants usually miss a target when they first try to reach it because of the immaturity of their proprioception, motors and sensory-motor mappings.

Using novelty and habituation as the driving forces in the early robot learning stage is much inspired by infant learning. This approach allows the system to develop sensory-motor coordination skills in the most important areas first, and later extend to other areas. This can help the system to gain enough experience before exploring the areas which may need high sensory-motor skills and have less frequent use, for example, the areas in figure 4 with most distortion.

8   Other Related Work

There have been many research projects that have explored the issues involved in creating truly autonomous embodied learning systems. These include behaviour-based architectures [6], hybrid approaches [4], evolutionary methods [15], and probabilistic techniques [16]. However, a missing element in all these is an explicitly developmental approach to learning, that is, a treatment of learning in the context of an epigenetic framework that allows environmental and internal constraints to shape increasing competence and the gradual consolidation of control, coordination and skill.

Our developmental approach to robotics is supported by much psychological work, e.g. [11], and recognises the characterisation of cognitive growth and expertise due to progressively better interpretation of input from the standpoint of background knowledge and expectation [7].

The most extensive work on computer based Piagetian modelling has been that of Drescher [9]. However, Drescher's system tries to cross-correlate all possible events and is computationally infeasible as a brain model. Maes showed how Drescher's approach can be improved by using focus of attention mechanisms, specifically sensory selection and cognitive constraints [10], but there remains much further work to be done to properly understand the value of these concepts for robotic devices. In particular, all the key previous work has been done in simulation and the concepts of constraint lifting and scaffolding have not been explored to any extent.

The computing framework used by others has been the sensory-motor *schema* drawn from Piaget's conception of schemas in human activity [18]. This was used as the fundamental representational element in Drescher's system [9], following early work by Becker [2]. The main proponent of schema theory has been Arbib [1]. We have also used the schema based approach in previous work; in order to capture spatial aspects we have developed a sensory-motor topological mapping scheme [12]. Our diagrams of sensory-motor spaces have similarities to those in Churchland [8].

Wang [21] extended Stanley's model by introducing long term habituation into the model using an inverse S-shaped curve. Habituation has been also used in mobile robots for novelty detection based on a self-organising feature mapping neural network [13].

9   Discussion

We assume that a rest position exists (equivalent to arm being in the lateral position) and the result of driving the motors "full on" brings the hand to the body centre-line (in a position equivalent to the mouth). This rather ballistic approach to motor action is widely reported in three month old infants. In experiments where kicking behaviour is able to disturb a stimulus, infants learn to adapt their kicking to achieve a desired visual change but they do this by altering the timing and frequency of their actions but not the duration of the basic motor pattern [20]. It seems that the neuronal burst duration is constant but the firing rate is modulated. This allows multiple muscles to be synchronised as they all have the same time-base while the amplitudes are varied to alter behaviour.

Interestingly, the apparently more "scientific" pattern of trying one motor alone and then the other does not seem to be present. It seems Nature has preferred algorithms for development, and these are likely to be well worth understanding and imitating.

Acknowledgments

We are very grateful for the support of EPSRC through grant GR/R69679/01 and for laboratory facilities provided by the Science Research Investment Fund.

References

1. Arbib M.A., *The Metaphorical Brain 2*, J. Wiley, NY, 1989.
2. Becker J., A Model for the Encoding of Experiential Information, in Shank R. and Colby K., (eds) *Computer Models of Thought and Language*, Freeman, pp396-435, 1973.
3. Bernstein Nikolai A., The Co-ordination and Regulation of Movements, Pergamon Press, Oxford, 1967.
4. Bonasso R. Peter, Firby R. James, Gat Erann, Kortenkamp David, Miller David P and Slack Mark G., Experiences with Architecture for Intelligent, Reactive Agents, *Journal of Experimental and Theoretical Artificial Intelligence*,Vol.9, No.2, pp237-256, 1997.
5. Bosco G., Poppele R. E., and Eian J., Reference Frames for Spinal Proprioception: Limb Endpoint Based or Joint-Level Based? J Neurophysiol, May 1, 2000; 83(5): 2931 - 2945.
6. Brooks R., Intelligence Without Representatio, *Artificial Intelligence*, 1991.
7. Bruner, J.S., *Acts of Meaning*, Harvard University Press, Cambridge, MA 1990.
8. Churchland P.S. and Sejnowski T.J., *The Computational Brain*, MIT Press, 1994.
9. Drescher, G. L., *Made up minds: a constructivist approach to artificial intelligence*, MIT Press, 1991.
10. Foner L.N. and Maes P., Paying Attention to What's Important: Using Focus of Attention to Improve Unsupervised Learning, Proc. 3rd Int. Conf. Simulation of Adaptive Behaviour, Cambridge MA, pp256-265, 1994.
11. Hendriks-Jansen H., *Catching Ourselves in the Act*, MIT Press, 1996.
12. Lee M.H., A Computational Study of Early Spatial Integration, Proc. 5th Int. Congress of Cybernetics and Systems, Mexico, 1981.
13. Marsland Stephen, Nehmzow Ulrich, and Shapiro Jonathan, Novelty Detection on a Mobile Robot Using Habituation,In From Animals to Animats, The Sixth International Conference on Simulation of Adaptive Behaviour, Paris, pp189-198, 2000.
14. Newcombe Nora S. and Huttenlocher Janellen, Making Space, MIT Press, 2000.
15. Nolfi, S. and Floreano, D., Evolutionary Robotics: The Biology, Intelligence, and Technology of Self-Organizing Machines, Cambridge, MA: MIT Press/Bradford Books,2000.
16. Perl J., Fusion, propagation, and structuring in belief networks, *Artificial Intelligence*, 29(3), pp241-288, 1986.

17. Pfeifer R., On the role of morphology and materials in adaptive bahavior. In Meyer, Berthoz, Floreano, Roitblat, and Wilson(eds.): *From animals to Animates 6. Proceedings of the sixth International Conference on Simulation of Adaptive Behavior 2000*, 23-32.
18. Piaget J., *The Child's Conception of the World*, translated by J. and A. Tomlinson, Paladin, London, 1973.
19. Stanley, J. C., Computer simulation of a model of habituation. Nature, 261, pp146-148, 1976.
20. Thelen, E. Grounded in the world: Developmental origins of the embodied mind, Infancy, 1(1), pp3-28, 2000.
21. D. Wang, Habituation. In M. A. Arbib, editor, The Handbook of Brain Theory and Neural Networks, pp441-444, MIT Press, 1995.

Modular Learning Schemes for Visual Robot Control

Gilles Hermann, Patrice Wira, and Jean-Philippe Urban

Université de Haute-Alsace, Laboratoire Mips,
4 rue de Frères Lumière, F-68093 Mulhouse Cedex, France
{patrice.wira, gilles.hermann, jean-philippe.urban}@uha.fr

Abstract. This chapter explores modular learning in artificial neural networks for intelligent robotics. Mainly inspired from neurobiological aspects, the modularity concept can be used to design artificial neural networks. The main theme of this chapter is to explore the organization, the complexity and the learning of modular artificial neural networks. A robust modular neural architecture is then developed for the position/orientation control of a robot manipulator with visual feedback. Simulations prove that the modular learning enhances the artificial neural networks capabilities to learn and approximate complex problems. The proposed bidirectional modular learning architecture avoids the neural networks well-known limitations. Simulation results on a 7 degrees of freedom robot-vision system are reported to show the performances of the modular approach to learn a high-dimensional nonlinear problem. Modular learning is thus an appropriate solution to robot learning complexity due to limitations on the amount of available training data, the real-time constraint, and the real-world environment.

1 Introduction

The artificial neural networks are a neurobiologically inspired paradigm that emulate the functioning of the vertebrate brain. The brain is a highly structured entity with localized regions of neurons specialized in performing specific tasks. On the other hand, the mainstream monolithic artificial neural networks are generally unstructured black boxes, which is their major performance limiting characteristic. The non explicit structure and monolithic nature of the artificial neural networks result in lack of capability to incorporate functional or task-specific a priori knowledge in the design process. Furthermore, if artificial neural networks often satisfy the requirements of nonlinear control and complex function estimation, it does not imply, however, that it is equally easy to learn to represent any function from a finite amount of data. Biologically inspired studies recently showed some very new interesting perspectives in control systems engineering. Various works demonstrated that modularity, present in biological organisms, would extend the capabilities of artificial neural networks.

For a robotic system to carry out complex tasks, taking its environment and several sensorial modalities into account, a single neural network may not be

S. Wermter et al. (Eds.): Biomimetic Neural Learning, LNAI 3575, pp. 333–348, 2005.
© Springer-Verlag Berlin Heidelberg 2005

sufficient. The modular approach we propose is a sequential and bidirectional structure of neural modules. It is difficult to train globally a chained functionality of networks. The training of such an architecture is achieved through the introduction of supplementary neural modules that are only active during the learning phase. These modules allow to reconstruct intermediate representations. The neural architecture is thus referred to as a bidirectional approach.

Only specific combination diagrams are suited for an efficient learning. In our case, the organization of the neural modules is based on a geometrical comprehension of the considered problem. The modular neural decomposition and its learning paradigm is validated with SOM-LLM-type networks. The objective is to learn and to perform nonlinear control tasks using visual feedback. The proposed architecture has been tested on a 7 degrees of freedoms (dof) robot-vision system and the modular bidirectional architecture turns out to be efficient. Indeed, the resulting proposed modular neural network models demonstrate greater accuracy, generalization capabilities, comprehensible simplified neural structure, ease of training and more user confidence. These benefits are obvious for certain problems, depending upon availability and usage of a priori knowledge about the problems.

This paper is organized as follows. Section 2 introduces the concept of modularity in term of a biological inspiration to develop new learning architectures. This section motivates the need of a modular neural architecture, presents existing modular networks and presents the bidirectional learning architecture. Section 3 briefly reviews the basics of the SOM-LLM network which will serve as a module in our approach. In section 4, we will present the adopted serial modular architecture for the specific task of robot control. Section 5 reports the results of training the architecture on the visual servoing task. Section 6 draws final conclusions and suggests possible extensions.

2   Modular Learning

2.1   Motivation and Biological Inspiration

The basic concept of artificial neural networks stems from the idea of modeling individual brain cells or neurons in a fairly simple way, and then connecting these models in a highly parallel fashion to offer a complex processing mechanism which exhibits learning in terms of its overall nonlinear characteristics. Because of the modeling operation, different model types have been derived, ranging widely in their complexity and operation, e.g., some are analog whereas others are binary. This means that some of the models vary considerably when their operation is compared to actual brain cells.

The overall architecture of a brain, which is not yet well understood, is highly complex in terms of connectivity and structure. The brains are not of one particular form, i.e. they are not all identical but they are composed of different regions which consist of different types of neuron. Artificial neural networks (used in conventional approaches) are generally well structured and simply coupled, thereby enabling the possibility of understanding their mode of operation. Based on this assumption and considerations, artificial neural networks can rarely be used in a black-box, stand-alone mode, in that they are generally not powerful or versatile enough to deal with more than two or three tasks, often being restricted in operation to one specific role. For an actual application, therefore, it should be remembered that, although a neural network may be employed as a key processing element, much interfacing, expert or rule-based systems, must also be employed.

If the nature or complexity of the problem is greater than that with which the neural network can deal adequately, poor overall performance will result. This could mean that the structural complexity of the neural network employed is not sufficient for the problem in hand. However, in some cases, neural networks may be too complex for the problem, i.e. the network is over-parameterized or functionally too powerful for the particular problem addressed.

By combining several networks, complex tasks, nonlinear problems, the environment and also several sensorial modalities can be taken into account. When multiple neural networks are integrated in a modular architecture, it is not always possible to define the learning. Indeed, only specific combination diagrams are suited for efficient learning. In particular, it appears almost impossible to create neural processing architectures, or to treat multiple functionalities. On the other hand, a modular neural network learning scheme with an adequate decomposition allows to learn and solve complex problems which cannot be treated or which are treated with difficulty with a single monolithic network. This is the reason for the employment of modular neural network architectures.

2.2   The Concept of Modular Learning

In a very general and abstract sense, modular systems can be defined as systems made up of structurally and/or functionally distinct parts. An overview of the modularity concept is proposed in [3]. While non-modular systems are internally homogeneous, modular systems are segmented into modules, i.e., portions of a system having a structure and/or function different from the structure or function of other portions of the system. Modularity can be found at many different levels in the organization of organisms.

The biological modularity concept has several largely independent roots [4]. In developmental biology the modularity concept is based on the discovery of semi-autonomous units of embryonic development. On the other hand, evolutionary modules are defined by their variational independence from each other and the integration among their parts, either in interspecific variation or in mutational variation. Functional modules, on the other hand, are parts of organisms that are independent units of physiological regulation. The precise definition of all these concepts is somewhat difficult and still controversial. Different approaches and architectures have been developed and proposed over the past years and the real challenge, however, is to determine how these different kinds of modules learn and relate to each other [4].

Whatever the decomposition, the learning of a modular neural scheme is crucial. Each module has to learn a portion of the problem and one should notice that the modules can also be trained using biologically inspired training rules [14].

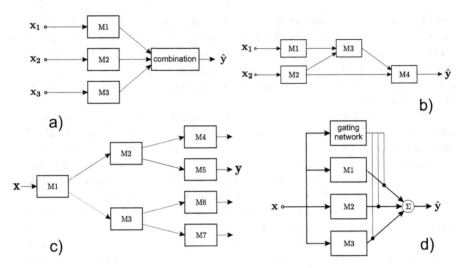

Fig. 1. Different modular neural architectures, a) a parallel architecture, b) a serial decomposition, c) a hierarchical structure, d) the Mixture of Experts model

2.3   Existing Modular Decompositions

The modularity concept is illustrated by numerous examples in [3]. These various architectures do no pose the same problem in terms of learning. The organization of the learning is more clearly presented in [2]. This work presents a classification of the architectures depending on the nature of the modules and on the elements that are combined by considering homogeneous architectures (where all modules are equivalent to an ideal neural network defined by its interface and its capability to approximate any arbitrary function). The authors distinguish Mixtures of Experts, hierarchical, parallel and serial structures. These modular neural structures are represented by figure 1.

Mixtures of Experts [10, 11] exploit the capabilities of the divide and conquer principle in the design and learning of the modular artificial neural networks. The strategy of divide and conquer solves a complex computational problem by dividing it into simpler sub-problems and then combining the individual solutions to the sub-problems into a solution to the original problem. The divisions of a task considered in this chapter are the automatic decomposition of the mappings to be learned, decompositions of the artificial neural networks to minimize harmful interaction during the learning process, and explicit decomposition of the application task into sub-tasks that are learned separately.

A typical hierarchical decomposition can be represented by a decision tree, where the output of a network in a given layer will select a network of the inferior layer. Numerous combination schemes of parallel decompositions have been proposed, in which several neural networks treat in parallel the same or similar information. Different approaches can be distinguished: data fusion architectures, ensemble-based approaches, local experts, etc. Serial architectures split a complex problem into successive partial tasks, the set of input variables are not presented to a single network anymore. Each module treats a few inputs and intermediate results appear between the modules.

More recently, functional networks have been developed in [5, 6]. Unlike other neural networks, in these networks there are no weights associated with the links connecting neurons, and the internal neuron functions are not fixed but learnable. These functions are not arbitrary, but subject to strong constraints to satisfy the compatibility conditions imposed by the existence of multiple links going from the last input layer to the same output units.

Moreover, modular architecture networks specially dedicated to motor learning and robot control have been proposed and experimented. Examples can be found in [7, 11, 13, 1, 17]. The idea of endowing robotic systems with learning capabilities is not new [18] and all these variants have been proposed to improve the performance over monolithic neural networks.

2.4 Our Modular Decomposition

The Serial Decomposition. In terms of a control systems environment, the majority of practical controllers actually in use are both simple and linear, and are directed toward the control of a plant which is either reasonably linear or at least linearizable. The use of artificial neural networks for such an environment is considered and summarized in [18], however it is sufficient here to state that an artificial neural network is usually a complex nonlinear mapping tool, and the use of such a device for relatively simple linear problems makes very little sense at all. The fact that modular neural networks have the capability of dealing with complex nonlinearities in a fairly general way is very interesting and extremely attractive. By their nature, nonlinear systems are nonuniform and invariably require custom designed control scheme/modular architecture to deal with individual characteristics. No general theory deals comprehensively with the modular decomposition, so we propose a serial approach to divide a problem in simpler sub-problems.

Because of the global nature of the approximations obtained in fully connected networks, it is generally difficult to train such networks when the data are sampled from an underlying function that has significant variation on a local or intermediate scale.

The advantage of modular networks is that they can be structured more easily then fully connected networks. A modular architecture may contain a variety of types of network modules that are more or less appropriate for particular tasks. Types of modules can be considered to be networks with different parameters, with different topologies, with different architectures, with different complexity, and with different learning rules. A priori knowledge can thus be used to choose particular classes of neural modules that are appropriate for particular tasks. Also, by partitioning a complex mapping, modular architectures tend to find representations that are more easily interpretable then those of fully connected networks.

Among basic architectures, one can find the parallel and the serial decompositions. A parallel architecture consists of a set of neural modules having the same inputs (i.e. trained for the same task) and whose estimations are combined.

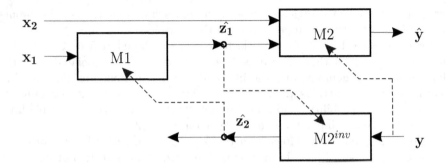

Fig. 2. Modular architecture with two serial modules, M1 and M2, and an inverse module M2^{inv} for the learning of M1

A serial architecture is a collection of individual neural modules trained on different sub-problems of a same complex problem. This architecture may be useful for partitioning the learning task with several neural networks of reduced sizes.

We have adopted the serial architecture to decompose the complexity of a robot-control problem. By choosing SOM-LLM neural modules of less than three inputs, the convergence of the self-organizing maps is presumed. In this modular architecture, the neural module can be individually or sequentially trained.

The decomposition of a complex problem in different sub-problems brings in some internal representations. These representations are often not available, and we thus need to estimate them by an inverse module [13].

Bidirectional Approach. The principle of the bidirectional architecture will be presented with an example. Consider two serial blocks, A and B (see figure 2). Each neural module is implemented with a SOM-LLM which requires a supervised learning scheme. The desired output of module A is not available. This internal representation must be estimated. We thus introduce an additional module, C, which has to learn the same internal representation. Z_k^A is the estimated internal representation made with module A, Z_k^C is the estimation of the internal representation made with module C. Z_k^A will serve to define the error signal for the learning of module C. On the other hand, Z_k^C will serve to determine the error signal for the learning of module A. Z_k^A is also the input of module B. One should note that module C represents the inverse of module B, and that module C does not participate to achieve the output of the whole modular architecture.

To enforce the convergence of the learning of the modular architecture, the estimations of the intermediate representations have to be constrained. In [2], the authors show that by imposing the statistical properties on the internal representations (i.e. mean and variance), the learning stability and convergence are then ensured.

The versatility and capabilities of the new proposed modular neural networks are demonstrated by simulation results. A comparison of the introduced modular neural network design techniques with a single network is also presented for reference. The results presented in this chapter lay a solid foundation for design and learning of the artificial neural networks that have a sound neurobiological basis that leads to superior design techniques.

3   The Extended Kohonen Maps (SOM-LLM)

The neural network we adopted for the implementation of a neural module is the SOM-LLM (Self-Organizing Map-Local Linear Map)[2]. This neural network is a combination of the extended self-organized Kohonen map [12, 15] and ADA-LINE networks [16]. The SOM-LLM has been chosen for its simplicity, for its implementation facilities, and for its topological properties (neighborhood and competition between neurons). The main objective of the SOM-LLM is to approximate any transformations by a linear local estimation.

The SOM-LLM is composed of a grid of neurons for the discretization of the input space, a grid of neurons for the discretization of the output space, and ADALINES associated to each neuron from the output map to compute a linear local response. The SOM-LLM architecture is represented by figure 3.

For each input vector (stimulus), the neurons of the input map compete to be the best representation of the input x_k. The weights of the winner, $w_s^{(in)}$, are those that minimize the distance with the input:

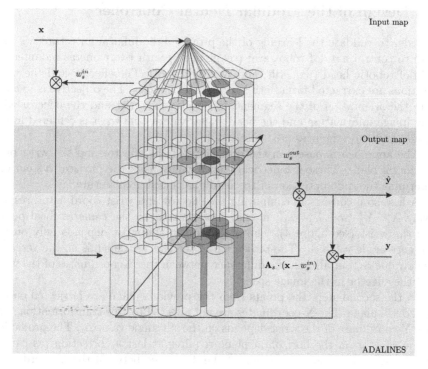

Fig. 3. SOM–LLM architecture

$$s = \arg\min \left\| x_k - w^{(in)} \right\| . \tag{1}$$

The weights $\Delta w^{(in)}$ and $\Delta w^{(out)}$, respectively in the input and in the output maps, are adapted using the two following equations:

$$\Delta w^{(in)} = w^{(in)} + \varepsilon h_s \left(x_k - w^{(in)} \right) , \tag{2}$$

$$\Delta w^{(out)} = w^{(out)} + \varepsilon' h'_s \left(u - w^{(out)} \right) , \tag{3}$$

where ε and ε' are the learning rates, h_s and h'_s the neighborhood functions, and u the desired output.

Thus, the output of the network is expressed by:

$$y = w_s^{(out)} u + A_s \left(x_k - w_s^{(in)} \right) , \tag{4}$$

where A is the ADALINE weight vector.

Each module of our learning architecture will be a SOM-LLM network.

4   Design of the Modular Neural Controller

In order to validate the learning of the proposed modular architecture, we propose to control a 4-dof robot-arm (see figure 6) with two cameras mounted on a 4-dof robotic head (pan, tilt, and two vergences). The effector and the wrist positions are extracted form both left and right images. The objective is to compute the orientation of the segment defined by the wrist and the effector with only image information and the head angles values (the wrist is centered in the images and the vergences are symmetric).

The approach is based on the projection of the effector and the wrist on a horizontal plane. These geometrical transformations of the physical system will determine the decomposition of the modular neural architecture.

A first step consists in computing the effector and wrist coordinates, respectively X''_E, Y''_E and X''_P, Y''_P, in the plane defined by the cameras focal points and the wrist (see figure 4). The projection of the wrist depends only on the vergence angle value α_v. The projection of the effector depends on the vergence angle value α_v but also on the difference between the X-coordinates of the wrist and the effector in the image space.

In the second step, the points from the previous plane are projected on the horizontal plane. The X-coordinates are not affected by this transformation. The new Y-coordinate of the wrist depends on the tilt angle value α_t. The projection of the effector on the horizontal plane requires at last a correction (as can be seen in figure 5) which is function of the difference between the Y-coordinates of the wrist and the effector in the image space.

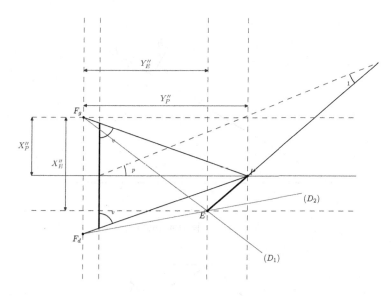

Fig. 4. Plane formed by the two focal points F_g and F_d and by the wrist P

The estimated orientation can then be expressed by the following equation:

$$\theta_4 = \cos^{-1}\left(\frac{\sqrt{(X_P - X_E)^2 + (Y_P - Y_E)^2}}{l_4}\right), \tag{5}$$

with l_4 the length of the effector of the robot.

5   Results

5.1   Application to a Visual Servoing Task

The performances of the proposed architecture are evaluated on a visual servoing task. The proposed architecture is used to control a 4-dof robot (see figure 6) through visual feedback. Indeed, the neural architecture estimates the orientation of an object in the scene which will serve as a control signal for the fourth axis of the robot-arm. The computation of the three first axis has been presented in [8], also with a modular architecture. The information used to control the robot-arm are the joint angle values of the stereoscopic head and the positions of the effector and of the wrist in both images. The principle of visual servoing is presented in [9]. Our scheme is depicted in figure 7.

The inputs of the modular neural architecture are defined as follow:

- α_p, α_t and α_v, the pan, tilt and vergence angle values of the stereoscopic head (the left and right vergences are kept symmetric),
- X_{gE}, Y_{gE}, X_{dE} and Y_{dE}, respectively the coordinates of the effector in the left and right image,

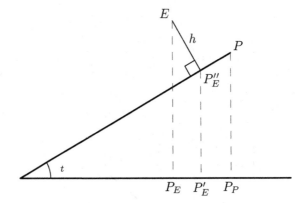

Fig. 5. Projection plane

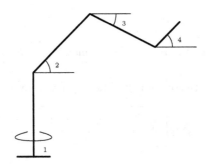

Fig. 6. The 4 degrees of freedom robot

The outputs of the modular neural architecture are:

- X_E, Y_E, the coordinates of the projected effector on the horizontal plane.
- X'_E, Y'_E, the coordinates of the projected effector on the horizontal plane before correction.
- X''_E, Y''_E, the coordinates of the effector in the plane defined by the cameras focal points and the wrist.

5.2   Wrist and Effector Projected Coordinates

The modular decomposition we propose is based on geometrical knowledge about the robot-vision system. The global architecture is showed by figure 8. One can see that the estimation of the wrist's projected coordinates is learned by only one module which is represented by figure 9. The learning of this module results in a mean error that is less than 1% and a maximum of 3% .

The estimation of the effector's projected coordinates requires three steps and thus three neural modules (see figure 10): a first step (P''_E) consists in computing the effector's coordinates in the plane defined by the cameras focal points and

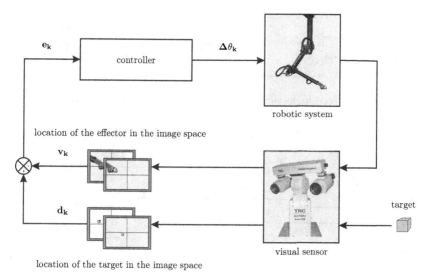

Fig. 7. Visual servoing principle

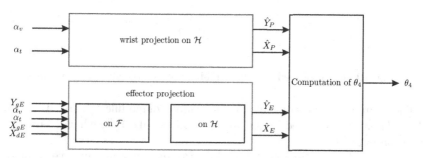

Fig. 8. The proposed decomposition based on significant values

the wrist, the two other steps are the projection on the horizontal plane and a correction (H and P_E).

The first module is function of the vergence angle and function of the coordinates of the wrist in the image. This relationship is expressed by the following equations:

$$X_E'' = \frac{-B.b.c}{a.d - b.c} \quad , \quad Y_E'' = \frac{-B.a.c}{a.d - b.c} \quad . \tag{6}$$

with (f is the focal distance) :

$$a = f\sin(\alpha_v') - X_{gE}dx\cos(\alpha_v') \ , \tag{7}$$

$$b = f\cos(\alpha_v') + X_{gE}dx\sin(\alpha_v') \ , \tag{8}$$

$$c = f\sin(\alpha_v') + X_{dE}dx\cos(\alpha_v') \ , \tag{9}$$

$$d = -f\cos(\alpha_v') + X_{dE}dx\sin(\alpha_v') \ . \tag{10}$$

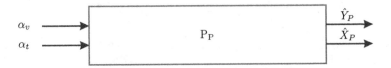

Fig. 9. The proposed single-module-network for implementing the wrist projection

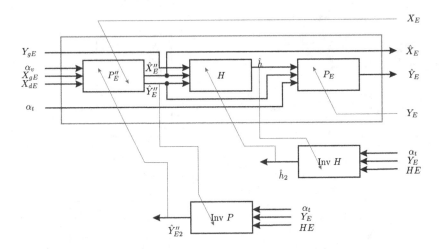

Fig. 10. The proposed modular architecture for the effector projection

The desired necessary outputs for the training of module P_E'', are either known (like X_E'') either estimated by an inverse module (like Y_E''), which is also represented on figure 10.

X_E'' and Y_E'' are estimated by module P_E and by using an inverse module. The results are presented by figures 11-13. The first figure, 11, shows $\hat{Y}''_E = f(Y_E'')$. It clearly shows the bijective relationship (close to linear) between the computed and the desired output. The rate is not one because the estimation of the intermediary representation is only constrained with a null average and a unit variance. These constraints are very interesting. Firstly, without these constraints, we cannot enforce the convergence to a representative estimation of the intermediate space. Secondly, the learning performances of Kohonen maps are better with input signals confined in a same range. In other words, the input signals of a map must have the same statistics. Thus, the online learning of an architecture composed of several cascaded modules is possible because all the input signals are constrained in the same manner. We know that other constraints can be used.

The second step is for estimating the internal representation h which allows to compute Y_E from Y_E'. H is a function of three parameters, which are X_E'', Y_E'' and Y_{gE}. The first two (X_E'' and Y_E'') are estimated by module P_E''. The learning of module H thus requires the outputs of module P_E'' associated with the output of another inverse module. The learning performances are validated

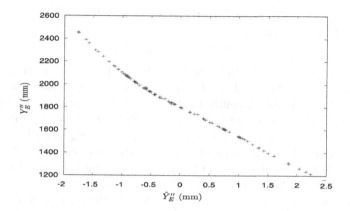

Fig. 11. $\hat{Y}''_E = f(Y''_E)$ (in mm)

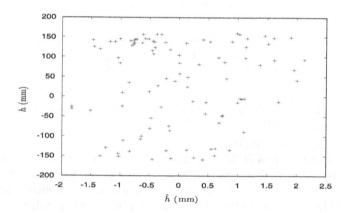

Fig. 12. $\hat{h} = f(h)$ (in mm)

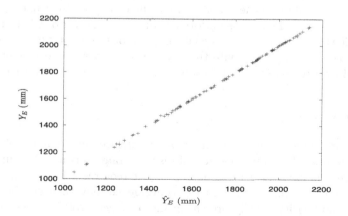

Fig. 13. $\hat{Y}_E = f(Y_E)$ (in mm)

by comparing the estimation $\hat{h}$ to the theoretical value given by the following geometrical relationship:

$$\hat{h} = a.h + b.Y_E'' + c \tag{11}$$

The set of data points represented by figure 12 and representing the intermediate space seems to show that the learning process failed. In fact, h is estimated by two different neural modules. Module H gives a first estimation $\hat{h} = f_1(X_E'', Y_E'', Y_{gE})$ while an inverse module delivers a second estimation of h depending from other values, $\hat{h} = f_2(Y_E, HE, \alpha_t)$, with HE the effector's height in the 3-D space. The only constraints used to force the outputs of f_1 and f_2 to converge to the same value are a null average and a unit variance. The resulting estimation is either h or every other solution of f_1 and f_2.

A more detailed analysis allows us to say that the estimation $\hat{h}$ is a linear combination of h and Y_E''.

$$\hat{h} = a.h + b.Y_E'' + c \ . \tag{12}$$

In this case:

$$f_1 = a\frac{\sqrt{X_E''^2 + Y_E''^2}Y_{gE}}{f} + bY_E'' + c \ , \tag{13}$$

$$f_2 = a\frac{HE - Y_E\tan(\alpha_t)}{\cos(\alpha_t)} + b\frac{Y_E\tan(\alpha_t)}{\sin(\alpha_t)} + HE - Y_E\tan(\alpha_t) + c \ , \tag{14}$$

with a, b and c, some fixed coefficients.

Having an output $\hat{h}$ of H different from h does not disturb the learning (12) of the whole modular architecture. The performance of the learning is represented by figure 13. A comparison is presented with a single monolithic network by figure 14. What we need is that h is contained in $\hat{h}$. The learning is good, the mean error is less than 1% and the maximum is about 4% for the estimation of h.

θ_4, the orientation of the robot, can now be computed by using X_P, Y_P and X_E, Y_E (5). This estimation is performed with a mean error of 1.5 degrees. This joint angle value is then used as a control signal for the robot. One can note that this approach allows to evaluate the orientation of every object defined in the scene and thus to control a 4-dof robot to reach its position and orientation.

6   Conclusion

We have presented a robust modular neural architecture that is able to learn complex systems. Learning robotic tasks is not straightforward. Nonlinear, often with a great number of degrees of freedom, it requires a lot of computational power, lots of training data, and the convergence may then not be enforced. We chose modularity, widely present in the biological world, to decompose the complexity of a problem. A high dimensional fully connected neural network is replaced by a set of well organized modules which can be more easily trained.

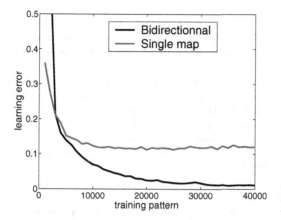

Fig. 14. The learning error of the global bidirectional modular learning architecture compared to a single network (a single map)

The main contribution of the work is a formulation that makes a set of neural modules able to converge and to learn any complex system.

Negative aspects of the use of artificial neural networks remain, however, in that they present problems in terms of stability and approximation analysis, and furthermore, it is often difficult to choose the modular network architecture. Even though, the proposed modular neural network is very well suited for online applications.

Future study will concern the possibility of using other constraints to enforce the convergence toward representative internal representations. Finally, further work is needed to understand the structural properties of the modular network and the plant under control.

References

1. J.-L. Buessler, and J.-P. Urban. *Visually guided movements: learning with modular neural maps in robotics.* Neural Networks, Special Issue on Neural Control and Robotics : Biology and Technology, 11(7-8): 1395–1415, 1998.
2. J.-L. Buessler and J.-P. Urban. Modular neural architectures for robotics. In R.J. Duro, J. Santos, and M.Grana, editors, *Biologically Inspired Robot Behavior Engineering*, pages 261–298. Springer Verlag, 2003. ISBN 3-7908-1513-6.
3. T. Caelli, L. Guan, and W. Wen. *Modularity in Neural Computing.* Proceedings of the IEEE, Special Issue on Computational Intelligence, 87(9): 1497–1518, 1999.
4. R. Calabretta, S. Nolfi, D. Parisi, and G. P. Wagner. *Duplication of modules facilitates the evolution of functional specialization.* Artificial Life, 6(1): 69–84, 2000.
5. E. Castillo. *Functional Networks.* Neural Processing Letters, 7(3): 151–159, 1998.
6. E. Castillo, A. Cobo, J. M. Gutirrez, and E. Pruneda. *Functional Networks With Applications. A Neural Based Paradigm.* Kluwer Academic, Norwell, Massachusetts, 1999.
7. H. Gomi, and M. Kawato. *Recognition of Manipulated Objects by Motor Learning with Modular Architecture Networks.* Neural Networks, 6(4): 485–497, 1993.

8. G. Hermann, P. Wira, and J.-P. Urban. Neural networks organizations to learn complex robotic functions. In Michel Verleysen, editor, *11th European Symposium on Artificial Neural Networks (ESANN'2003)*, pages 33–38, Bruges, Belgium, 2003. D-Facto.

9. S. Hutchinson, G. Hager, and P. Corke. *A Tutorial on Visual Servo Control*. IEEE Transactions on Robotics and Automation, 12(5): 651–670, 1996.

10. R. A. Jacobs, M. I. Jordan, S. J. Nowlan, and G. E. Hinton. *Adaptive Mixture of Local Experts*. Neural Computation, 3(1): 79–87, 1991.

11. R. A. Jacobs, and M. I. Jordan. *Learning Piecewise Control Strategies in a Modular Neural Network Architecture*. IEEE Transactions on Systems, Man, and Cybernetics, 23(2): 337–345, 1993.

12. T. Kohonen. *Self-Organizing Maps*, volume 30 of *Information Sciences*. Springer-Verlag, Berlin, 1995.

13. H. Miyamoto, S. Schaal, F. Gandolfo, H. Gomi, Y. Koike, R. Osu, E. Nakano, Y. Wada, and M. Kawato. A kendama learning robot based on bi-directional theory. *Neural Networks*, 9(8):1281–1302, 1996.

14. P. Poirazi, C. Neocleous, C. S. Pattichis, and C. N. Schizas. *Classification Capacity of a Modular Neural Network Implementing Neurally Inspired Architecture and Training Rules*. IEEE Transactions on Neural Networks, 15(3): 597–612, 2004.

15. H. J. Ritter, T. M. Martinetz, and K. J. Schulten. *Neural Computation and Self-Organizing Maps*. Computation and Neural Systems. Addison-Wesley, Reading, MA, 1992.

16. B. Widrow, and E. Wallach. *Adaptive Inverse Control*. Prentice Hall Press, Upper Saddle River, 1996.

17. D. M. Wolpert, and M. Kawato. *Multiple Paired Forward and Inverse Modeles for Motor Control*. Neural Networks, Special Issue on Neural Control and Robotics : Biology and Technology, 11(7-8): 1317–1329, 1998.

18. A. M. S. Zalzala and A. S. Morris. *Neural Networks for Robotic Control*. Ellis Horwood, Norwell, New York, 1996.

Neural Robot Detection in RoboCup

Gerd Mayer, Ulrich Kaufmann, Gerhard Kraetzschmar, and Günther Palm

University of Ulm,
Department of Neural Information Processing,
D-89069 Ulm, Germany

Abstract. Improving the game play in ROBOCUP middle size league requires a very robust visual opponent and teammate detection system. Because ROBOCUP games are highly dynamic, the detection system also has to be very fast. That both conditions are not necessarily contradictory is shown in this paper. The described multilevel approach documents, that the combination of a simple color based attention control and a subsequent neural object classification can be applied successfully in real world scenarios. The presented results indicate a very good overall performance regarding robustness, flexibility and computational needs.

1 Introduction

In ROBOCUP middle size league, teams consisting of four robots (three field players plus one goal keeper) play soccer against each other. The game itself is highly dynamic. Some of the robots can drive up to 3 meters per second and accelerate the ball even more. Cameras are the main sensor used here (and often the only one). To play reasonably well within this environment, at least 10–15 frames per second must be processed.

Instead of grasping the ball and running towards the opponents' goal, team play and thus recognizing teammates and opponent robots is necessary to further improve robot soccer games. Because of noisy self localization and an unreliable communication between the robots, it is somewhat risky to rely only on the robots' positions communicated between the teammembers. In addition, opponent robots do not give away their positions voluntarily. Hence, there is a need for a visual robot detection system.

A method used for this task has to be very specific to avoid detecting too many uninteresting image parts, yet flexible enough to even detect robots that were never seen before. A robot recognition approach used in a highly dynamic environment like ROBOCUP also has to be very fast to be able to process all images on time.

In contrast to the scenarios where lot of other object detection methods on mobile robots are used, the setting described here is a lot more natural. Objects are in front of different, cluttered backgrounds, are often partially occluded, blurred by motion and are extremely variable in sizes between different images. Besides that, the shapes of robots from different teams vary highly, only limited by a couple of ROBOCUP regulations and some physical and practical

S. Wermter et al. (Eds.): Biomimetic Neural Learning, LNAI 3575, pp. 349–361, 2005.
© Springer-Verlag Berlin Heidelberg 2005

constraints. Additional difficulties arise because the robots' vision system, which includes a very large aperture angle, causes extreme lens distortions.

The approach presented here is a multilevel architecture. This includes one level for the selection of interesting regions within the image that may contain a robot using simple and fast heuristics. The next level then classifies these regions with more complex and costly methods to decide if there is in fact a robot within the selected regions or not. This classification step is realized using artificial neural networks.

In this paper we focus more on the later steps: the choice of the used features, their parameterization and the subsequent neural decision making. We discuss in detail the performance of the neural networks on robots from the training image set, on robot views previously unseen and even on totally unknown robot types. The test cases used for these results contain various kinds of perceptual difficulties as described above. Considering this, the presented results indicate a very good overall performance of the presented approach.

In the next Section the problem is explained. In Section 4 our method is described in detail, including all separate steps. Results are presented in Section 5. In Section 3 our results are discussed in the context of related work. Finally, Section 6 draws conclusions.

2   The Problem

As already mentioned, ROBOCUP is a quite realistic testbed. In contrast to other object recognition problems implemented on mobile robots, a lot more problems

Fig. 1. Example images, taken from THE ULM SPARROWS, showing different types of perceptual difficulties. In the first two pictures (upper row) the variability in sizes of robots from the same type can be seen nicely. Recordings three and four showing occlusions, image five illustrates the lens distortion, which let the robot tilt to the left, and finally picture six is blurred by the robots own movement

arise here that may also occur in real world scenarios: partial occlusions between objects or at the image borders, vast size variations, huge variations in (robot) shapes, motion blur through own or object motion, cluttered, unstructured and often unknown background and even unknown objects (e.g. robot types) that were never seen before. Figure 1 illustrates some of these problems, showing images recorded from one of our robots during the last tournament. In the first two pictures the variability in sizes of robots from the same type can be seen nicely. Recordings three and four showing occlusions, image five illustrates the lens distortion, which let the robot tilt to the left, and finally picture six is blurred by the robots own movement (the used pictures are admittedly not that badly blurred). Please note also, how different the individual robots look like.

3   Related Work

Object detection is a well known problem in current literature. There are many approaches to find and classify objects within an image, e.g. from Kestler [1][2], Simon [3] or Fay [4][5] to name just a few that are developed and investigated within our department.

Within RoboCup the problems are rather less well defined then in their scenarios and real-time performance is not an absolute prerequisite for them, which may be the main reason that up to now there are only few workings published about more complex object detection methods in RoboCup. Most of the participants in the RoboCup middle size league use color based approaches, like e.g. in [6][7][8]. One interesting exception is presented by Zagal et. al. [9]. Although they still use color-blob information, they let the robot learn different parameters for the blob evaluation, like e.g. the width or the height of the blob using genetic algorithms. Thereby they are able to even train the robot to recognize multi-colored objects as used for the beacons on both sides of the playing field (as used in the Sony legged league, which is well comparable to the middle size league). Another method used in this league is introduced by Wilking et. al. [10]. They are using a decision tree learning algorithm to estimate the pose of opponent robots using color areas, its aspect ratio, angles between line segments and others.

One attempt to overcome the limitations of color based algorithms is presented by Treptow et. al. [11] where an algorithm called Adaboost uses small wavelet like feature detectors. Another approach, that does not even need a training phase at all, is presented by Hanek et. al. [12]. They use deformable models (snakes), which are fitted to known objects within the images by an iterative refining process based on local image statistics to find the ball.

As you will see, the presented solution approach has several advantages over these methods: First of all, the system is able to work in real time. As real time is a very burdened item, it is meant here as being able to process as much images as necessary to fulfill a specific task, in this case at least around 15 images per second. Another advantage is, that the system do not need an explicit, a prior known model. The model is learned automatically from the example data.

And because of the layered architecture, individual components of the system can be exchanged easily without the need of readjusting the remaining parts. In contrast, it can be easily extended with e.g. a tracking module to further stabilize the object detection task.

4   Solution Approach

The robot recognition method we present in this paper can be roughly divided into the following individual steps:

1. Detect regions of interest.
2. Extract features from these regions.
3. Classify them by two neural networks.
4. Arbitrate the classification results.

The final result is a decision, if a robot is seen or not, and if it is part of the own or the opponent team. The whole process is illustrated in Figure 2 to further clarify the individual steps.

In the first step, potential robot positions are searched within the recorded images to direct the robots attention to possibly interesting places (1). In the next step different features are calculated for each of the detected regions of interest (ROI). They describe different attributes of the robot as general as possible to be sensitive to the different robot shapes yet specific enough to avoid false

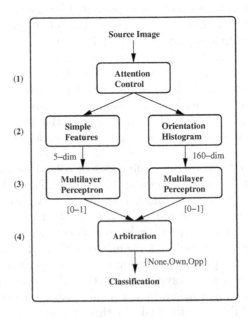

Fig. 2. Data flow from the source image, passing through the various processing stages, up to the recognition result

positives (2). Used features are for example the width of the robot, a percentage value of the robots color and orientation histograms (and others described later in detail). All these features (i.e. the vector describing the features) are then passed over to two multilayer perceptron networks which vote for each feature vector if it belong to a robot or not (3). A final arbitration instance then decides wehther we assume a teammate, an opponent robot or no robot at all (4).

Regions of interest are by the way rectangular bounding boxes around found image areas. Such rectangular boxes are considerably faster to process than pixel-precise shapes around the object. The drawbacks of such simplified areas can be mitigated by the below described process using binary masks (Section 4.2).

4.1   Region of Interest Detection

In ROBOCUP, all relevant objects on the playing field are color coded. The floor is green, the ball is orange, the goals are blue and yellow and the robots are mainly black with a cyan or magenta color marker on top of them. So a simple attention control can be realized using a color based mechanism.

Consequently, region of interests are, in principle, calculated by a blob search within a segmented and color-indexed image (for a detailed explanation see e.g. [13] or [14]). Additional model based constraints reduce the number of found ROIs. Different possible methods are already explained in detail in [15], so we will not detail this here.

Of course, such a simple method cannot be perfect. There is always a tradeoff between how many robots are focused and how many regions not containing a robot are found. Because all subsequent processing steps only consider these found regions, it is important not to miss too many candidates here. On the other hand, too many non-sense regions increases the computational cost of the whole process.

4.2   Feature Calculation

The following features are finally used:

- perceived width of the robot on the image,
- percentage amount of black color within the whole ROI,
- percentage amount of (cyan or magenta) label color, separated into left, middle and right part of the ROI,
- an orientation histogram.

The first three items (respectively five values) are so called simple features (see Figure 2). The black and label color percentages are determined using the already mentioned color-indexed image.

The orientation histogram, the right process sequence in Figure 2, is a method to describe the overall shape of an object in a very general and flexible way. First of all, the found ROI is subdivided into several subwindows as illustrated in Figure 3. For each subwindow we then calculate independently the direction and the strength of the gradient in x- and y-direction (using a Sobel operator on

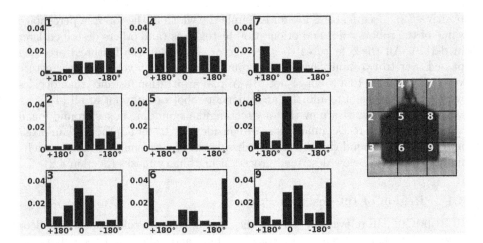

Fig. 3. Orientation histogram using 3 × 3 subwindows and 8 discretisation bins. The individual histograms corresponds with the particular subwindows from the right, original image. For example a significant peak at direction zero degree, like e.g. in histogram eight, indicate a sole vertical line in the original image at the corresponding position

a grayscale image). The directions of the gradients are discretized into a specific number of bins (e.g. eight in Figure 3). Finally we sum up all occurrences of the specific direction weighted by their strengths. To be even more flexible, the individual subwindows may overlap to a specified degree (see Section 5).

Because of extreme lens distortions, hollow robot bodies or partial occlusions with other objects, there is a high chance that there is also a lot of noise inside the selected regions of interest. To minimize the influence of this, a binary mask is calculated from the robot and label colors (again with respect to the color-indexed image). The mask is afterwards dilated several times to fill small holes. The orientation histogram is then only calculated, where the binary mask contains a positive value.

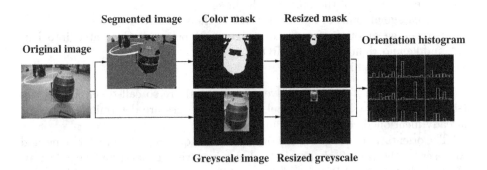

Fig. 4. Sketch of the various steps to calculate the histogram features including the robot mask and the maximal size reduction

Another problem arises due to very different robot sizes. A close-up of a robot may contain a lot more details compared to a wide-distance image. So if a ROI is larger than a specific threshold, the ROI is scaled down to this upper size limit. For a more clear and concise explanation of the various preprocessing steps, please have a look at Figure 4.

4.3   Neuronal Networks Classification

After calculating the different features, they are passed to two artificial neural networks as can be seen in Figure 2. One network only processes the simple features, the input for the second one are the orientation histogram values. The breakdown into two networks turned out to be necessary in order not to let the pure mass of data of the orientation histogram overvote the few simpler features.

Both neural networks are standard multilayer perceptron networks containing only one hidden layer and one single output neuron trained with a normal backpropagation algorithm. Both networks produce a probability value that describes the certainty of seeing a robot regarding the given input vector.

The input layer of the first network consists of 5 input neurons according to the five values described above. The number of neurons for the input layer of the second network depends on the parameterization of the orientation histogram. In Section 5 we present different parameterizations, but in general the size is the product of the number of subwindows in vertical and horizontal direction multiplied with the number of discretisation steps for the gradient direction. The number of neurons in the hidden layers is up to the network designer and is also evaluated in detail in Section 5.

4.4   Arbitration

The final classification decision is made from a combination of the outputs of the two neural networks. Because every network only uses a subset of the features, it is important to get an assessment as high as possible from each individual network. Of course a positive feedback is easy, if both networks deliver an assessment of nearly 100%, but in real life, this is only rarely the case. So the network outputs are rated that way, that only if both networks give a probability value above 75%, it is assumed that a robot is found within the region of interest. The membership of the robot to the own or the opponent team is determined using the robots color marker. If the ROI contains a lot more cyan pixels than magenta pixels or vice versa, the membership can be assumed for sure.

5   Experimental Results

As usual in a robot vision system, especially in combination with artificial neural networks, there are a lot of parameters that can and must be adjusted properly. The next sections discusses all these possible parameters followed by some results how the complete permutation of all possible values and the selected best combination perform during different tests.

5.1   Parameterization

Having a closer look at the orientation histogram: The larger the overlap between the individual subwindows, the less problematic are ROIs that do not match a robot exactly (in size or in position), but on the other hand, the result is getting less specific. The number of subwindows in both directions controls how exact the shape representation will be. This is again a decision between flexibility and specificity. It can be wise to choose a larger number here if all robot types are known already, but disadvantageous, if more flexibility is needed. Exactly the same applies for the direction discretisation, i.e. the number of bins used in the histogram.

Another starting point for parameterization are the perceptron networks. The number of input neurons is determined by the parameters of the orientation histogram or the number of simple features and the number of output neurons is fixed to only one (the probability for seeing a robot). But there is still the hidden layer which can be adjusted. The more neurons a network has in the hidden layer, the better it may learn the presented examples, but the higher is the risk to overfit the network and to lose plasticity and the ability to generalize the learned things. Another drawback of a large hidden layer is the increasing computational cost caused by the full connectivity between the neurons.

To be more concrete, the following values have been tested for the respective parameters:

- Subwindow overlap: 30%, 20%, 10% or 0% (no overlap).
- Number of subwindows: 2, 3 or 4 (in both directions).
- Number of histogram bins: 6, 8, 10, 12.
- Number of hidden neurons: identical to the number of input neurons or half, one third or one quarter of it (applies for the perceptron network which deals with the orientation histogram data, for the other network, see below).

5.2   Training and Classification

This paper focuses on the evaluation of the object recognition performance and not the performance of the attention control or the overall process like in [15]. So we labeled all the images used in this section by hand to ensure, that the performance is only determined by the classification instance and not by inaccuracies from the attention control.

We used images from six different types of robots with varying shapes collected during several tournaments (examples of the different robot types are already shown in Figure 1, Figure 5 shows some of the Ulm Sparrows with ROI drawn around them). The collection contains 682 robots from:

- AIS/BIT-Robots (84 ROIs)
- Osaka University Trackies (138)

Fig. 5. Some of the labeled Ulm Sparrows images from varying distances and viewing directions with the respective ROIs around them

- Attempto Tübingen (116)
- Clockwork Orange Delft (148)
- Mostly Harmless Graz (102)
- Ulm Sparrows (94)

Additionally, we used 656 ROIs as negative examples:

- 164 carefully selected ones, like e.g. black trousers or boxes
- 492 randomly chosen areas

The learning phase is done by using five out of the six robot types, choosing 90% of the ROIs of each robot type, mix them with again 90% of the negative examples and train the artificial neural networks with them. The training phase is done three times for each combination of robot types and parameters to choose the network with the lowest mean square error regarding the training material to avoid unfortunate values of the (random) initialization.

After that, we analyzed a complete confusion matrix on how the network recognizes the training data set itself or the remaining 10% of the robot and negative images which it has never seen before. Additionally, we calculated such a confusion matrix for a totally new (the sixth) robot type. Because the neural network emits a probability value between zero and one, a positive classification answer is assumed if the probability value is above 75% like it is handled for the final arbitration step (see Section 4.4).

5.3   Orientation Histogram

First results are provided for the neural network which processes the orientation histogram vector. In Table 1 the average results over all possible parameters for

Table 1. Resulting confusion matrices when presenting different data sets to the neural network responsible for the orientation histogram values. The results are averaged over all possible parameter permutations

Training data	Positive	Negative		Evaluation data	Positive	Negative
Correct decision	99.2%	99.8%		Correct decision	91.9%	97.1%
False decision	0.8%	0.2%		False decision	8.1%	2.9%

Table 2. Resulting confusion matrices when using the best parameter settings for the orientation histograms. The results are averaged over all possible robot type permutations

Evaluation data	Positive	Negative		Unknown robot type	Positive
Correct decision	100%	99.5%		Correct decision	96.4%
False decision	0.0%	0.5%		False decision	3.6%

the different data sets (the training and the evaluation set) are listed. The upper left cell displays the percentage of robots that are correctly identified as such, whereas the lower right cell represents the remaining percentage of regions that are assumed as not being a robot while in fact are a robot. The same applies for the other cells for regions not containing a robot.

First we test, how well the artificial networks perform if confronted again with the data set used during the training phase (the first part of Table 1). Note that the results are almost perfect, showing nearly 100% in the upper row. This means, that the networks are able to memorize the training set over a wide range of used parameters. The second part of the table is calculated using the evaluation data set. Remember that these values are made using images from the same robot types which are not presented to the network during the training phase. So even when averaging over all possible parameter values, the networks classified already over 90% correct.

Remember, that we trained the networks alternately only with five robot types. So now we can select the parameter combination, for which the networks perform best if confronted with the sixth type. Table 2 shows the performance for these settings one time for the evaluation data set and the other time for the sixth robot type (without negative examples respectively). The parameters proven to perform best are the following: Subwindow overlap 20%, number of subwindows 4, number of histogram bins 10, number of hidden neurons: 160 (identical to the number of input neurons). There are a couple of other parameter settings that perform only slightly worse.

5.4   Simple Features

Next, results are shown for the artificial neural networks dealing with the simple features. The size of the hidden layer is the only variable parameter here. After testing all networks with a number of 3, 5, 10 and 15 hidden neurons, differences

Table 3. Resulting confusion matrices for the network dealing with the simple features. The results are averaged over all possible robot type permutations

Evaluation data	Positive	Negative
Correct decision	96.7%	100%
False decision	3.3%	0.0%

Unknown robot type	Positive
Correct decision	94.6%
False decision	5.4%

Table 4. Resulting overall confusion matrices for one parameter set and one robot type combination

Evaluation data	Positive	Negative
Correct decision	96.0%	100%
False decision	4.0%	0.0%

Unknown robot type	Positive
Correct decision	100%
False decision	0.0%

seems to be marginal, so we finally choose 3 neurons for this layer. Whereas computing time can be disregarded here, the fewer neurons should increase the ability to generalize the learned things. Table 3 again presents the resulting value.

5.5   Overall Classification Results

Finally we want to present an overall result. The parameters are chosen identically to the above ones. The network is trained with the following robot types: AIS/BIT-Robots, Osaka University Trackies, Attempto Tübingen, Clockwork Orange Delft and Mostly Harmless Graz (and the usual negative examples). As you can see in Table 4, the unknown robot type (THE ULM SPARROWS) is classified perfectly. The worst values are for the AIS/BIT-Robots where the network classifies 88% correctly and Tuebingen, where 9% of the robots are not recognized.

5.6   Computing Time

The vision system currently used in THE ULM SPARROWS provides 30 frames per second. Although it is not necessary to detect the robots in each single frame, a fast processing is an absolute prerequisite to be able to use the method in real tournaments. As absolute timing values should always been taken with a grain of salt, they may nevertheless give an impression on how fast the method can be. The values are measured on a 2.6GHz Pentium4 Processor, the image sizes are 640×480 Pixels. The various image preprocessing steps, like color indexing, mask building, observance of the upper size limit etc. need around 5ms, processing the orientation histogram averages 10ms and the final classification and arbitration step approximately 2ms. Note that some of the preprocessing steps like the color indexing need to be done anyway in the vision system for attention control of a lot of other objects.

6   Conclusions and Future Work

The presented results indicate a very good overall performance of our approach considering all the problematic circumstances mentioned above. We showed that splitting the problem into several subtasks (i.e. simple color based preprocessing in combination with neural network classification) made the problem manageable. The method appears to be a good basis for further improvements.

Even though the images are taken during real tournaments, there are always surprises in real robotics. So additional, more specific features, or enhancements like temporal integration may further help to stabilize the overall detection rate (e.g. occluded robots can be tracked even if the robot is not observable or detected in every single image).

Acknowledgment. The work described in this paper was partially funded by the DFG SPP-1125 in the project *Adaptivity and Learning in Teams of Cooperating Mobile Robots* and by the MirrorBot project, EU FET-IST program grant IST-2001-35282.

References

1. Kestler, H.A., Simon, S., Baune, A., Schwenker, F., Palm, G.: Object Classification Using Simple, Colour Based Visual Attention and a Hierarchical Neural Network for Neuro-Symbolic Integration. In Burgard, W., Christaller, T., Cremers, A., eds.: Advances in Artificial Intelligence. Springer (1999) 267–279
2. Kestler, H., Sablatnög, S., Simon, S., Enderle, S., Kraetzschmar, G.K., Schwenker, F., Palm, G.: Concurrent Object identification and Localization for a Mobile Robot. KI-Zeitschrift (2000) 23–29
3. Simon, S., Kestler, H., Baune, A., Schwenker, F., Palm, G.: Object Classification with Simple Visual Attention and a Hierarchical Neural Network for Subsymbolic-Symbolic Integration. In: Proceedings of IEEE International Symposium on Computational Intelligence in Robotics and Automation. (1999) 244–249
4. Fay, R., Kaufmann, U., Schwenker, F., Palm, G.: Learning object recognition in a neurobotic system. In Horst-Michael Groß, Klaus Debes, H.J.B., ed.: 3rd Workshop on SelfOrganization of AdaptiVE Behavior SOAVE 2004, Düsseldorf, VDI (2004) 198–209
5. Fay, R., Kaufmann, U., Knoblauch, A., Markert, H., Palm, G.: Integrating object recognition, visual attention, language and action processing on a robot using a neurobiologically plaussible associative architecture. In: 27th German Conference on Artificial Intelligence (KI2004), NeuroBotics Workshop, University of Ulm (2004)
6. Jamzad, M., Sadjad, B., Mirrokni, V., Kazemi, M., Chitsaz, H., Heydarnoori, A., Hajiaghai, M., Chiniforooshan, E.: A fast vision system for middle size robots in robocup. In Birk, A., Coradeschi, S., Tadokoro, S., eds.: RoboCup 2001: Robot Soccer World Cup V. Volume 2377 / 2002 of Lecture Notes in Computer Science., Springer-Verlag Heidelberg (2003)

7. Simon, M., Behnke, S., Rojas, R.: Robust real time color tracking. In Stone, P., Balch, T., Kraetzschmar, G., eds.: RoboCup 2000: Robot Soccer. World Cup IV. Volume 2019 / 2001 of Lecture Notes in Computer Science., Springer-Verlag Heidelberg (2003)

8. Jonker, P., Caarls, J., Bokhove, W.: Fast and accurate robot vision for vision based motion. In Stone, P., Balch, T., Kraetzschmar, G., eds.: RoboCup 2000: Robot Soccer. World Cup IV. Volume 2019 / 2001 of Lecture Notes in Computer Science., Springer-Verlag Heidelberg (2003)

9. Zagal, J.C., del Solar, J.R., Guerrero, P., Palma, R.: Evolving visual object recognition for legged robots. In: RoboCup 2003 International Symposium Padua (to appear). (2004)

10. Wilking, D., Röfer, T.: Real-time object recognition using decision tree learning. In: Proceedings of RoboCup-2004 Symposium (to appear). Lecture Notes in Artificial Intelligence, Berlin, Heidelberg, Germany, Springer-Verlag (2004)

11. Treptow, A., Masselli, A., Zell, A.: Real-time object tracking for soccer-robots without color information. In: Proceedings of the European Conference on Mobile Robotics (ECMR 2003). (2003)

12. Hanek, R., Schmitt, T., Buck, S., Beetz, M.: Towards robocup without color labeling. In: RoboCup 2002: Robot Soccer World Cup VI. Volume 2752 / 2003 of Lecture Notes in Computer Science., Springer-Verlag Heidelberg (2003) 179–194

13. Mayer, G., Utz, H., Kraetzschmar, G.K.: Towards autonomous vision self-calibration for soccer robots. Proceeding of the IEEE/RSJ International Conference on Intelligent Robots and Systems (IROS-2002) 1 (2002) 214–219

14. Mayer, G., Utz, H., Kraetzschmar, G.: Playing robot soccer under natural light: A case study. In: Proceedings of RoboCup-2003 Symposium. Lecture Notes in Artificial Intelligence, Berlin, Heidelberg, Germany, Springer-Verlag (2004)

15. Kaufmann, U., Mayer, G., Kraetzschmar, G.K., Palm, G.: Visual robot detection in robocup using neural networks. In: Proceedings of RoboCup-2004 Symposium (to appear). Lecture Notes in Artificial Intelligence, Berlin, Heidelberg, Germany, Springer-Verlag (2004)

A Scale Invariant Local Image Descriptor for Visual Homing

Andrew Vardy and Franz Oppacher

School of Computer Science,
Carleton University,
Ottawa, K1S 5B6, Canada
Fax: +1 (613) 520-4334
avardy@scs.carleton.ca
http://www.scs.carleton.ca/~avardy

Abstract. A descriptor is presented for characterizing local image patches in a scale invariant manner. The descriptor is biologically-plausible in that the necessary computations are simple and local. Two different methods for robot visual homing based on this descriptor are also presented and tested. The first method utilizes the common technique of corresponding descriptors between images. The second method determines a home vector more directly by finding the stationary local image patch most similar between the two images. We find that the first method exceeds the performance of Franz et. al's *warping method*. No statistically significant difference was found between the second method and the warping method.

1 Introduction

Visual homing is the act of returning to a place by comparing the image currently viewed with an image taken when at the goal (the *snapshot image*). While this ability is certainly of interest for mobile robotics, it also appears to be a crucial component in the behavioural repertoire of insects such as bees and ants [1]. We present here two methods for visual homing which employ a novel image descriptor that characterizes a small patch of an image such that the descriptor is invariant to scale changes. Scale change is a prevalent source of image distortion in visual homing where viewed landmarks generally appear larger or smaller than in the snapshot image. The image descriptor developed here has a simple structure which might plausibly be implemented in the limited hardware of the insect brain.

Approaches to visual homing range from those purely interested in robotic implementation (e.g. [2]) to those concerned with fidelity to biological homing (e.g. [3]). Both camps have proposed methods which find correspondences between image features and use these to compute a home vector. These feature-based methods rely on visual features such as regions in 1-D (one-dimensional) images [4,5], edges in 1-D images [3], image windows around distinctive points

S. Wermter et al. (Eds.): Biomimetic Neural Learning, LNAI 3575, pp. 362–381, 2005.
© Springer-Verlag Berlin Heidelberg 2005

in 1-D images [2], coloured regions in 2-D images [6], and Harris corners in 2-D images [7, 8]. Any visual feature is subject to distortions in scale, illumination, and perspective, as well as distortions from occlusion. The ability to correspond features in the presence of these distortions is critical for feature-based homing. Scale invariant schemes do exist. Notable examples include Lowe's scale invariant keypoints [9], and a visual homing method using scale invariant features based on the Fourier-Mellin transform [10]. However, it is currently unclear how complex these schemes might be for implementation in the neural hardware of an insect. The descriptor presented here is partially invariant to scale changes and has a direct and simple neural implementation.

The first of our two homing methods operates in a manner quite similar to that described above in that it searches for correspondences between descriptors in the snapshot image and descriptors in the currently viewed image. However, the second method takes advantage of the structure of the motion field for pure translation to avoid this search process. This method only pairs descriptors at the same image position. Very similar pairs ideally correspond to one of two stationary points in the motion field, known as the focus of contraction and focus of expansion. Finding either of these foci is equivalent to solving the visual homing problem.

An alternate approach to visual homing is Franz et. al's *warping method* [11]. This method warps 1-D images of the environment according to parameters specifying displacement of the agent. The parameters of the warp generating the image most similar to the snapshot image specify an approximate home vector. As the warping method is known for its excellent performance (see reports in [12, 13]) we use it here for comparison with our methods.

The images used in this paper are panoramic and were taken from a robot equipped with a panoramic imaging system. The results we present were obtained on a database of images collected within an office environment. We compare the performance of our two methods with the warping method on these images. Note that we make the assumption that all images were captured at the same compass orientation. A robot homing by one of our methods would require a compass to allow the differing orientation of images to be corrected. The warping method does not share this requirement. However, it has been found that the warping method performs better when it can be assumed that all images are taken from the same orientation [14].

In the next section we define a model of image scaling which is employed in the subsequent section on the development of our scale invariant image descriptor. We then present the two homing methods based on this descriptor. Next is a results section which shows the performance of these two homing methods and the warping method on a database of panoramic images. This is followed by a discussion section. The main content of the chapter ends with concluding remarks and references. An appendix includes a derivation of one of the principles underlying the image descriptor.

2   Scaling Model

We define a model of image scaling applicable to a local image patch. Let **p** be the coordinates of the centre of an image patch. The effect of scaling is to change the distance of image features to **p** by a factor k. Features nearby to **p** will shift by a smaller amount then distant features, yet the same scaling factor of k is applied for all image features. Hence, we refer to this as linear scaling.

Assume we have an image I which has been subject to linear scaling about point **p** by factor k. A point **a** in the original image I now corresponds to a point **a**$'$ in the scaled image I'. That is, a pixel in the original image at **a** will have the same value as a pixel in the scaled image at **a**$'$.

$$I(\mathbf{a}) = I'(\mathbf{a}') \tag{1}$$

Note that $I(\mathbf{a})$ is shorthand for the value of the pixel in image I with coordinates (a_x, a_y). Also, for simplicity we ignore pixel discretization and treat **a** as real-valued.

We now formulate an expression for **a** which involves the centre of scaling **p**. The following parametric equation of a line represents **a** with respect to its distance l from **p**, and with respect to the direction from **p** to **a** indicated by the unit vector **v**.

$$\mathbf{a} = \mathbf{p} + l\mathbf{v} \tag{2}$$

$$\mathbf{v} = \frac{\mathbf{a} - \mathbf{p}}{||\mathbf{a} - \mathbf{p}||} \tag{3}$$

The point **a**$'$ corresponding to **a** after linear scaling is similarly represented.

$$\mathbf{a}' = \mathbf{p} + kl\mathbf{v} \tag{4}$$

Note that this scaling model assumes the scaling factor k to be constant across the whole image. This is generally not true for the panoramic images employed here. However, linear scaling is a reasonable model for the scaling that occurs within local image patches of a panoramic image.

3   The Scale Invariant Descriptor

In this section we develop a local image descriptor which is partially invariant to scale changes. Figure 1 shows an image I and two increasingly scaled variants I' and I''. The figure also plots the value of each image along the ray $\mathbf{p} + l\mathbf{v}$ where $l > 0$ and **v** is arbitrarily set on a diagonal. We refer to this ray as a *channel*. The house image consist only of edges so the plots show isolated pulses where the channel crosses an edge.

It can be observed that while the positions of edge pulses along the channel have changed between I and I', the same two pulses are still found. Hence, the

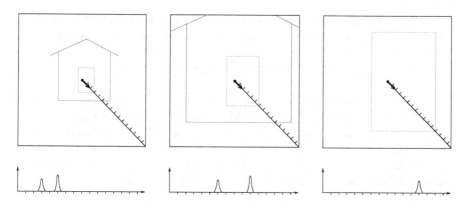

Fig. 1. Scaling of an image I and the value of I along a channel. Three images are shown with scale doubling incrementally from left to right. Beneath each image is a plot of the image values along the indicated channel. The images consist only of edges with darker edges having a higher value than lighter edges

area underneath these two pulses is the same. This observation prompts our first proposal for an invariant measure, which is the sum of image values along the channel

$$f^{\mathbf{p},\mathbf{v},I} = \int_0^{l_{\max}} I(\mathbf{p} + l\mathbf{v})\, dl \tag{5}$$

If indeed the same pulses are found along the same channel of I and I' then the following is true

$$f^{\mathbf{p},\mathbf{v},I'} = f^{\mathbf{p},\mathbf{v},I} \tag{6}$$

However, if the scaling factor k is too large then this condition will not hold. For example, in image I'' of figure 1 the outside edge of the house has been scaled entirely out of the frame. The channel now shows only a single pulse. Thus, $f^{\mathbf{p},\mathbf{v},I} \neq f^{\mathbf{p},\mathbf{v},I''}$. The same problem occurs for contraction ($k < 1$). If I'' was the original image and had been scaled down to I, the pulse representing the outside of the house would have appeared—and again $f^{\mathbf{p},\mathbf{v},I} \neq f^{\mathbf{p},\mathbf{v},I''}$. To mitigate the problem of appearance/disappearance of image features we propose a new invariant measure which includes a decay function

$$g^{\mathbf{p},\mathbf{v},I} = \int_0^{l_{\max}} w(l) I(\mathbf{p} + l\mathbf{v})\, dl \tag{7}$$

The purpose of the decay function $w()$ is to reduce the impact of outlying features on g. The appendix includes a derivation which places some constraints on $w()$. One obvious function which satisfies these constraints is

$$w(l) = \frac{1}{l^\zeta} \tag{8}$$

where $\zeta < 1$.

The objective now is to determine the relationship between $g^{\mathbf{p},\mathbf{v},I}$ and $g^{\mathbf{p},\mathbf{v},I'}$. This relationship is explored in the appendix and found to be as follows:

$$g^{\mathbf{p},\mathbf{v},I'} \approx kw(k)g^{\mathbf{p},\mathbf{v},I} \tag{9}$$

The presence of the factor $kw(k)$ implies that g is not scale invariant. We will deal with this problem momentarily. More fundamentally, however, is the fact that a scalar quantity such as $g^{\mathbf{p},\mathbf{v},I}$ is likely insufficient to describe a local image patch robustly. A richer descriptor is required to allow image patches to be disambiguated. We obtain such a descriptor by forming a vector $\mathbf{g}$ of g values computed from the same point $\mathbf{p}$ but at different directions

$$\mathbf{g}^{\mathbf{p},I} = \begin{pmatrix} g^{\mathbf{p},\mathbf{v}_0,I} \\ g^{\mathbf{p},\mathbf{v}_1,I} \\ \vdots \\ g^{\mathbf{p},\mathbf{v}_n,I} \end{pmatrix} \tag{10}$$

The length of the vector $\mathbf{g}$ is n. An obvious choice for the channel direction vectors $\mathbf{v}_i$ is to arrange them evenly in a radial pattern. For example, if $n = 4$ we would choose left, up, right, and down. If $n = 8$ we would add the four diagonals as well.

For the first algorithm presented below we will not be concerned with the length of $\mathbf{g}$, but only its direction. Therefore we define a normalized vector $\mathbf{h}$

$$\mathbf{h}^{\mathbf{p},I} = \frac{\mathbf{g}^{\mathbf{p},I}}{||\mathbf{g}^{\mathbf{p},I}||} \tag{11}$$

By normalizing we remove the factor $kw(k)$, hence

$$\mathbf{h}^{\mathbf{p},I'} \approx \mathbf{h}^{\mathbf{p},I} \tag{12}$$

and we can say that $\mathbf{h}$ is a scale invariant image descriptor. For the second algorithm it will be necessary to know whether k is greater or less than one. Thus, in the description for this algorithm we will also make reference to $\mathbf{g}$.

3.1   Conditions

The image descriptor $\mathbf{h}$ is invariant to scale changes given the following qualitative conditions:

1. The scale factor k is neither too great nor too small. The decay function can offset the impact of edge pulses being scaled in and out of range, but the scaling of outlying edge pulses will still generally distort the direction of $\mathbf{h}$.
2. If image edges are particularly dense then the edge pulses along a channel may interfere with each other in the summation of equation (7). Thus, it is advantageous for image edges to be relatively sparse.

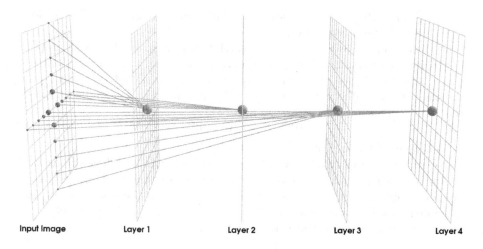

Input Image   Layer 1   Layer 2   Layer 3   Layer 4

Fig. 2. Structure to compute **g**. This descriptor has $n = 4$ and $l_{max} = 5$. The grid on the far left represents the input image. Large spheres indicate elements of the descriptor which sum their weighted inputs. Small spheres show connections to the input image. The radii of the small spheres are proportional to weights given by the decay function in equation (8)

3.2   Structure

The computation of **g** involves only the weighted summation of input image values. Figure 2 illustrates the structure that would be required to compute **g**.

This structure is repeated across all image positions that we wish to characterize. Such repetition of structure is similar to the retinotopic arrangement of columns of neurons in the visual systems of insects such as the honeybee [15, 16] and vertebrates such as cats [17]. Further, the computation for **g** consists only of local weighted sums. This style of processing is characteristic of artificial neural networks and is generally believed to be within the space of the processing operations that biological neural networks are capable of. Thus, while our image descriptor is not a model of any known neural structure in the animal kingdom, it is at least *plausible* that this descriptor could be implemented in an animal's brain.

4   1:N Matching Method

We present here the first of two homing methods which use the image descriptor developed above. This method is based on matching each descriptor in the snapshot image to N descriptors in the current image at neighbouring image positions. The coordinates of the best matches are then used to generate correspondence vectors. These correspondence vectors are then mapped to home vectors using the method described in [14]. The average home vector is the final output from this method.

We refer to the positions of descriptors in the snapshot image S as *source positions*. Each source position is matched with descriptors in the current image

C at N *candidate positions*. These candidate positions are located in a block surrounding the source position.

For each source position $\mathbf{p}$ in S we search to find the candidate position $\mathbf{p}'$ in C which is at the centre of the image patch most similar to the image patch at $\mathbf{p}$. To judge the degree of match between these two image patches we compute the scale invariant descriptors $\mathbf{h}^{\mathbf{p},S}$ and $\mathbf{h}^{\mathbf{p}',C}$ and find the dot product between them:

$$DP(\mathbf{p}, \mathbf{p}') = \mathbf{h}^{\mathbf{p},S} \cdot \mathbf{h}^{\mathbf{p}',C} \qquad (13)$$

A high value of DP indicates a good match.

To reduce computational complexity we do not consider all positions in S as source positions, but only a sampling of positions at integer multiples of the horizontal step size m_x and the vertical step size m_y, where m_x and m_y are also integers. Given images of width w pixels and height h pixels we define the number of horizontal and vertical sampling points

$$n_x = \lfloor w/m_x \rfloor \qquad (14)$$
$$n_y = \lfloor h/m_y \rfloor \qquad (15)$$

The total number of source positions is $n_x n_y$. Each source position requires a search for the best candidate position. This search involves computing DP for N candidate positions. The candidate positions are located within a radius of q pixels from the source position $\mathbf{p}$. Hence, $N = (2q + 1)^2$.

We select $\check{\mathbf{p}}$ as the candidate position with the highest DP:

$$\check{\mathbf{p}} = \arg\max_{\mathbf{p}' \in E_q(\mathbf{p})} DP(\mathbf{p}, \mathbf{p}') \qquad (16)$$

$$E_q([p_x, p_y]) = \{(p_x + i, p_y + j) \mid i, j \in \mathbb{Z}, |i| \le q \wedge |j| \le q\} \qquad (17)$$

There is an additional constraint made on the correspondence search whereby source positions in the snapshot image will only be paired with candidate positions which are on the same side of the horizon. The horizon of the panoramic image is the line which does not undergo vertical translations under movements of the robot in the plane. As long as the robot moves purely within a single plane, no image features should cross the horizon. Therefore we constrain our search to avoid any such spurious matches.

The candidate position $\check{\mathbf{p}}$ with the highest DP is used to compute the correspondence vector δ

$$\delta = \begin{pmatrix} \delta_x \\ \delta_y \end{pmatrix} = \begin{pmatrix} \Delta_x(\check{p}_x - p_x) \\ \Delta_y(\check{p}_y - p_y) \end{pmatrix} \qquad (18)$$

where Δ_x represents the inter-pixel angle in the horizontal direction and Δ_y represents vertical inter-pixel angle. These multipliers are required so that δ is expressed as a pair of angles.

We now have a set of correspondence vectors which ideally describe the movement of image features in S to their new positions in C. From each of these correspondence vectors we can determine an individual home vector. We use the 'vector mapping' method presented in [14] for this purpose. Finally, the average of these home vectors is computed, normalized, and used as the final home vector.

5 1:1 Pairing Method

We now present our second homing method based on the scale invariant image descriptor. While the method above repeatedly searches for correspondences between source positions in S and candidate positions in C, the method below considers only one candidate position for each source position. Only positions at the same image position are compared and the best matching pair is used to compute a final home vector directly.

In general, a landmark seen in the snapshot image will either move to a new position in the current image, or will disappear. However, there is an important exception to this rule. Landmarks at the focus of contraction (FOC) or focus of expansion (FOE) will maintain the same image position if the displacement of the agent from the goal consists of pure translation. For pure non-zero translation the flow field (field of correspondence vectors) exhibits two foci separated by 180°. We assume here that the world is imaged onto the unit sphere, hence both foci are always visible. All correspondence vectors are parallel to great circles passing through the foci. Correspondence vectors are oriented from the FOE to the focus of contraction FOC [1]. Figure 3 shows an ideal flow field for an agent within a simulated environment where all surfaces are equidistant from the agent. It can be observed that the amplitude of flow (the length of correspondence vectors) approaches zero at the foci.

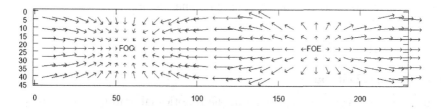

Fig. 3. An ideal flow field for pure translation. Vectors were generated by tracking the displacement of unique markers on the surface of a sphere, where the sphere was centred on the agent for the snapshot image and then shifted to the right for the current image

The 1:1 method computes descriptors $\mathbf{g}$ and $\mathbf{h}$ for positions along the horizon of the snapshot image S and the current image C. Descriptors at the same image position in both images are then compared by computing the dot product between them. The coordinates $\hat{\mathbf{p}}$ of the pairing with the highest DP value is determined

$$\hat{\mathbf{p}} = \arg\max_{\mathbf{p}} \mathrm{DP}(\mathbf{p}, \mathbf{p}) \tag{19}$$

If our descriptor truly provides a unique means of characterizing local image patches then $\hat{\mathbf{p}}$ represents a local image patch that is stationary between the

---

[1] See [18] for a more thorough discussion.

snapshot image and current image. Such an image patch could either represent a very distant landmark, or else it could represent one of the foci. Here we assume the latter. In the experiments described below the range of distances to objects remains rather small. However, if distant landmarks were present then some sort of filtering scheme might be employed to remove them from the image [19].

We determine which of the two foci $\hat{\mathbf{p}}$ corresponds to by comparing the length of the vector $\mathbf{g}^{\hat{p},S}$ with $\mathbf{g}^{\hat{p},C}$. Growth of the descriptor vector $\mathbf{g}$ from the snapshot to the current image occurs in the neighbourhood of the FOC. By definition, image features move in towards the FOC, and as they do, become weighted more heavily by the decay function $w()$. The opposite situation occurs at the FOE where features become weighted less heavily as they expand away from the FOE. The quantity b equals 1 for the case of contraction and -1 for expansion

$$b = \begin{cases} 1 & \text{if } ||\mathbf{g}^{\hat{p},S}|| < ||\mathbf{g}^{\hat{p},C}|| \\ -1 & \text{if } ||\mathbf{g}^{\hat{p},S}|| > ||\mathbf{g}^{\hat{p},C}|| \\ 0 & \text{otherwise} \end{cases} \tag{20}$$

Finally, the computed home vector is given by converting the image coordinate $\hat{\mathbf{p}}$ into a vector and using b to reverse that vector if appropriate

$$\mathbf{w} = b \begin{pmatrix} \cos(\Delta_x \hat{p}_x) \\ \sin(\Delta_y \hat{p}_y) \end{pmatrix} \tag{21}$$

The vector $\mathbf{w}$ above is the final estimated home vector.

6   Results

6.1   Image Database

A database of images was collected in the robotics laboratory of the Computer Engineering Group of Bielefeld University. Images were collected by a camera mounted on the robot and pointed upwards at a hyperbolic mirror[2]. The room was unmodified except to clear the floor. The capture grid had dimensions 2.7 m by 4.8 m, which covered nearly all of the floor's free space. Further details on the collection and format of these images has been reported in [14].

The images used for homing are low-resolution (206×46) panoramic images. Figure 4 shows sample images along a line from position (6,4) to position (0,4).

6.2   Methods

Both homing methods require edges to be extracted from input images. A Sobel filter is applied for this purpose. Parameters described below control the level of low-pass filtering applied prior to the Sobel filter.

Some parameters of the homing methods are method-specific while others are shared by both methods. The method-specific parameters were set to values

---

[2] The camera was an ImagingSource DFK 4303. The robot was an ActivMedia Pioneer 3-DX. The mirror was a large wide-view hyperbolic mirror from Accowle Ltd.

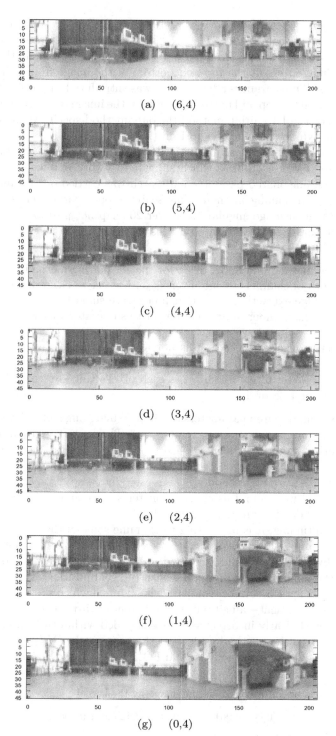

Fig. 4. Sample images from image database along a line of positions from (6,4) to (0,4)

which generally appeared to provide good results. For the 1:N matching method
these included the search radius q (set to 30), the horizontal step size m_x (4), and
the vertical step size m_y (4). Another parameter excluded points in the specified
number of image rows at the top and bottom of the image from being used as
source points. This parameter (set to 10) was introduced upon observing that
image patches in the top and bottom portions of the image tended to be relatively
indistinct. For the 1:1 pairing method the only method-specific parameter is the
height of the window around the horizon (9).

One important shared parameter is the exponent of the decay function, ζ.
This parameter was set to 0.75 which appeared to work well for both methods.
For the remaining shared parameters a search was carried out to find the best
settings for each homing method. This search scored parameter combinations
according to the average angular error over 20 snapshot positions, as described
below. Four parameters were varied in this search process:

- The length of descriptor vectors, n, was set to either 8 or 32 for the 1:N
  matching method and 8, 32, or 64 for the 1:1 pairing method.
- The length of channels to sum over, l_{max}, was set to either 20 or 50.
- Prior to edge extraction, the input images are smoothed by a Gaussian operator [3]. The number of applications of this operator was set to either 0 or
  4.
- As described in section 3.1, it is not advantageous for image edges to be
  excessively dense. The density of edges can be reduced by passing the image
  through a power filter, which raises each pixel's value to exponent τ. τ was
  set to either 1, 2, or 4.

The best found shared parameters for the 1:N matching method were: $n = 32$,
$l_{max} = 50$, 0 Gaussian operations, and $\tau = 4$. The best shared parameters for
the 1:1 pairing method were: $n = 64$, $l_{max} = 50$, 4 Gaussian operations, and
$\tau = 4$.

Franz et. al's warping method was also tested for comparison [11]. Parameters
for this method were found using a parameter search similar to that described
above. Further details can be found in [14].

Before continuing, it is interesting to examine some of the internal workings
of our two homing methods. We begin by examining the correspondence vectors
generated by the 1:N matching method. Figure 5 shows these vectors as computed for the images shown in figure 4 with the goal position at (6,4). The flow
fields generally appear correct (compare with figure 3 which shows the ideal flow
for the same movement—albeit within a different environment). However, there
are a number of clearly incorrect vectors embedded within these flow fields.

For the 1:1 pairing method we look at the variation in $DP(\mathbf{p}, \mathbf{p})$. This quantity should show peaks at the FOC and FOE. Figure 6 shows $DP(\mathbf{p}, \mathbf{p})$ for the

---

[3] The Gaussian operator convolves the image by the kernel

$$[0.005 \ 0.061 \ 0.242 \ 0.383 \ 0.242 \ 0.061 \ 0.005]$$

applied separately in the x and y directions.

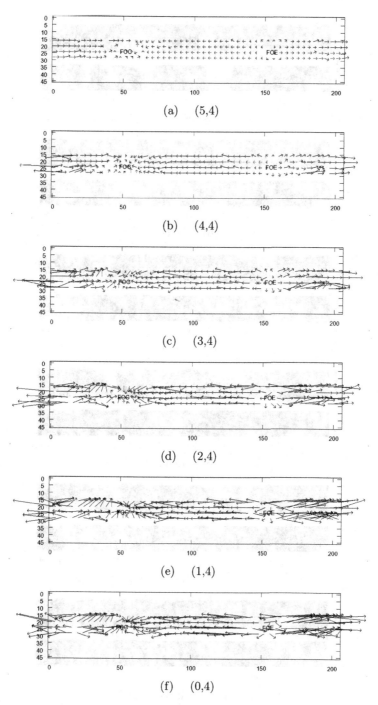

(a)   (5,4)

(b)   (4,4)

(c)   (3,4)

(d)   (2,4)

(e)   (1,4)

(f)   (0,4)

Fig. 5. Correspondence vectors for the 1:N matching method. The snapshot image was captured at (6,4), which is to the right of the positions indicated above. Hence, the correct FOC should be around (52,23) while the correct FOE should be near (155,23)

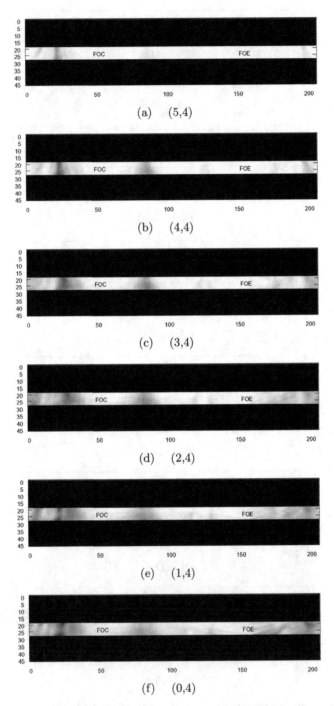

Fig. 6. Variation in $DP(\mathbf{p}, \mathbf{p})$ for the 1:1 pairing method. Snapshot position and labels are as in figure 5

images shown in figure 4 with the goal again at (6,4). Indeed, two major peaks near the ideal locations of the FOC and FOE are found.

In the first set of experiments, a single image is selected from the capture grid as the snapshot image. The method in question then computes home vectors for all other images. Figure 7 shows the home vectors generated by our two methods and the warping method for snapshot positions (6,4) and (0,16).

Both the 1:N matching method and the warping method perform quite well at position (6,4). It is evident for both of these methods that the home vectors for all positions would tend to lead the robot to the goal from all start positions, although the paths taken by the warping method would be somewhat shorter. For the 1:1 matching method, however, there are a number of incorrect vectors embedded within the otherwise correct vector field. At position (0,16) it is apparent that the 1:N matching methods yields the best results of the three methods. The 1:1 pairing method exhibits appropriate home vectors for some positions, but also generates vectors which are directed 180° away from the correct direction, as well as others which point in a variety of incorrect directions. The warping method generates appropriate vectors only within a small neighbourhood around the goal position.

For a more qualitative determination of the success of homing we compute the *average angular error* (AAE) which is the average angular deviation of the computed home vector from the true home vector. We indicate the average angular error for snapshot position (x, y) as $AAE_{(x,y)}$. Values for AAE are shown in the captions for figure 7. These values generally reflect the qualitative discussion above.

It is clear from figure 7 that homing performance is dependent on the chosen snapshot position. To assess this dependence we tested all homing methods on a sampling of 20 snapshot positions and computed AAE for each position. Figure 8 shows these snapshot positions which were chosen to evenly sample the capture grid. Figure 9 shows the computed AAE for all methods over these 20 snapshot positions. All methods exhibit higher error for snapshot positions near the fringes of the capture grid. The captions in this figure show the angular error averaged over all test positions, and all snapshot positions.

To obtain a more quantitative understanding of the difference between these methods we performed statistical tests on AAE_*. A repeated measures ANOVA with Tukey-Kramer multiple comparisons test was carried out between all three methods. Table 1 presents the results of this test. The test indicates that the 1:N matching method exhibits a significantly lower error than the warping method. No significant difference was found between the error of 1:N matching and 1:1 pairing. Nor was a significant difference found between the error of 1:1 pairing and the warping method.

While it is interesting to compare the performance of these homing methods against each other, it is useful also to compare them to an absolute standard. As described in [11], a homing method with an angular error that is always less than $\pi/2$ will yield homing paths that converge to the goal—perhaps taking a very inefficient route, but arriving eventually. Having an average error below $\pi/2$ does

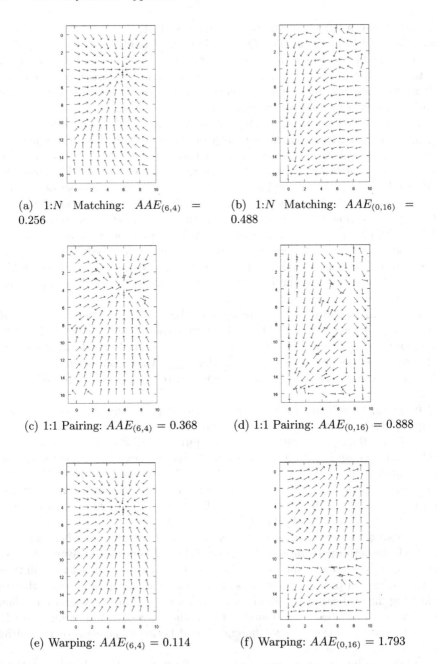

(a) 1:N Matching: $AAE_{(6,4)} =$ 0.256

(b) 1:N Matching: $AAE_{(0,16)} =$ 0.488

(c) 1:1 Pairing: $AAE_{(6,4)} = 0.368$

(d) 1:1 Pairing: $AAE_{(0,16)} = 0.888$

(e) Warping: $AAE_{(6,4)} = 0.114$

(f) Warping: $AAE_{(0,16)} = 1.793$

Fig. 7. Home vector fields for 1:N matching (a,b), 1:1 pairing (c,d), and warping (e,f), for snapshot positions (6,4) (a,c,e) and (0,16) (b,d,f)

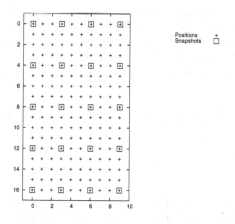

Fig. 8. Positions of images and snapshots within the capture grid

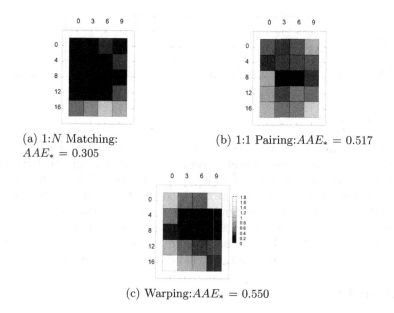

(a) 1:N Matching:
$AAE_* = 0.305$

(b) 1:1 Pairing:$AAE_* = 0.517$

(c) Warping:$AAE_* = 0.550$

Fig. 9. AAE for the twenty snapshot positions shown in figure 8 for all methods on image collection **original**

not imply convergent homing but it is a useful threshold. We have performed a statistical analysis of the difference between the angular error and $\pi/2$ using the Wilcoxon rank sum test. Table 2 presents the results of this test for $\pi/2$, and also for increasingly small angles $\pi/4$, $\pi/6$, and $\pi/8$. The test indicates whether each method exhibits an AAE smaller than the threshold. All three methods exhibit error significantly less than both $\pi/2$ and $\pi/4$. However, only the 1:N matching method exhibits an error significantly less than $\pi/6$.

Table 1. Statistical significance of the difference in AAE_* between homing methods. Significance for each cell is indicated if the method on the vertical axis is significantly better than the method on the horizontal axis. Empty fields indicate no significant difference. *Legend:* $*$ = $(p < 0.05)$, $**$ = $(p < 0.01)$, $***$ = $(p < 0.001)$, $****$ = $(p < 0.0001)$, X = self-match. *Test:* Repeated measures ANOVA with Tukey-Kramer multiple comparisons

	1:N Matching	1:1 Pairing	Warping
1:N Matching	X		*
1:1 Pairing		X	
Warping			X

Table 2. Statistical significance of AAE_* being less than the reference angles $\pi/2$, $\pi/4$, $\pi/6$, and $\pi/8$. Significance for each cell is indicated if the method on the vertical axis has an angular error significantly less than the threshold on the horizontal axis. See table 1 for legend. *Test:* Wilcoxon rank sum test

	$\pi/2$	$\pi/4$	$\pi/6$	$\pi/8$
1:N Matching	****	****	**	
1:1 Pairing	****	***		
Warping	****	*		

7 Discussion

Of the three homing methods tested above, the 1:N matching method exhibits the lowest error and overall best results. According to our statistical tests the 1:1 pairing method performs equivalently to the warping method. The pairing method is of interest because it does not require repeated searching to find correspondences between images. In theory, the computational complexity of the 1:1 pairing method should be considerably lower than the 1:N matching method. However, the pairing method appears to be less robust to parameter settings and requires the most expensive parameters (high n, and $l_{\max}$) in order to perform well.

8 Conclusions

This chapter introduced a new descriptor for local image patches which is partially invariant to scale changes. The descriptor has a simple structure that is suitable for neural implementation. Two homing methods based on this descriptor were presented. The first method employed the standard technique of matching descriptors between images. The second method, however, employed the novel notion of extracting one of the foci of motion, and using the position of that focus to compute the home vector directly. The performance of the 1:N matching method was found to exceed that of the warping method. No statistically significant difference was found between the 1:1 pairing method and the warping method. Future work will look at improvements to our descriptor as well as possibilities for using other scale invariant descriptors for the 1:1 pairing method.

Acknowledgments

Many thanks to Ralf Möller for the use of his lab, equipment, and software, and particularly for helpful reviews and discussion. Thanks also to his students Frank Röben, Wolfram Schenck, and Tim Köhler for discussion and technical support. This work has been supported by scholarships from NSERC (PGS-B 232621 - 2002), and the DAAD (A/04/13058).

References

1. Collett, T.: Insect navigation en route to the goal: Multiple strategies for the use of landmarks. Journal of Experimental Biology **199** (1996) 227–235
2. Hong, J., Tan, X., Pinette, B., Weiss, R., Riseman, E.: Image-based homing. In: Proceedings of the 1991 IEEE International Conference on Robotics and Automation, Sacramento, CA. (1991) 620–625
3. Möller, R.: Insect visual homing strategies in a robot with analog processing. Biological Cybernetics **83** (2000) 231–243
4. Cartwright, B., Collett, T.: Landmark learning in bees. Journal of Comparative Physiology A **151** (1983) 521–543
5. Lambrinos, D., Möller, R., Labhart, T., Pfeifer, R., Wehner, R.: A mobile robot employing insect strategies for navigation. Robotics and Autonomous Systems, Special Issue: Biomimetic Robots **30** (2000) 39–64
6. Gourichon, S., Meyer, J.A., Pirim, P.: Using colored snapshots for short-range guidance in mobile robots. International Journal of Robotics and Automation, Special Issue: Biologically Inspired Robotics **17** (2002) 154–162
7. Vardy, A., Oppacher, F.: Low-level visual homing. In Banzhaf, W., Christaller, T., Dittrich, P., Kim, J.T., Ziegler, J., eds.: Advances in Artificial Life - Proceedings of the 7th European Conference on Artificial Life (ECAL), Springer Verlag Berlin, Heidelberg (2003) 875–884
8. Vardy, A., Oppacher, F.: Anatomy and physiology of an artificial vision matrix. In Ijspreet, A., Murata, M., eds.: Proceedings of the First International Workshop on Biologically Inspired Approaches to Advanced Information Technology. (2004) (to appear)
9. Lowe, D.: Distinctive image features from scale-invariant keypoints. International Journal of Computer Vision **60** (2004) 91–110
10. Rizzi, A., Duina, D., Inelli, S., Cassinis, R.: A novel visual landmark matching for a biologically inspired homing. Pattern Recognition Letters **22** (2001) 1371–1378
11. Franz, M., Schölkopf, B., Mallot, H., Bülthoff, H.: Where did I take that snapshot? Scene-based homing by image matching. Biological Cybernetics **79** (1998) 191–202
12. Weber, K., Venkatesh, S., Srinivasan, M.: Insect-inspired robotic homing. Adaptive Behavior **7** (1999) 65–97
13. Möller, R.: A biorobotics approach to the study of insect visual homing strategies, Habilitationsschrift, Universität Zürich (2002)
14. Vardy, A., Möller, R.: Biologically plausible visual homing methods based on optical flow techniques. Connection Science, Special Issue: Navigation (2005 (to appear))
15. Hertel, H.: Processing of visual information in the honeybee brain. In Menzel, R., Mercer, A., eds.: Neurobiology and Behavior of Honeybees. Springer Verlag (1987) 141–157

16. Ribi, W.: The structural basis of information processing in the visual system of the bee. In Menzel, R., Mercer, A., eds.: Neurobiology and Behavior of Honeybees. Springer Verlag (1987) 130–140
17. Hubel, D., Wiesel, T.: Receptive fields, binocular interaction and functional architecture in the cat's visual cortex. Journal of Physiology **160** (1962) 106–154
18. Nelson, R., Aloimonons, A.: Finding motion parameters from spherical motion fields (or the advantage of having eyes in the back of your head). Biological Cybernetics **58** (1988) 261–273
19. Cartwright, B., Collett, T.: Landmark maps for honeybees. Biological Cybernetics **57** (1987) 85–93

Appendix

The goal of this appendix is to determine the relationship of $g^{\mathbf{p},\mathbf{v},I'}$ to $g^{\mathbf{p},\mathbf{v},I}$. We begin with $g^{\mathbf{p},\mathbf{v},I'}$

$$g^{\mathbf{p},\mathbf{v},I'} = \int_0^{l_{\max}} w(l)I'(\mathbf{p}+l\mathbf{v})\,dl \tag{22}$$

From equations (1), (2), and (4) we obtain

$$I(\mathbf{p}+l\mathbf{v}) = I'(\mathbf{p}+kl\mathbf{v}) \tag{23}$$

With a change of variables we have the following:

$$I(\mathbf{p}+\frac{l}{k}\mathbf{v}) = I'(\mathbf{p}+l\mathbf{v}) \tag{24}$$

We insert the above into the right hand side of equation (22) to obtain

$$g^{\mathbf{p},\mathbf{v},I'} = \int_0^{l_{\max}} w(l)I(\mathbf{p}+\frac{l}{k}\mathbf{v})\,dl \tag{25}$$

Next we replace the integration variable l with $j = \frac{l}{k}$

$$g^{\mathbf{p},\mathbf{v},I'} = \int_0^{l_{\max}/k} w(jk)I(\mathbf{p}+j\mathbf{v})k\,dj \tag{26}$$

Now we place our first assumption on $w()$. We assume this function has the property

$$w(xy) = w(x)w(y) \tag{27}$$

Utilizing this property on expression (26) and renaming the integration variable j back to l gives

$$g^{\mathbf{p},\mathbf{v},I'} = kw(k)\int_0^{l_{\max}/k} w(l)I(\mathbf{p}+l\mathbf{v})\,dl \tag{28}$$

To proceed further we must place another constraint on $w()$. The intention of this decay function is to reduce the impact of outlying features on g. Therefore it makes sense that $w(l)$ should be small for large values of l. We first define a new constant $l^*_{\max} = \min(l_{\max}, l_{\max}/k)$. The second constraint on $w()$ is as follows:

$$w(l) \approx 0 \quad \text{for } l > l^*_{\max} \tag{29}$$

Therefore

$$g^{\mathbf{p},\mathbf{v},I} \approx \int_0^{l^*_{\max}} w(l)I(\mathbf{p} + l\mathbf{v})\, dl \tag{30}$$

and

$$g^{\mathbf{p},\mathbf{v},I'} \approx kw(k) \int_0^{l^*_{\max}} w(l)I(\mathbf{p} + l\mathbf{v})\, dl \tag{31}$$

Combining these two approximations gives us the desired relationship

$$g^{\mathbf{p},\mathbf{v},I'} \approx kw(k)g^{\mathbf{p},\mathbf{v},I} \tag{32}$$

Author Index

Lecture Notes in Artificial Intelligence (LNAI)

Vol. 3336: D. Karagiannis, U. Reimer (Eds.), Practical Aspects of Knowledge Management. X, 523 pages. 2004.

Vol. 3327: Y. Shi, W. Xu, Z. Chen (Eds.), Data Mining and Knowledge Management. XIII, 263 pages. 2005.

Vol. 3315: C. Lemaître, C.A. Reyes, J.A. González (Eds.), Advances in Artificial Intelligence – IBERAMIA 2004. XX, 987 pages. 2004.

Vol. 3303: J.A. López, E. Benfenati, W. Dubitzky (Eds.), Knowledge Exploration in Life Science Informatics. X, 249 pages. 2004.

Vol. 3301: G. Kern-Isberner, W. Rödder, F. Kulmann (Eds.), Conditionals, Information, and Inference. XII, 219 pages. 2005.

Vol. 3276: D. Nardi, M. Riedmiller, C. Sammut, J. Santos-Victor (Eds.), RoboCup 2004: Robot Soccer World Cup VIII. XVIII, 678 pages. 2005.

Vol. 3275: P. Perner (Ed.), Advances in Data Mining. VIII, 173 pages. 2004.

Vol. 3265: R.E. Frederking, K.B. Taylor (Eds.), Machine Translation: From Real Users to Research. XI, 392 pages. 2004.

Vol. 3264: G. Paliouras, Y. Sakakibara (Eds.), Grammatical Inference: Algorithms and Applications. XI, 291 pages. 2004.

Vol. 3259: J. Dix, J. Leite (Eds.), Computational Logic in Multi-Agent Systems. XII, 251 pages. 2004.

Vol. 3257: E. Motta, N.R. Shadbolt, A. Stutt, N. Gibbins (Eds.), Engineering Knowledge in the Age of the Semantic Web. XVII, 517 pages. 2004.

Vol. 3249: B. Buchberger, J.A. Campbell (Eds.), Artificial Intelligence and Symbolic Computation. X, 285 pages. 2004.

Vol. 3248: K.-Y. Su, J. Tsujii, J.-H. Lee, O.Y. Kwong (Eds.), Natural Language Processing – IJCNLP 2004. XVIII, 817 pages. 2005.

Vol. 3245: E. Suzuki, S. Arikawa (Eds.), Discovery Science. XIV, 430 pages. 2004.

Vol. 3244: S. Ben-David, J. Case, A. Maruoka (Eds.), Algorithmic Learning Theory. XIV, 505 pages. 2004.

Vol. 3238: S. Biundo, T. Frühwirth, G. Palm (Eds.), KI 2004: Advances in Artificial Intelligence. XI, 467 pages. 2004.

Vol. 3230: J.L. Vicedo, P. Martínez-Barco, R. Muñoz, M. Saiz Noeda (Eds.), Advances in Natural Language Processing. XII, 488 pages. 2004.

Vol. 3229: J.J. Alferes, J. Leite (Eds.), Logics in Artificial Intelligence. XIV, 744 pages. 2004.

Vol. 3228: M.G. Hinchey, J.L. Rash, W.F. Truszkowski, C.A. Rouff (Eds.), Formal Approaches to Agent-Based Systems. VIII, 290 pages. 2004.

Vol. 3215: M.G.. Negoita, R.J. Howlett, L.C. Jain (Eds.), Knowledge-Based Intelligent Information and Engineering Systems, Part III. LVII, 906 pages. 2004.

Vol. 3214: M.G.. Negoita, R.J. Howlett, L.C. Jain (Eds.), Knowledge-Based Intelligent Information and Engineering Systems, Part II. LVIII, 1302 pages. 2004.

Vol. 3213: M.G.. Negoita, R.J. Howlett, L.C. Jain (Eds.), Knowledge-Based Intelligent Information and Engineering Systems, Part I. LVIII, 1280 pages. 2004.

Vol. 3209: B. Berendt, A. Hotho, D. Mladenic, M. van Someren, M. Spiliopoulou, G. Stumme (Eds.), Web Mining: From Web to Semantic Web. IX, 201 pages. 2004.

Vol. 3206: P. Sojka, I. Kopecek, K. Pala (Eds.), Text, Speech and Dialogue. XIII, 667 pages. 2004.

Vol. 3202: J.-F. Boulicaut, F. Esposito, F. Giannotti, D. Pedreschi (Eds.), Knowledge Discovery in Databases: PKDD 2004. XIX, 560 pages. 2004.

Vol. 3201: J.-F. Boulicaut, F. Esposito, F. Giannotti, D. Pedreschi (Eds.), Machine Learning: ECML 2004. XVIII, 580 pages. 2004.

Vol. 3194: R. Camacho, R. King, A. Srinivasan (Eds.), Inductive Logic Programming. XI, 361 pages. 2004.

Vol. 3192: C. Bussler, D. Fensel (Eds.), Artificial Intelligence: Methodology, Systems, and Applications. XIII, 522 pages. 2004.

Vol. 3191: M. Klusch, S. Ossowski, V. Kashyap, R. Unland (Eds.), Cooperative Information Agents VIII. XI, 303 pages. 2004.

Vol. 3187: G. Lindemann, J. Denzinger, I.J. Timm, R. Unland (Eds.), Multiagent System Technologies. XIII, 341 pages. 2004.

Vol. 3176: O. Bousquet, U. von Luxburg, G. Rätsch (Eds.), Advanced Lectures on Machine Learning. IX, 241 pages. 2004.

Vol. 3171: A.L.C. Bazzan, S. Labidi (Eds.), Advances in Artificial Intelligence – SBIA 2004. XVII, 548 pages. 2004.

Vol. 3159: U. Visser, Intelligent Information Integration for the Semantic Web. XIV, 150 pages. 2004.

Vol. 3157: C. Zhang, H. W. Guesgen, W.K. Yeap (Eds.), PRICAI 2004: Trends in Artificial Intelligence. XX, 1023 pages. 2004.

Vol. 3155: P. Funk, P.A. González Calero (Eds.), Advances in Case-Based Reasoning. XIII, 822 pages. 2004.

Vol. 3139: F. Iida, R. Pfeifer, L. Steels, Y. Kuniyoshi (Eds.), Embodied Artificial Intelligence. IX, 331 pages. 2004.

Vol. 3131: V. Torra, Y. Narukawa (Eds.), Modeling Decisions for Artificial Intelligence. XI, 327 pages. 2004.

Vol. 3127: K.E. Wolff, H.D. Pfeiffer, H.S. Delugach (Eds.), Conceptual Structures at Work. XI, 403 pages. 2004.

Vol. 3123: A. Belz, R. Evans, P. Piwek (Eds.), Natural Language Generation. X, 219 pages. 2004.

Vol. 3120: J. Shawe-Taylor, Y. Singer (Eds.), Learning Theory. X, 648 pages. 2004.

Vol. 3097: D. Basin, M. Rusinowitch (Eds.), Automated Reasoning. XII, 493 pages. 2004.

Vol. 3071: A. Omicini, P. Petta, J. Pitt (Eds.), Engineering Societies in the Agents World. XIII, 409 pages. 2004.

Vol. 3070: L. Rutkowski, J. Siekmann, R. Tadeusiewicz, L.A. Zadeh (Eds.), Artificial Intelligence and Soft Computing - ICAISC 2004. XXV, 1208 pages. 2004.

Vol. 3068: E. André, L. Dybkjær, W. Minker, P. Heisterkamp (Eds.), Affective Dialogue Systems. XII, 324 pages. 2004.

Vol. 3067: M. Dastani, J. Dix, A. El Fallah-Seghrouchni (Eds.), Programming Multi-Agent Systems. X, 221 pages. 2004.